Autologous Fat Transplantation

Autologous Fat Transplantation

edited by
Melvin A. Shiffman
Tustin, California

MARCEL DEKKER, INC. NEW YORK • BASEL

ISBN: 0-8247-0484-3

This book is printed on acid-free paper.

Headquarters
Marcel Dekker, Inc.
270 Madison Avenue, New York, NY 10016
tel: 212-696-9000; fax: 212-685-4540

Eastern Hemisphere Distribution
Marcel Dekker AG
Hutgasse 4, Postfach 812, CH-4001 Basel, Switzerland
tel: 41-61-261-8482; fax: 41-61-261-8896

World Wide Web
http://www.dekker.com

The publisher offers discounts on this book when ordered in bulk quantities. For more information, write to Special Sales/Professional Marketing at the headquarters address above.

Current printing (last digit):
10 9 8 7 6 5 4 3 2 1

PRINTED IN THE UNITED STATES OF AMERICA

Preface

The transplantation of fat from one area of the body to another is a safe and effective procedure. Autologous fat will vascularize without rejection or allergic reaction if injected in small aliquots. There is no cost for the fat and it is available in sufficient quantities from almost any patient. The convenience is thus inestimable.

Autologous fat transplantation is operator-dependent and requires gentle handling, proper instruments, and careful selection of recipient sites. Despite all precautions, the transplant will often resorb either partially or completely. In most cases, at least a portion of the fat will remain viable. Repeat transplants may be necessary to obtain the desired results.

Fat can also be frozen and used at a later date for refinement of a particular problem or for cosmetic use in other areas. The proper method of freezing and thawing has not been adequately researched and it is not known if the fat survives or the residual collagen results in filling of defects.

There is little doubt that autologous fat transplantation has become a recognized and established tissue filler. This book attempts to give the cosmetic surgeon an understanding of the techniques that are being utilized by surgeons around the world. It is hoped that there will be continued research concerning fat transplants so that even better results can be obtained in the future.

Melvin A. Shiffman

Introduction

I began the modern technique of fat transfer after my first ten years of liposculpture when confronted with the problem of creating better anatomical form in areas of the body that had depressions from the suction of excessive fat. The early techniques not only entailed feathering the margins close to the entrance incisions but also consisted of replacing the fat with autologous fat transfer. This evolved into transferring fat to other body parts. In 1988, fat transfer to the breast was initiated with the intention to improve consistency and contour with augmentation rather than the insertion of breast prostheses. This method included placement of modest amounts of adipose above the pectoral muscle and around the mammary gland but not in the gland proper.

Confirmation of this technique came about ten years ago when Pierre Fournier and I came across *The Face and Its Improvement by Aesthetic Plastic Surgery,* written in 1926 by Charles H. Willi. In the book, Willi describes the technique and advantages of syringe fat transfer. In 1986, I had begun to augment the zygomatic and malar region with autologous fat and to fill deep nasolabial folds. At that time, fairly large quantities of fat were deposited in these anatomical subunits, which caused overfilling to compensate for anticipated graft loss. As a result, 50% of the transferred fat was lost. This was consistent with the literature but was not acceptable. By 1992, there was a better understanding of grafting techniques, biology, and physiology of adipose tissue and closer attention was being paid to harvest techniques and processing. I began to harvest the adipose tissue with small syringes and a sharp needle. Processing of fat was abandoned and the adipose tissue was allowed to settle into layers. In 1992 and 1993, the technique focused on graft survival rather than anticipated graft loss. Because less fat was transferred, there was a great increase in long-term survival and a decrease in early graft loss.

In 1997, I developed a technique that consisted of the transfer of rice grain–sized fat parcels for total facial rejuvenation. This technique focuses on multiple transfers to the subdermal, interdermal, muscular, and fat layers using a microcannula to deposit tiny parcels of adipose tissue. The intention was early and reliable neovascularization. The microcannula has also allowed correction of problems that were once difficult or impossible to manage. Signs of aging in the infraorbital region caused by the orbiculais oculi muscles shining through the delicate translucent infraorbital skin could be restored by depositing approximately 4 cc of adipose tissue on one side of the face from forehead to chin. The age of the patient will often determine the number of transfers performed—this technique is ideal for patients 25–30 years old. The rice grain–sized implants stimulate new vessel formation and gives the skin improved color, consistency, and texture.

These early successes led me to correct other deformities such as acne scarring using fat transfer. The pathophysiology of acne scarring is complicated and acne-related deformities are difficult to manage as acne lesions affect the epidermis, dermis, and subdermal fat. The current method is used to restore the atrophied subcutaneous tissues with subsequent dermabrasion of the acne scars.

The next step in the evolution of fat transfer focused on lines of force and strength in the face. These are the lines that act against gravity and are responsible for the typical aged appearance of the aging face. These factors were used to determine the vector of placement of the fat. Fat is transferred in an oblique manner to support the skin and is deposited from the chin to forehead in the shape of a herringbone, the head of the herringbone at the chin and the tail toward the forehead.

The clinical scientific aspects of fat transfer are exciting and challenging areas of cosmetic surgery. All factors must be taken into consideration and addressed to maintain and improve outcomes. As the technology of medicine and surgery continues to evolve so do the techniques and our understanding. The future of fat transfer is in our hands as scientists and surgeons.

Giorgio Fischer, M.D.

Contents

Preface iii

Introduction Giorgio Fischer v

Contributors xi

1. History of Autologous Fat Transfer 1
Melvin A. Shiffman

2. Principles of Autologous Fat Transplantation 5
Melvin A. Shiffman

3. Guidelines for Autologous Fat Transfer 23
Robert W. Alexander

4. Autologous Fat Volume Retention Evaluation by Magnetic Resonance Imaging 31
Hans Wolfgang Hoerl and Axel-Mario Feller

5. History of Autologous Fat Transplant Survival 43
Melvin A. Shiffman

6. Complications of Fat Augmentation 53
Melvin A. Shiffman

7. Fat Transfer with Rice Grain–Size Fat Parcels 55
Giorgio Fischer

8. Preoperative Consultation 65
Melvin A. Shiffman

9. Fat Transplant to the Glabella and Forehead 69
Felix Rüdiger G. Giebler

10. Multilayered Malar and Mental Lipotransplant Surgery 77
Julius Newman

11. Lipotransfer to Nasolabial Folds and Marionette Lines 95
David S. Alkek

12. Autologous Fat Transfer to the Lips 113
Steven B. Hopping

13. Eyebrow Lift with Fat Transfer 127
Giorgio Fischer

14. Carbon Dioxide Laser Resurfacing and Autologous Fat Injection 131
Steven L. Harlan

15. Rejuvenation of the Hand: The Hand Fill 169
Pierre Fournier

16. Fat Transplant to the Buttocks and Legs for Enhancement and Deformities—Large Volumes of Fat Transplant 187
Lina Valero de Pedroza

17. History of Breast Augmentation with Fat 199
Melvin A. Shiffman

18. Stabilization of Fat Transfer to the Breast with Autologous Platelet Gel 207
James E. Fulton

19. Fat Transplants in Male and Female Genitals 219
Enrique Hernandez-Perez and Filadelfo Venadero-Albarran

20. Autologous Fat for Liposuction Defects During and After Surgery 233
Pierre Fournier

21. Lipofilling Acne Scars 243
David S. Alkek

22. Autologous Frozen Fat for Transplant 261
Ziya Saylan

23. Long-Term Frozen Fat Transplant 267
Katsuya Takasu and Shizu Takasu

24. Autologous Fat Mixed with Polymethylmethacrylate 279
Marcelo Benedetti and Adriana Ponti

25. Autologous Collagen and Fat Transfer 287
Julio A. Ferreira

26. New Consideration in Fat Transfer: A Possible Role for Maintaining Interstitial Protein to Reduce Shrinkage of Transferred Volume 299
Mitchell V. Kaminski, Jr., James E. Fulton, and John J. Wolosewick

27. Medical Legal Aspects of Autologous Fat Transplantation 311
Melvin A. Shiffman

28. Editor's Analysis of Fat Transfer Techniques 313
Melvin A. Shiffman

Index 317

Contributors

Robert W. Alexander, M.D. Alderwood SurgiCenter, Lynnwood, Washington

David S. Alkek, M.D. Dermatology Department, University of Texas, Southwestern Medical School at Dallas, Dallas, Texas

Marcelo Benedetti, M.D. Clinica Estetica Charcas, Buenos Aires, Argentina

Axel-Mario Feller, M.D. Department of Plastic Surgery, Behandlungszentrum, Vogtareuth, Germany

Julio A. Ferreira, M.D. South American Academy of Cosmetic Surgery and Aesthetic Medicine and Aesthetic Surgery Department, Gendarmeria Hospital, Buenos Aires, Argentina

Giorgio Fischer, M.D. Fischer Institute, Rome, Italy

Pierre Fournier, M.D. Private Practice, Aesthetic Plastic Surgery, Paris, France

James E. Fulton, M.D. Newport Point Medical Center, Newport Beach, California

Felix Rüdiger G. Giebler, M.D. Vincemus Klinik, Eider, Germany

Steven L. Harlan, M.D. Laser Resurfacing and Vein Center, Des Moines, Iowa

Enrique Hernandez-Perez, M.D. Center for Dermatology and Cosmetic Surgery, San Salvador, El Salvador

Hans Wolfgang Hoerl, M.D. Private Practice, Plastic Surgery, Munich, Germany

Steven B. Hopping, M.D. George Washington University and Center for Cosmetic Surgery, Washington, D.C.

Mitchell V. Kaminski, Jr., M.D. North Shore Cosmetic Surgery and Wellness Center, Northfield, Illinois

Julius Newman, M.D. Department of Cosmetic Surgery, Graduate Hospital, Philadelphia, Pennsylvania

Adriana Ponti, M.D. Clinica Estetica Charcas, Buenos Aires, Argentina

Ziya Saylan, M.D. Saylan Comestic Surgery Center, Düsseldorf, Germany

Melvin A. Shiffman, M.D., J.D. Private Practice, Cosmetic and Reconstructive Surgery, Tustin, California

Katsuya Takasu, M.D., Ph.D. Takasu Clinic, Nagoya, Japan

Shizu Takasu, M.D. Takasu Clinic, Nagoya, Japan

Lina Valero de Pedroza, M.D. Nubell Clinic for Aesthetic Surgery and Unidad Medica Clinica Del Country, Bogota, Colombia

Filadelfo Venadero-Albarran, M.D. Center for Dermatology and Cosmetic Surgery, San Salvador, El Salvador

John J. Wolosewick, Ph.D. Department of Anatomy and Cell Biology, University of Illinois, Chicago, Illinois

Autologous Fat Transplantation

1

History of Autologous Fat Transfer

Melvin A. Shiffman
Private Practice, Cosmetic and Reconstructive Surgery, Tustin, California

Neuber (1893)[1] reported the use of small pieces of fat from the upper arm to reconstruct a depressed area of the face resulting from tuberculosis osteitis. He concluded that small pieces of fat, bean or almond size, appeared to have a good chance of survival. Czerny (1895)[2] used a large lipoma to fill a defect in the breast following resection of a benign mass. The transplanted breast, however, appeared darker in color and smaller in volume than the opposite breast. Verderame (1909)[4] observed that fat transplants solved the problem of shrinkage at the transplant site. Lexer (1910)[3] reported personal experience with fat transplants and found that larger pieces of fat gave better results. Bruning (1911)[5] used fat grafts to fill a postrhinoplasty deformity by placing fat in a syringe and injecting the tissue through a needle. Tuffier (1911)[6] inserted fat into the extrapleural space to treat pulmonary conditions. Biopsy of the fat 4 months posttransplant showed most of the fat was resorbed and replaced by fibrous tissue.

Straatsma and Peer (1932)[7] used free fat grafts to repair postauricular fistulas and depressions or fistulas resulting from frontal sinus operations. Cotton (1934)[8] used a technique of broad undercutting and insertion of finely cut fat which was molded to fill defects.

Peer (1956)[9] noted that grafts the size of a walnut appear to lose less bulk after transplanting than do smaller multiple grafts. He also found that free fat grafts lose about 45% of their weight and volume 1 year or more following transplantation owing to the failure of some fat cells to survive the trauma of grafting as well as the new environment. Fat grafts are affected by trauma, exposure, infection, and excessive pressure from dressings.[10] Peer (1959)[11] stated that microscopically grafts appear like normal adipose tissue 8 months after transplantation.

Liposuction was conceived by A. Fischer and G. Fischer in 1974[12] and put into practice in 1975.[13]

Fischer (1976)[14] first reported removal of fat by means of 5-mm incisions using a rotating, alternating instrument electrically and air powered. This allowed aspiration of fat through a cannula. Saline solution was injected through a separate incision to dilute the fat. In 1977, Fischer and Fischer[15] reviewed 245 cases of liposuction with the planotome for treatment of cellulite in the lateral trochanteric areas. There was a 4.9% incidence of seromas despite wound suction catheters and compression dressings. Pseudocyst formation, which required removal of the capsule through a wider incision and the use of the planotome, occurred in 2% of cases.

The advent of liposuction spurred a move toward using the liposuctioned fat for reinjecting areas of the body for filling defects or augmentation. Bircoll (1982)[16] first reported the use of autologous fat from liposuction for contouring and filling defects. Illouz (1986)[17] claimed he began to inject aspirated fat in 1983. Johnson[18] stated that he began in 1983 to use autologous fat injection for contouring defects of the buttocks, anterior tibial area, lateral thighs, coccygeal area, breasts, and face. Bircoll (1984)[19] presented the method of injecting fat that had been removed by liposuction. Krulig (1987)[20] asserted that he began to use fat grafts by means of a needle and syringe. He called the procedure lipoinjection. He began to use a disposable fat trap to facilitate the collection process and to ensure the fat's sterility. Newman (1987)[21] stated that he began reinjecting fat in 1985. The idea of utilizing the aspirated fat, which was otherwise wasted, was an attractive idea and other surgeons began to make use of the aspirate to augment defects and other abnormalities.

The American Society of Plastic and Reconstructive Surgery (ASPRS) Ad-Hoc Committee on New Procedures produced a report on September 30, 1987, regarding autologous fat transplantation.[22] The conclusions were:

1. Autologous fat injection has a historical and scientific basis.
2. It is still an experimental procedure.
3. Fat injection has achieved varied results, and long-term, controlled clinical studies are needed before firm conclusions can be made regarding its validity.
4. Fat transplant for breast augmentation can inhibit early detection of breast carcinoma and is hazardous to public health.

REFERENCES

1. Neuber F. Fettransplantation. Chir Kongr Verhandl Dtsch Ges Chir 1893;22:66
2. Czerny M. Plastischer Ersatz der brusterlruse durch ein lipom. Verhandl Dtsch Ges Chirurg
3. Lexer E. Freie Fettransplantation. Dtsch Med Wochenschr 1910;36:640
4. Verderame P. Ueber fettransplantation bei adharenten knochennarben am orbitalrand. Klin Monatsbl Augenheilkd 1909;47:433–442
5. Bruning P. Cited by Broeckaert, TJ, Steinhaus, J.: Contribution e l'etude des greffes adipueses. Bull Acad R Med Belg 1914;28:440
6. Tuffier T. Abces gangreneux du pouman ouvert dans les bronches: hemoptysies repetee operation par decollement pleuro-parietal; gueriscn. Bull Mem Soc Chir Paris 1911;37:134
7. Straatsma C.R., Peer L.A. Repair of postauricular fistula by means of a free fat graft. Arch Otolaryngol 1932;15:620–621
8. Cotton F.J. Contribution to technique of fat grafts. N Engl J Med 1934;211: 1051–1053
9. Peer L.A. The neglected free fat graft. Plast Reconstr Surg 1956;18:233
10. Peer L.A. Loss of weight and volume in human fat grafts. Plast Reconstr Surg 1950;5:217
11. Peer L.A. Transplantation of Tissues, Transplantation of Fat. Baltimore: Williams & Wilkins, 1959
12. Fischer G. The evolution of liposculpture. Am J Cosm Surg 1997;14(3):231–239
13. Fischer G. Surgical treatment of cellulitis. IIIrd International Congress of the Acad Cosmit Surgery, Rome, May 31, 1975
14. Fischer G. First surgical treatment for modeling body's cellulite with three 5 mm incisions. Bull Int Acad Cosm Surg 1976;2:35–37
15. Fischer A, Fischer G. Revised technique for cellulitis fat reduction in riding breeches deformity. Bull Int Acad Cosm Surg 1977;2(4):40–43
16. Bircoll M. Autologous fat transplantation. The Asian Congress of Plastic Surgery, February, 1982
17. Illouz Y.G. The fat cell graft: A new technique to fill depressions. Plast Reconstr Surg 1986;78(1):122–123
18. Johnson G.W. Body contouring by macroinjection of autologous fat. Am J Cosm Surg 1987;4(2):103–109
19. Bircoll M.J. New frontiers in suction lipectomy. Second Asian Congress of Plastic Surgery, Pattiyua, Thailand, February, 1984
20. Krulig E. Lipo-injection. Am J Cosm Surg 1987;4(2):123–129
21. Newman J., Levin J. Facial lipo-transplant surgery. Am J Cosm Surg 1987;4(2): 131–140
22. American Society of Plastic and Reconstructive Surgery: Report in Autologous Fat Transplantation. Plastic Surgery Nursing, Winter 1987, pp 140–141

2
Principles of Autologous Fat Transplantation

Melvin A. Shiffman
Private Practice, Cosmetic and Reconstructive Surgery, Tustin, California

The introduction of liposuction for fat reduction and body contouring has developed into transplantation of the extracted fat for augmentation of defects or for cosmetic purposes. There has been controversy concerning the manner of collecting fat, injecting the fat, cleansing the fat, and the effectiveness of the fat transfer. Some physicians have been disappointed with the long-term results of fat transplantation.

The process of fat transplantation has not been standardized, and there is a need to analyze some of the methods and results.

I. FAT TRANSPLANT SURVIVAL

Vitamin E is a necessary factor in fat tissue maintenance,[1] whereas insulin increases the metabolic activity of fat cells[2] and retards lipolysis.[3–7]Hiragun[8] theorized that insulin may induce fibroblasts to pick up lipid lost from lipolysis and become adipocytes.

Skouge[3] felt that fat cells from an area of relatively poor vascularity will be more hardy, have decreased metabolic needs, and increase survival. Asken,[9] however, stated that more fibrous areas, such as the upper abdomen, are not ideal for donor sites.

Fat characteristics may be helpful in determining which area of fat is more likely to be retained. Those adipocytes with alpha 2 receptors are antilipolytic with a poor response to diet and appear more likely to survive with little change from weight loss or weight gain in comparison to adipocytes with beta 1 receptors.

	Fat Characteristics[8]	
	Alpha 2 receptors	Beta 1 receptors
Lipolysis	Antilypolytic	Lipolytic
Response to diet	Poor	Good
Region of fat	Abdominal, trochanteric (genetic fat)	Facial, arms Upper torso

Survival of adipocytes depends on the instrumentation used for harvesting and injecting the fat. Damage is inversely related to the diameter of the instrument to extract and inject.[10] The pressure generated in injecting fat increases as a function of decreasing needle diameter (from 16 to 22 gauge). There is some decrease in the metabolic activity of fragments passed through 20-gauge needles or smaller.

	Needle Size and Cell Survival[11] Needle gauge			
	16	18	20	22
Pad integrity	+	+	0	–
Cell morphology	+	+	0	–
Nuclear morphology	+	+	0	–
Fat globule	+	+	+	+

+ = 75% or more without cell damage.
0 = 25–75% cell damage.
– = >75% cell damage.

The presence of blood in the fat injected stimulates macrophage activity to remove the cells. Washing the cells in a physiological solution prior to injection will solve the problem.[12–14] Skouge[3] raised the question of whether washing decreases the viability of fragile adipocytes.

Campbell[11] concluded that adipocyte integrity and metabolism of fat fragments subjected to mechanical manipulation by liposuction using wall suction remain intact. Illouz[8] biopsied the areas of fat injection and found normal fat cells.

McCurdy[15] analyzed fat cell survival and concluded that the technical factors to accomplish the goal of 40–50% transplanted adipocyte survival includes:

1. Low vascularity of donor site
2. High vascularity for recipient site
3. Low-pressure technique of aspiration of fat
4. Filtering and washing harvested adipocytes
5. Use of >2-mm cannula for injection to minimize adipocyte injury
6. Multilayered deposition of fat
7. Overcorrection of recipient site

Because of the problem of resorption of fat with fat transplantation, 30–50% overinjection is ordinarily used.[16–21] Asadi[16] determined that subdermal injection is important for long-term results.

II. INDICATIONS FOR FAT TRANSPLANTATION

There have been two papers which relate the indications for autologous fat transplantation.

A. Skouge[22]: Indications for Fat Transplantation

1. Facial

Aging changes
Melolabial grooves
Central cheek depressions
Subcommissural depressions
Flattened upper lip
Glabella
Diffuse age-related lipoatrophy

2. Cosmetic

Lip augmentation
Chin augmentation
Malar augmentation

3. Scars

Traumatic
Lipoatrophy, acne
Idiopathic lipodystrophy
Facial hemiatrophy

4. Nonfacial

Rejuvenation of the hands
Body contour defects
Depressions, liposuction induced
Breast enlargement
Traumatic scars

B. Berdeguer[23]: Clinical Indications for Fat Transplantation

1. Depressed Scars—Face and Body

Postsurgical
Posttraumatic

2. Aging Skin with Loss of Supportive Tissue

Glabellar furrows
Upper lip
Melolabial folds
Hollow cheeks
Dorsal hands

3. Esthetic Enhancement

Cheek augmentation
Chin augmentation
Breast augmentation
Leg contour surgery

4. Congenital Defects

Hemifacial atrophy
Soft-tissue defects of the body

In analyzing these lists, of indications for autologous fat transplantation, a simpler and more useful classification can be devised:

1. Fill defects
   Congenital
   Traumatic
   Disease (acne)
   Iatrogenic

2. Cosmetic
   Furrows (wrinkles)
   Refill Lost Supportive Tissue (aging)
   Enhancement

III. COMPLICATIONS OF FAT TRANSPLANTATION

Injection of small globules of fat will prevent cyst formation. Johnson[24] showed that 1-, 3-, and 5-mL injections resulted in small cysts but 10 mL injections had macroscopic cyst formation. Oil cysts develop through confluence of necrotic fat cells, have a lining of macrophages, and resorption may take years, thus giving a false impression of a successful transplantation.[25]

Sterility of fat retrieval and injection must be maintained. Infection has not been reported.[22]

Bruising, temporary swelling, and tenderness may result from fat transplantation.[22]

Teimourian[26] reported a case of injection of fat into the glabellar frown lines with the onset of pain and loss of vision in one eye while receiving the injection. There was central retinal artery thrombosis, probably secondary to fat particle embolism.

Calcifications have only been reported in fat transplantation to the breast for augmentation. This does not appear to be a significant risk, since the timing of the appearance, the position, and the character of the calcifications will indicate the etiology.

The most important problem encountered is fat resorption. Trauma to the cells, desiccation during transfer, and the presence of blood are contributing factors. At least an 18-gauge needle should be used to reinject fat. Ersek[27] reported that very little autologous fat survives, but his use of a whisk in the cleansing process probably destroyed most of the fat cells.

IV. TECHNIQUE OF AUTOLOGOUS FAT TRANSPLANTATION

The lack of standardization of fat collection and transplantation allows a wide range of methods with varied results.

Following are methods utilized by certain cosmetic surgeons the author obtained by personal communication.

A. Billie[28]

I do have patients who go back over 10 years at this point. I have had good fortune overall with nonsmokers and actually a moderate amount of success in smokers. I have found that the younger the patients are the better they seem to do. We have placed the fat in multiple sites including defects in the legs from traumatic events such as automobile accidents or recluse spider bites and in the nasal labial furrows in aging patients.

Over the years, I have washed the fat, not washed the fat, added insulin, not added insulin, and tried everything. Currently, I am utilizing a 4-mm cannula to remove the fat, catching it in the sterile in-line trap, not washing it, and reapplying it utilizing a 16-gauge fat-grafting needle with a 10-mL syringe apparatus.

B. Fragen[28]

I have found that autologous fat transplantation is a very effective part of my facial rejuvenation surgery, providing I give the patients a detailed explanation of the limited nature of the procedure and the fact that it is always somewhat temporary. Depending on the patient and the location to which the fat is transferred, the fat survives for a variable period. I have found that transferring fat under skin grafts and scars and on top of semirigid or rigid surfaces improves the viability of the fat transfer. For example, if one transfers fat under the skin postmastectomy, it seems to stay there and offers some padding. Putting it under burn scars will help increase the padding of the burn scar and make the skin grafts over it more pliable and flexible. If fat is transferred to a lip, it seems to survive there least well. I think because of the active nature of the lip.

My method of transfer is very simple. I like to call it a closed system. Essentially what is done is the area for fat harvesting is prepped and draped and infiltrated with a Klein solution. The fat is then harvested with a 14-gauge blunt cannula on a 10-mL syringe. If there appears to be excess saline, the excess saline is decanted. If there is excess bloody tissue, then the specimen is washed in saline and again decanted. If, as is usual, essentially pure fat is removed from the donor site, then it is maintained within the syringe with the blunt 14-gauge cannula.

A small stab incision is made near the site for fat transfer, and the blunt cannula is then placed into the donor area. Several tunnels are made with the blunt cannula so that the fat is not squeezed into the area but rather easily injected into the donor site. Then the fat is transferred to the donor site. Both sites are prepared sterilely. If the patient is under general anesthesia, then usually no anesthesia is used for the recipient site. If this is done under local anesthesia, a small amount of 1% Xylocaine (lidocaine) with adrenaline (epinephrine) is infiltrated into either the lips or the glabella or to whatever site we are transferring the fat.

I usually overfill the graft site by approximately 50%, and I tell patients that the swelling will last 3–5 days. I routinely do fat transferring on face lift patients. I do in the neighborhood of 10 or more face lifts per month, and I would guess that 80% of those are associated with a fat transfer.

Untoward effects include bruising, short-term swelling, occasional lumpiness, and stimulation of fever blisters. The lumping has never been a problem, in that the fat can easily be compressed even months later. One time a patient had a small fat cyst which was easily evacuated.

It is my feeling that fat injected into the lower portion of the nasolabial fold, the lips, and the droll lines has a relatively short life span, with the ideal results being reached in approximately 2 weeks and slow disappearance over 2–4 months. In the glabella, I believe the fat will last 6 or more months and, in many cases, over a year. I think the area nearer the nose in the nasolabial fold will retain fat a little better. In that area also the fat will last 6–8 months. I think fat injected under graft sites and scars and over other hard prominences lasts for many months, and I have several cases where the fat has laster several years. The primary advantage of fat transfer is that it can very effectively camouflage cosmetic defects (such, e.g., as thin upper lips with wrinkling, glabellar frown lines, drool line) which are difficult to correct without other extensive procedures. In Palm Springs, we find many people who do not want to restrict their outdoor activities, such as tennis and golf. These patients accept the safe, although temporary, correction by fat transfer. Their biggest complaint is that the wonderful result they get is short-lived, but until we find a safe, nonresorbable filler which the FDA (US Food and Drug Administration) will approve, we do not have a better alternative.

C. Tobin[30]

About 10 years ago, when liposuction surgery was first introduced, we began hearing recommendations for reinjection of fat. My initial experience with this procedure was to attempt to refine breast reconstruction cases by injecting small amounts of fat adjacent to implants or in patients on whom other surgeons had carried out flap reconstructions. We initially harvested the fat with a syringe and reinjected it using an old, mechanical injector that was designed initially to inject Teflon into the vocal cords. In essence, we were injecting it through an

18-gauge needle with a very precise ratchet mechanism. Our results were discouraging because of rapid reabsorption. We felt that perhaps this was related to the fact that we were injecting into a scarred area.

At about the same time, we began injecting fat into the face. My first experience with this procedure was to attempt to correct grooving in the cheeks that was caused by facial liposuction. We did not understand the risks that were involved when liposuction was carried out in this area. Many of us ended up with patients who had skin irregularities or waviness. Again, we used the same technique—namely, aspiration with a syringe and reinjection through the Teflon gun. Again, out results were discouraging.

Because of these failures, we essentially abandoned the technique. Sometime later, we heard about successes with injection of fat into the back of the hand, and we attempted a few cases. By this time, we had stopped using the Teflon gun and were simply aspirating the fat with the syringe and transferring it to smaller syringes through a small transfer tube after which the fat was injected into the back of the hand. Our technique included aspiration of fat with a syringe, rinsing and straining with saline, and then reinjection. Again, both we and our patients were disappointed with the results.

About 3 years ago, after hearing of successes with the injection of separated fat, we were tempted to try again. Several surgeons had various techniques of morselizing the fat and injecting the fibrous portion. Often this material was called autologous collagen, although I am not aware of any confirmation that the material was in any way similar to the bovine collagen that had become so popular under the trade name Zyderm.

We utilized the technique (Becker Co., Palm Beach, FL) with kits that were available which included syringes for transferring the material through a progressively smaller orifice. This resulted in the morselization of the material. Following this, the material was centrifuged and the collagenous component was obtained to be used for reinjection. The material was supposedly capable of being preserved by freezing, and we attempted this as well. We probably treated about 25 patients with this process, carrying out multiple injections over a period of several months. As far as I can determine, we did not have one patient who was really pleased with the results, and we have since abandoned it.

At the present time, our use of injectable fat is uncommon. When patients request it, we explain the fact that out experience has not been very positive, but we do offer it as an option. Occasionally, patients request it, but once again, I have not seen any convincing evidence that there is any permanent augmentation.

Obviously, I am perplexed by the reports and the literature by reputable surgeons who claim they see permanent results. Until I see a series of consecutive cases presented over a relatively long period of time, I will remain unconvinced but will attempt to be open minded.

The results of literature review and personal communications are in Table 1. Analysis of the data shows that of 21 surgeons, the following techniques were utilized.

1. Harvesting Fat

Syringe vs Machine	No. of Surgeons: 21
Syringe	8
Machine	7
Unstated	8

There was a tendency to use the syringe to collect a smaller amount of fat and the suction machine to retrieve larger amounts of fat, with some surgeons using both methods.

Needle (Gauge)	No. of Surgeons: 7
13	2
14	5

Cannula (mm)	No. of Surgeons: 12
2	3
3	7
4	4
>4	2

When a needle was utilized to obtain fat, the 14-gauge the one was most often the choice. When a cannula was utilized to obtain fat, the 3 mm one was most often the choice. Some surgeons used multiple sizes.

2. Treatment of the Harvested Fat

Ten surgeons washed the fat, five strained the fat, and six decanted the excess blood and fluid. Some surgeons used more than one method.

Table 1 Methods of Autologous Fat Transplantation

Reference (Year)	No. of patients	Harvesting: Syringe (mL)	Harvesting: Machine (mL)	Harvesting: Needle gauge	Fat: Cannula (mm)	Fat: Treatment	Reinjection: Insulin	Reinjection: Needle gauge
Agrus (1987)[31]	157	+50	+50	14		Drained through cheesecloth for microdeposition		14
Asadi (1993)[1]	1		+ (low)		3 and 6	Irrigated with lactated Ringer's microdeposition	+	15
Asken (1987)[9]		+20	+20cc		2–3 3–4	Excess fluid drained		
Berleguer (1995)[23]	70	+			4	Strained		
Billie[28]					4	No washing		16
Bircoll (1987)[32]			+			Strained in coarse guaze pad	+	16
Carraway (1990)[33]			+	14		Liquid absorbed with a cottonoid		18
Chang (1994)[34]	24				3	Decanted		14 or 16
Ersek (1991)[27]	15				2–3	Agitated with wire wisk; cleansed with Eagle's medium suspended in physiological solution (Tsol)		16 or 18

Fournier (1986)[35]				13				13
Fragen[29]				14		Washed with saline		14
Hin (1988)[36]		+		14	4			14 or 16
Illouz (1986)[37]	37					Removed blood		
Johnson (1987)[24]	50		+			Sifted and washed		
Krulig[38]	520		+			Caught in trap; strained and washed with physiological solution		13
Lewis (1992)[3]	12	+			3	Cleansed with saline		
Newman (1987)[40]	96		+		4	Luken's strap irrigated with lactated Ringer's	+	18 or 20
Pinski (1992)[41]	43				3	Sterile trap decanted		15–18
Skouge (1992)[22]				13		Decanted, washed with lactated Ringer's if necessary		16–18
Toledo (1991)[42]					1.8–3, 3–6	Spun and decanted or washed with lactated Ringer's		18
Shiffman (1997)[43]		+	+	14	3	Washed with saline		14 or 18

3. Reinjection of Fat

No surgeon used a cannula to reinject the fat. When using a needle to reinject, six surgeons used more than one gauge.

Reinjection of Fat	
Needle (gauge)	No. of Surgeons: 14
14	5
15	2
16	7
18	7

The 16- and 18-gauge needles were most often used for reinjection of fat.

At the present time, small cannulas have been devised with relatively blunt tips which can be used for reinjection without the problem of bleeding in the recipient area.

D. Insulin

Some physicians have added insulin to the fat in preparation for transplantation.[12,44,45] The theory is that insulin inhibits lipolysis.

Sidman (1956)[45] found that insulin decreases lipolysis. Hiragun et al. (1980)[47] stated that theoretically insulin may induce fibroblasts to pick up the lipid lost and become adipocytes.

Chajchir et al. (1993)[48] found that the use of insulin did not show any positive effect on adipocyte survival during transplantation compared to fat not prepared with insulin.

E. Centrifugation

Some physicians centrifuge the adipose tissue to remove blood products and free lipids to improve the quality of the fat to be injected.[13,49,50]

Asken (1987)[9] stated that his "method of reducing the material to be injected to practically pure fat is to place the fat-filled syringe with a rubber cap (the plunger having been previously removed and kept in a sterile environment) into a centrifuge. The syringe is then spun for a few seconds at the desired rpm and the serum, blood, and liquefied fat collects in the dependent part of the syringe. . . ."

Toledo (1991)[42] reported that "for facial injection we spin the full syringes for 1 minute. . . . . in a manual centrifuge (about 2000 rpm), eject the unwanted solution, and transfer the fat. . . ."

Uebel (1992)[51] centrifuged autologous fat at 10,000 rpm for 10 min in order to obtain a "fat-collagen graft." The centrifuged material on histological examination showed cell residues, collagen fibers, and 5% intact fat cells. The material is absorbed at a slow rate and maintains contour and volume for 18–24 months. A new graft procedure is always performed to achieve a more permanent result.

Chajchir et al. (1993)[48] centrifuged 1 mL of bladder fat pad from mince (both at 1000 rpm for 5 min and 5000 rpm for 5 min) and injected into malar area subdermis. Microscopically, after 1–2 months, there were macrophages filled with lipid droplets, giant cells, focal necrosis of adipocytes, and cyst-like cavities of irregular sizes and shapes. After 112 months following injection, no recognized adipocytes could be found. Total cellular damage was present in both groups.

Brandow and Newman (1996)[52] found that centrifugation of harvested fat did not alter the microscopic structural integrity of cells. Spun and unspun samples were examined and were similar.

Fulton et al. (1998)[53] found that fat centrifuged for 3 min at 3400 rpm works well for small-volume transfers but not for large volume transfers into breasts, biceps, or buttocks.

F. Ratchet Gun for Injection

Neuman and Levin (1987)[40] designed a lipoinjector with a gear-driven plunger to inject fat tissue evenly into desired sites. Fat injected with excessive pressure in the barrel of a syringe can cause sudden injections of undesired quantities of fat which will pour into recipient sites.

Agris (1987)[31] stated that a rachet-type gun allows controlled accurate deposition of autologous fat. Each time the trigger is pulled, 0.1 mL is deposited.

Neichajev (1992)[54] used a ratchet gun for free transplantation of fat harvested at –0.5 atm pressure. He noted only partial resorption of the fat but with significant improvement of contour.

Asaadi and Haramis (1993)[55] described the use of a gun with a disposable 10-mL syringe for fat injection.

Niechajev and Sevcuk (1994)[56] utilized a special pistol and blunt cannula with 2.3-mm internal diameter to inject the fat.

Berdeguer (1995)[23] used a lipotransplant gun to inject fat into areas to be enhanced.

Fulton et al. (1998)[53] stated that it is beneficial for a beginning surgeon to use a ratcheted pistol for injection, as this gives a more uniform injection volume.

G. Severing Tethering Bands

Several authors have sugested severing of tethering bands, usually with a needle paddle-shaped needle or a V-shaped ("pickle-fork") instrument, to allow the skinto lift more easily with injection of fat.[42,57–59] Recurrence of depressions was thought to be less likely.

H. Machine Liposuction

"Liposuction harvesting of fat is traumatic and results in a graft composed of intact cells combined with cellular debris and free lipid."[60] Liposuction removal of autologous fat by –1 atm suction was reported by Nguyen et al. (1990)[61] as showing microscopically 90% elongated, irregularly shaped, and ruptured adipocytes and only 10% unchanged, normal-appearing adipocytes. With use of a 10-mL syringe for aspiration of fat, they found 95% unchanged adipocytes. May (1990),[62] in commenting of Nguyen's study stated, ". . . one would have thought that aspiration could produce nearly the same degree of suction (1 atm) as formal suctioning. If the degrees of negative pressure produced by these two techniques are similar, and if the cannulae are similar then the degree of cell damage should have been similar."

Niechajev (1992)[63] obtained fat for grafting using a vacuum pump with –0.5 atm pressure. Using a ratchet gun for injection into the cheek, he noted only partial resorption of the fat over 1.5–4.0 years (mean 3 years).

Niechajev and Sevchuk (1994)[64] reported 50% fat survival over 3.5 years after single fat transplantation with 50% overcorrection. They found that fat obtained under maximum negative pressure (–0.95 atm) results in partial breakage and vaporization of the fatty tissue. About two thirds of the fat withstood the trauma of aspiration. Low pressure (–0.5 atm) results in smaller cell size (29% smaller than with aspiration at –0.95% atm), and they assumed that high pressure causes mechanical distention of the adiopocytes which increases the risk of and sometimes causes cell breakage.

Elam et al. (1997)[65] noted more effective fat removal by lowering the negative suction pressure during liposuction. Negative pressures varied from 15 in of mercury (–375 mm mercury) to 30 in (–750 mm) (–760 mm = 1 atm). Above 25 in of mercury (–625 mm), an obvious amount of blood appears in the aspiration along with air bubbles. At maximum vacuum (–750 mm), the aspirate is a blood-tinged mixture of fatty globules with significant amounts of dark venous blood. The ideal liposuction vacuum pressure at sea level was felt to be a negative 20 in of mercury (–500 mm).

REFERENCES

1. Meschik Z. Vitamin E and adipose tissue. Edinburgh Med J 1944;51:486
2. Katoes AS Jr., Blalock K. Perfused fat cells: effects of lipolytic agents. J Biol Chem 1933;248:5089
3. Skouge JW. Autologous fat transplantation in facial surgery. Presented at meeting of American Academy of Cosmetic Surgery: Controversies in Breast and Facial Augmentation, Philadelphia, August 7–9, 1992
4. Sidman RL. The direct effect of insulin on organ cultures of brown fat. Anat Rec 1956;124:723
5. Smith U. Human adipose tissue in culture studies on the metabolic effect of insulin. Diabetologia 1976;12:137
6. Solomon SS. Comparative studies of the antilipolytic effect of insulin and adenosine in the perfused fat cell. Horm Metab Res 1980;12:601
7. Solomon SS, Duckworth WC. Effect of antecedent hormone administration on lipolysis in the perfused isolated fat cell. J Lab Clin Med 1976;88:984
8. Hiragun A, Sato M, Mitsui H. Establishment of a clonal cell line that differentiated into adipose cells in vitro. In Vitro 1980;16:685
9. Asken S. Autologous fat transplantation: Micro and macro techniques. Am J Cosm Surg 1987;4(2):111–121
10. Dolsky RL, Weber D Adipocyte survival. Presented at the third Annual Scientific Meeting of the American Academy of Cosmetic Surgery and The American Society of Lipo-Suction Surgery, Los Angeles, February 2-5, 1987
11. Campbell GL, Laudenslager N, Newman J. The effect of mechanical stress on adipocyte morphology and metabolism. Am J Cosm Surg 1987;4(2):89–94
12. Illouz YG. The fat cell "graft": a new technique to fill depressions. Plast Reconstr Surg 1986;78(1):122–123
13. Krulig E. Lipo-injection. Am J Cosm Surg 1987;4(2):123–129
14. Lewis CM. Correction of deep gluteal depression by autologous fat grafting. Aesthetic Plast Surg 1992;16:247–250
15. McCurdy JA Jr. Five years of experience using fat for leg contouring (Commentary). Am J Cosm Surg 1995;12(3):228
16. Asaadi M, Haramis HT. Successful autologous fat injection at 5-year follow-up. Plast Reconstr Surg 1993;91(4):755–756
17. Chajchir A, Benzaquen I. Liposuction fat grafts in face wrinkles and hemifacial atrophy. Aesthetic Plast Surg 1986;10(2):115–117
18. Chajchir A, Benzaquen I. Fat-grafting injection for soft tissue augmentation. Plast Reconstr Surg 1989;84(6):921–934

19. Chiu DTW, Edgerton BW. Repair and grafting of dermis, fat, and fascia. In: Plastic Surgery. J McCarthy, ed. Philadelphia: Saunders, 1990:515
20. Illouz YG. De l'utilization de la graisse aspiree pour combler les defects cutanes. Rev Chir Esth Langue Franc. 1985;10(40):13
21. Matsudo PKR, Toledo LS. Experience of injected fat grafting. Aesthetic Plast Surg 1988;12(1):35–38
22. Skouge J. The effectiveness and long term survival of transplanted fat. Presented at American Academy of Cosmetic Surgery, Philadelphia, August 7–9, 1992
23. Berdeguer P. Five years of experience using fat for leg contouring. Am J Cosm Surg 1995;12(3):221–229
24. Johnson GW. Body contouring by macroinjection of autogenous fat. Am J Cosm Surg 1987;492:103–109
25. Smahel J. Fat cylinder transplantation: An experimental study of three different kinds of fat transplants. Plast Reconstr Surg 1996;98(1):97–98
26. Teimourian B. Blindness following fat injections. Plast Reconstr Surg 1988; 82:361
27. Ersek RA. Transplantation of purified autologous fat: a 3-year follow-up is disappointing. Plast Reconstr Surg 1991;87(2):219–227
28. Billie JD. Autologous fat transplantation. Personal communication, 3/6/96
29. Fragen R. Autologous fat transplantation. Personal communication, 3/19/96
30. Tobin H. Fat transfer. Personal communication, 3//5/96
31. Agris J. Autologous fat transplantation: A 3-year study. Am J Cosm Surg 1987; 4(2):95–102
32. Bircoll M. Autologous fat tissue augmentation. Am J Cosm Surg 1987;4(2): 141–149
33. Carraway JH, Mellow CG. Syringe aspiration and fat concentration. A simple technique for autologous fat injection. Ann Plast Surg 1990;24(3):293–296
34. Chang KN. Surgical correction of postliposuction contour irregularities. Plast Reconstr Surg 1994;94(1):126–134
35. Fournier PF. Microlipoextraction and injection. Presented at Annual Meeting of American Society of Dermatology, Palm Springs, California, May 11, 1986
36. Hin LC. Syringe liposuction with immediate lipotransplantation. Am J Cosm Surg 1988;5(4):243–248
37. Illouz YG. The fat cell "graft": a new technique to fill depressions. Plast Reconstr Surg 1986;78(1):122–123
38. Krulig E. Lipo-injection. Am J Cosm Surg 1987;4(2):123–129
39. Lewis CM. Correction of deep gluteal depression by autologous fat grafting. Aesthetic Plast Surg 1992;16:247–250
40. Newman J, Levin J. Facial lipo-transplant surgery. Am J Cosm Surg 1987;4(2): 131–140
41. Pinski KS, Roenigk HH Jr. Autologous fat transplantation. J Dermatol Surg Oncol 1992;18:179–184

42. Toledo LS. Syringe liposculpture: A two-year experience. Aesthetic Plast Surg 1991;15:321–326
43. Shiffman MA. Autologous fat transplantion. Am J Cosm Surg 1997;14(4): 433–442
44. Ellenbogen R. Free autogenous pearl fat grafts in the face—a preliminary report of a rediscovered technique. Ann Plast Surg 1986:16:179–194
45. Newman J. Preliminary report on "fat recycling"—liposuction fat transfer for facial defects. Am J Cosm Surg 1986;3:67–69
46. Sidman RL. The direct effect of insulin on organ cultures of brown fat. Anat Rec 1956;124:723
47. Hiragun A, Sato M, Mitsui H. Establishment of a clonal line that differentiated into adipose cells in vitro. In Vitro 1980;16:685
48. Chajchir A, Benzaquen I, Moretti E. Comparative experimental study of autologous adipose tissue processed by different techniques. Aesthetic Plast Surg 1993;17:113–115
49. Toledo L. Syringe liposculpture: a two-year experience. Aesthetic Plast Surg 1991; 15:321–326
50. Zocchi M. Produccion y utilizacion de Colegeno Autologo para el remodelaje facial. II Congreso Chileno de Cirugia Plastica, Santiago, Chile, October 4-6, 1991
51. Uebel CO. Facial sculpture with centrifuged fat-collagen. In VT Hinderer, ed. Plastic Surgery 1992. Vol II. Amsterdam: Excerpta Medica, 1992:749–752
52. Brandow K, Newman J. Facial multilayered micro lipo-augmentation. Int J Aesthetic Restor Surg 1996;4(2):95–110
53. Fulton JE, Suarez M, Silverton K, Barnes T. Small volume fat transfer. Dermatol Surg 1998;24:1–9
54. Niechajev I. Autologous transplantation of fat (lipo-filling) for the improvement of the cheek contour, long-term results. In VT Hinderer, ed. Plastic Surgery 1992. Vol II. Amsterdam: Excerpta Medica, 1992:747–748
55. Asaadi M, Haramis HT. Successful autologous fat injection at 5-year follow-up. Plast Reconstr Surg 1993;91(4):755–756
56. Niechajev I, Sevcuk O. Long term results of fat transplantation: clinical and histologic studies. Plast Reconstr Surg 1994;94:496–506
57. Fournier PF. Liposculpture: The Syringe Technique. Paris: Arnette, 1991
58. Gasparotti M. Superficial liposuction: A new application of the technique for aged and flaccid skin. Aesthetic Plast Surg 1992;16:141–153
59. Grazer FM. Cellulite lysing. Aesthetic Surg 1991;11:11
60. Eppley BL, Sidner RA, Platis JM, Savove AM. Bioactivation of free-fat transfers: a potential new approach to improving graft survival. Plast Reconstr Surg 1991; 90:1022–1030
61. Nguyen A, Pasyk KA, Bouvier TN, Hassett CA, Argenta LC. Comparative study of survival of autologous adipose tissue taken and transplanted by different techniques. Plast Reconstr Surg 1990;85:378–386

62. May JW Jr. Comparative study of survival of autologous adipose tissue taken and transplanted by different techniques (Discussion). Plast Reconstr Surg 1990; 85:387–389
63. Niechajev I. Autologous transplantation of fat (lipo-filling) for the improvement of the check contour, long-term results. In VT Hinderer, ed. Plastic Surgery 1992. Vol II. Amsterdam: Excerpta Medica, 1992:747–748
64. Niechajev I, Sevchuk O. Long-term results of fat transplantation: clinical and histologic studies. Plast Reconstr Surg, 1994;94:496–506
65. Elam MV, Packer D, Schwab J. Reduced negative pressure liposuction (RNPL): could less be more? Int J Aesthetic Restor Surg 1997;5:101–104

3
Guidelines for Autologous Fat Transfer

Robert W. Alexander
Alderwood SurgiCenter, Lynnwood, Washington

With the evolution of technical lipocontouring techniques, interest in transfer of autologous fat from undesired areas to those areas with defects or contour deficiencies has greatly increased. Controversy remains relative to the ideal manner of harvesting, preparing, and transferring. Although the process of fat transplantation has not been standardized, many surgeons acknowledge and report very significant degrees of effectiveness and safety for the procedures. Fat transfer is no longer considered to be a matter of theory, and therefore it can now be accepted as a valid modality in the armamentarium of cosmetic surgeons around the world.

The author has formulated a guideline of the considerations and techniques which is efficacious, safe, predictable, and has had clinical success.

I. SELECTION OF APPROPRIATE RECIPIENT SITES

Candidates for autologous fat transfer would be systemically healthy individuals without specific medical contraindications to surgery or anesthesia.

1. Primary recipient sites are locations where fat normally resides. Examples include:

Facial and body depressions due to loss of subdermal fat deposits
Malar
Submalar cheeks
Glabella

Chin
Lipodystrophic areas
Lateral aspect of nasolabial folds
Retroglandular breast
Postliposuction irregularities

2. Secondary recipient sites are locations with adequate vascularity and limited density of tissues. Examples include:

Labial aspect of nasolabial folds
Submucosal (lingual) aspect of lips
Mucocutaneous area of lips
Various depressions associated with scarring or trauma

3. Recipient site considerations

Sites should be gently pretunneled prior to injection of fat to recipient bed.
Sites should not be infiltrated with fluids prior to fat introduction, particularly with solutions containing vasoconstrictors.
Should avoid fat transplantation into areas of preexisting inflammation or compromised circulation.
Should seek existing subdermal fat deposit plane and avoid attempts to place grafted tissue directly into the dermal plane or intralesional dense scar sites.
Selection of access opening for introduction of fat grafts should be made to ensure maximal cosmesis and still afford direct access to the areas to be augmented.

II. SELECTION OF APPROPRIATE DONOR SITES

Autologous fat appears to display "donor site memory"; that is, it retains the metabolic and storage characteristics of the cells common to the sites selected for harvest. This has significant clinical importance in determination of ideal donor site areas.

1. Screening for potential candidates and donor areas:
    a. Familial history of primary fat deposit sites and general body configuration tendencies appear strongly to follow female to female traits and locations and vice versa in males.
    b. Clear identification of the individual patient's areas of long-term difficulties with fat deposition and maintained storage locations (usually reported as diet and exercise resistant sites).

2. Ideal donor areas are locations whose cells are genetically designated as active in storage, metabolically resistant locations (commonly referred to as primary deposit sites). Grafts taken from these sites appear to have better long-term survival with storage capabilities. The exact locations are specific to each individual patient and represent the areas where they first put weight on and last areas to take weight off. It is believed that these areas should supersede all other areas for harvest regardless of convenience of other locations convenient for surgeon.

3. Sites should be accessible to allow for harvesting procedures and result in an acceptable esthetic scar.

4. Most common donor sites:

Abdomen
Flanks
Thighs (lateral, medial, and anterior)
Buttocks
Knees (medial aspect)

5. Less common donor sites:

Calves
Upper arms
Precostal and lateral thoracic deposits
Any location with an adequate palpable volume of fat deposition where safe access can be afforded and meets the first on, last off criteria.

III. LOW-PRESSURE HARVESTING OF AUTOLOGOUS FAT

1. Utilize appropriate sterile surgical technique for donor and recipient site preparation.

2. Preoperative broad-spectrum systemic antibiotics may be appropriate prior to initiation of any harvest.

3. Selected donor and recipient sites should be marked with the patient in an upright position, including identification and marking of approach openings for harvest and placement of grafts.

4. Photodocumentation of both donor and recipient sites is recommended.

5. Fat harvesting appears to have high cell viability and predictability of result when grafts are removed with low-pressure, closed-syringe techniques using a sterile saline or balanced salt solution buffer within the harvesting syringe. Excessive hypertonicity or hypotonicity of the buffer solutions is considered to be damaging to the cells to be grafted.

6. Extraction should be performed with superpolished interior and tip edge, polished edge, blunt needles or cannulas ranging in size from 2.1 to 3.7 mm in diameter. The use of needles smaller than these dimensions will result in higher extraction pressures and may cause a large amount of cellular damage. The use of larger diameter extraction devices may yield a higher proportion of collagenous matrix which may not be desirable for transfer.

7. Documentation of actual locations and specific volumes extracted should be written during the procedure and a record retained in the patient chart.

8. Donor sites should be gently injected utilizing multiport infiltrators or polished edged, blunt needles of diameter 2.1–3.0 mm. Tumescence is provided by volumes of saline with either epinephrine only (1:100,000) or saline with dilute local plus epinephrine (0.05–0.1% lidocaine with 1:100,000 epinephrine) is introduced at a ratio of two to four times the anticipated volume of fat to be harvested in preparation for transfer. Based on recent findings, multiple rinses of plain saline solution may not be able effectively to remove all remnants of the intracellular lidocaine. Effects of lidocaine on fat metabolism is currently being studied.

9. It is recommended to wait 15 min following instillation of tumescent solution into the donor areas to afford maximal hemostatic effects.

10. Limited, gentle pretunneling with the extraction cannula or needle is recommended. Overuse of pretunneling may result in more cellular damage and potential hemorrhage or inflammation postoperatively.

11. Pretunneling is considered to be valuable in order to create recipient channels into which columns of fat can be placed with minimal resistance and which can provide maximal margin contact between the grafted cells and the nutrient bed of the recipient site.

12. Attempts should be made to pass the harvesting cannula or ported needle gently through undisturbed fat deposits and use minimal repetitive passes through the areas being harvested. The concept of passing the harvesting device through fresh, undisturbed fat areas is consistent with minimal trauma and more clean fat cell removal.

13. Following aspiration of a small volume of normal saline into the harvesting syringe, the syringe plunger should be pulled back approximately one fourth to one third the volume of the syringe barrel. Approximate recommended volumes and vacuum applied are as follows:

a. 10-mL syringe. Use 2–5 mL normal saline (NS); draw plunger in 2:4-mL increments.
b. 35-mL syringe. Use 5–10 mL normal saline (NS); draw plunger in 5- to 10-mL increments.

c. 60 mL syringe. Use 10–15 mL normal saline (NS); draw plunger in 10- to 20 mL increments.

14. Attempt should be made to extract as pure and clean a cellular return as possible. In the event of excessive collagen matrix or hemorrhage, it is recommended to redirect the harvesting cannula or needle into adjacent, less disturbed areas or simply move to an alternative donor site.

15. In patients seeking lipocontouring changes at the same time, it is recommended that all fat harvesting for transfer be complete *prior to* providing extensive cell population reductions required for larger contour changes.

16. Harvest of volumes ranging from two to four times the anticipated volumes intended for graft purposes is recommended. Upon completion of the transfer, the remaining washed cellular fat can be discarded in accordance with the standard toxic waste guideline or set aside for storage according to the delayed transfer protocol.

IV. HANDLING AND PREPARATION OF HARVESTED FAT

1. Fat harvested should be allowed to decant thoroughly in order to separate the fat to be transferred from the infranatant layer containing saline and blood elements. There are three recognized means to accomplish this goal:

a. Decanting in the harvesting syringe by placement of the syringe in a vertical position, plunger up.
b. Placement of the harvested fat from the syringes directly into a large-mouth container with sterile normal saline where it is allowed to decant and rinse. Extraction of the infranatant layer and subsequent addition of an equal volume of normal saline, coupled with a gentle stirring, will facilitate the cleansing (and lidocaine reduction if used) process. Removal of the fat cellular components with a device capable of atraumatic loading of the syringes which will be used for actual transfer is accomplished with syringe tip closure and barrel removal.
c. Low-speed, hand-generated centrifugation may be utilized to facilitate identification of infranatant components and facilitate loading of the graft syringes.

2. Infranatant fluid levels should be clearly identified and carefully expressed from the syringe used to transfer the fat cells. This fluid should be relatively clear in appearance to ensure complete washing of the cellular elements desired for grafting.

3. Clean, sterile normal saline should be used to provide complete washing of the potential graft materials. If washing is to be accomplished inside the syringe used for harvesting, it is recommended that normal saline be drawn through a transfer base of no less than 2.1 mm diameter to ensure low-pressure exposure of the graft cells.

4. Rinsing should be continued until essentially clear infranatant lay can be produced. Depending on the purity of the harvest, this may require multiple (four to five) rinsing efforts to accomplish.

5. The supranatant layer is compromised of fatty oils and debris. This material should not be included in the actual grafting procedures. Failure to avoid injection of these materials may lead to a greater tissue inflammatory response and thereby lessen viability of grafted cells.

6. To the extent possible, avoid selection of fat cell elements extracted that contain extensive collagenous matrix unless the specific intent and purpose of the surgeon is to provide a fibrous fat– or dermal fat–type graft for augmentation. Clinical evidence suggests that such grafts do not provide improved cellular element survival, and to the extent that they produce increased inflammatory response, may actually result in less predictability of augmentation achieved and may result in more contracture of the recipient bed during the healing process.

V. TRANSFER TECHNIQUES INTO RECIPIENT SITES

1. The use of large-bore, blunt-edged cannulae or needles (2.1–3.7 mm) is recommended for fat injection into limited, pretunneled channels. The use of injector guns or devices with injector ends of diameters of less than 2.1 mm may result in exposure of the graft cells to high pressures and increased potential for cellular damage. This may result in less successful augmentation and predictability.

2. Pretunneling with small, blunt-edged cannulae or needles gently utilized in the exact desired recipient sites is recommended with the intent to create appropriate spaces to receive the graft cells under minimal resistance (therefore also minimal pressure during placement). Too aggressive pretunneling may disturb the recipient site integrity and vascular supply and may allow excessive clumping or shifting of the grafted cells in the developed sites.

3. It is considered to be important that the pretunneled areas afford maximal contact of grafted fat cells with the recipient bed cells necessary to provide vascularity, nutrition, oxygenation, and waste removal. Ideally, placement of tunnels should be in the areas normally occupied by fat cells.

4. Placement of fat grafts should be made in small aliquots (0.5–2.0 mL) into the tunnels utilizing a minimal amount of pressure. The use of mechanical injectors delivering exact, measured volumes (usually 0.5 mL) are both convenient and accurate and should subject fat cells to a minimal amount of pressure when used with larger injector tips (2.1–3 mm). It is acceptable to inject without such devices, with effort being made to place small volumes into the developed tunnels, to avoid "pooling or clumping" of injected graft materials, and use minimal pressure.

5. Overfilling of approximately 20–30% is usually recommended to compensate for the volume of normal saline fluid used to suspend the fat cells during their transfer.

VI. PROTOCOL FOR STORAGE OF FAT FOR DELAYED TRANSFER

1. Tissue culture evidence has demonstrated the viability of adipocytes following freezing and storage.

2. Specimens that have been harvested according to the above protocol, rinsed thoroughly, and placed in the injection syringes to be used should be capped, sealed, accurately labeled (according to patient, date of harvest, expiration date, and other information desired by the surgeon).

3. These isolated syringes are placed in individual freezer zip-lock bags and allowed gradually to be reduced in temperature to a −4°C and held at that temperature throughout the storage period. There is histological evidence that rapid freezing protocols (e.g., liquid nitrogen) lead to cellular crystallization and disruption, making such treatments not being recommended for subsequent adipocyte transfer.

4. At this time, volumes of fat stored have been limited to small-volume transfer syringes (i.e., 1-, 3-, 5-, and 12-mL syringes), where effort is made to store a volume of one to two times the amount transferred in freshly harvested cells. To date, large-volume transfers are limited to utilization of freshly harvested and prepared transfers. There is no known contraindications of larger volume storage units.

5. Specimens to be prepared for transfer are removed approximately 2–3 h prior to actual appointment time. They are placed in sterile saline solution at room temperature and gradually allowed to thaw and return to ambient temperatures. It is commonly noted that the fat tissues after freezing and thawing have an altered texture which does not seem to change the cellular integrity on histological examination of injectability. This texture is best described as "rubbery" and slightly more resistant to injection.

6. Placement of such storaged fat grafts is carried out under sterile conditions following the same protocol as the primary injection. It is believed that pretunneling is important in the development of a graft recipient bed, as in the primary transfer efforts.

7. At this time, it is customary to place patients receiving stored fat grafts on a 5-day oral antibiotic course (such as cephalexin [Keflex] or other appropriate medication). No cases using this protocol have been reported to have resulted in unfavorable clinical acceptance or infections.

8. Observation of patients receiving stored grafts suggests that they retain volumes at least equal to those seen with the primary, freshly harvested fat cells. To date, we have no evidence that they are less filling or effective than primary grafts.

9. It is recommended that the storage freezing device used for maintaining the stored grafts be clearly marked as bearing hazardous biological materials and should be monitored with a thermometer for daily log recordings of temperature (with initials of examining person).

10. Storage has been effective at 1 year from the time of harvest. Additional tissue culture evidence should be developed to determine efficacy and safety of longer storage intervals.

4

Autologous Fat Volume Retention Evaluation by Magnetic Resonance Imaging

Hans Wolfgang Hoerl
Private Practice, Plastic Surgery, Munich, Germany
Axel-Mario Feller
Behandlungszentrum, Vogtareuth, Germany

Today, the fat used in transplantation procedures is predominantly obtained by liposuction. Tiny particles of fat are transplanted after liposuction, which results in a larger area of contact for diffusion and neovascularization, reducing the danger of necrosis by transplantation of smaller, singular deposits. Despite the advantage of improved neovascularization by using small particles, there exists, however, the disadvantage that considerable tissue trauma is caused by the liposuction procedure.

I. TECHNIQUE FOR FAT ASPIRATION AND TRANSPLANTATION

Fat tissue is aspirated through a minor skin incision from such areas as the abdominal panniculus below the umbilicus, the buttocks, or the inner portion of the upper thigh. The suction site as well as the point of transplantation are locally anesthetized before the procedure. The aspiration system used consists of a suction cannula (Fig. 1), a connective hose, and a pressure pump that is fitted with a tissue filter, which traps any fat cell formation.

After injection of Klein's solution with hyaluronidase, a 30- to 45-min delay is taken to wait until the intercellularly binding connective tissues between

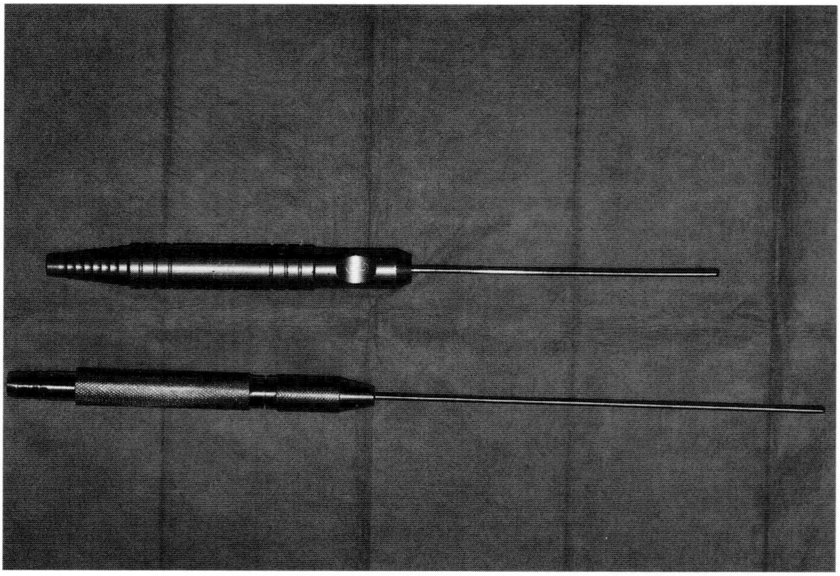

Figure 1 Suction cannula.

the fat cells are depolymerized. The hyaluronidase solution is given simultaneously with a vasoconstrictor to decrease bleeding.

Liposuction is performed through a suction cannula at a pressure of 0.8–1.0 atm. Blunt suction cannulae with three ovals bored on the side and half-rounded ends with diameters of 4 and 6 mm are used. The aspirated cells are then removed from the reservoir of the tissue filter, placed onto gauze, and rinsed with Ringer's solution to remove any cellular debris and blood constituents, as well as any local anesthetic, vasoconstrictor, and hyaluronidase solution (Fig. 2). The fat particles are then placed into a 2-ml syringe with a locking mechanism. For larger amounts, we use a gun injector (Fig. 3). The advantage of using a finely calibrated syringe is its built-in locking mechanism, which ensures accurate administration of a defined dose (Fig. 4). Injection is performed by using a wide luminar cannula (Fig. 5). The cannula is inserted into the previously anesthetized recipient site, and the fat tissue is injected as the cannula is gently pulled back. It is important to avoid any intracutaneous injection, which would lead to an inflammatory reaction and possible skin loss. The single deposits should not be any larger than 1 mL in size to ensure adequate

Figure 2 Prepared fatty tissue after washing procedure.

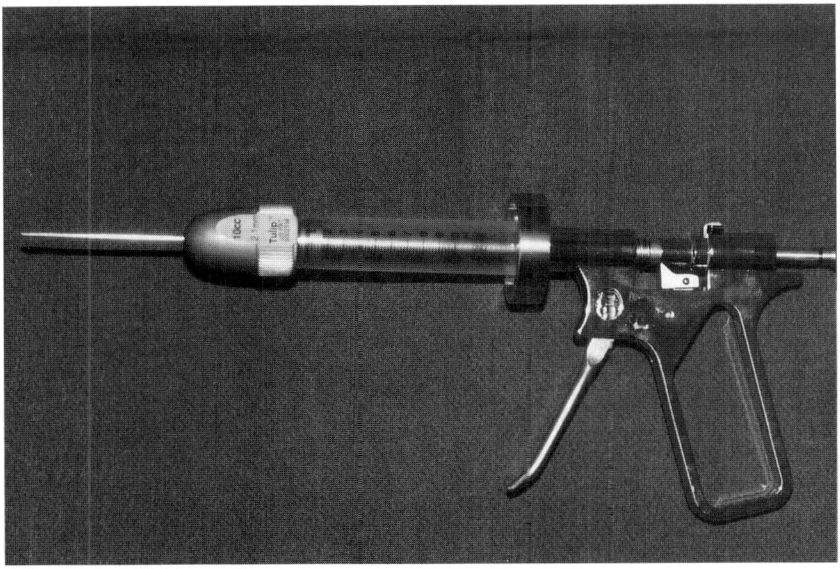

Figure 3 Gun injector for larger amounts of fatty tissue.

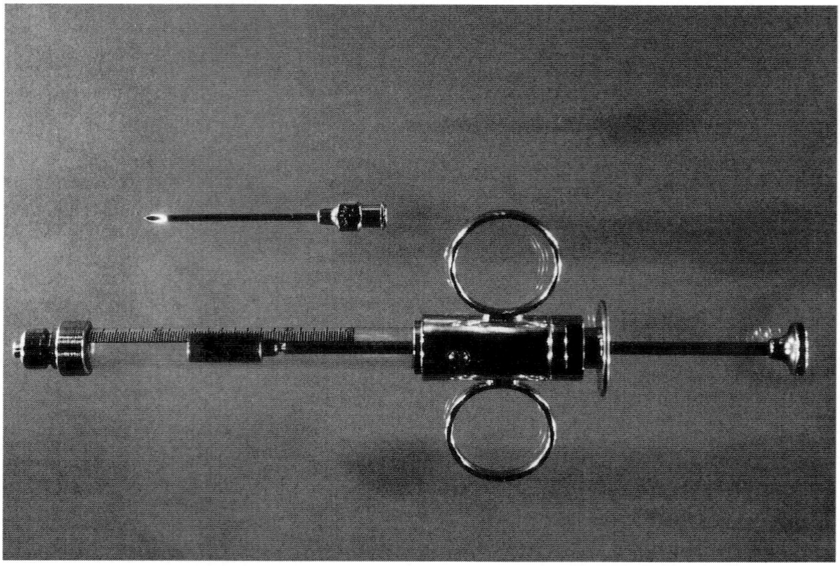

Figure 4 Injection syringe and cannula for transplantation. (Courtesy of Lippincott.)

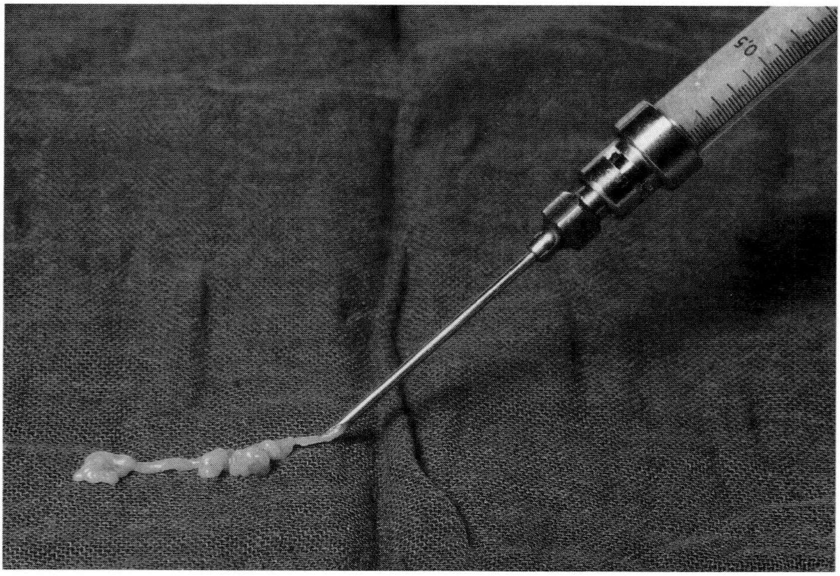

Figure 5 Prepared fatty tissue ready for transplantation.

diffusion within the first few days and subsequent neovascularization without necrosis.[1] Larger defects must have multiple small deposits injected at several visits to achieve augmentation over an extended period of time. This is especially important in view of the fact that there is some resorption, and the correction of the necessary fat volume cannot be achieved in one sitting.

II. POSTOPERATIVE VOLUME CONTROL

A resorption rate of approximately 50% can be expected within the first 6 months. Resorption rates between 30 and 60% have been reported in the literature.[2–5] To evaluate objectively this well-recognized volume loss, postoperative volume evaluation was performed for an area of the face in 10 patients by using magnetic resonance imaging (MRI) (Gyroscan, S15, Philips Co., Eindhoven, the Netherlands) and the results analyzed by computer. These patients achieved fat deposits of 1.2–3.2 ml. MRI proved to be a most suitable method for volume evaluation; providing good imaging of the soft tissues without the need for isotopes or contrast media. The methodology used for volume evaluation is derived from that used for heart volume determinations.[6,7] The volumes evaluated in cardiac investigations are much larger than those undertaken to measure in the fat tissues of the face.

A (T1) spin-lattice relaxation time was used to show the injected fat, which appears as light areas on the images. Connective tissue as well as scars, however, appear dark with lower intensity. Therefore, fat can be differentiated from fibrous replacement tissue. In contrast to larger fat sections, however, the vascular pattern cannot be used definitively to determine the viability of small fat deposits. It is also impossible to differentiate between revascularized fat and oil cysts. The survival of the transplanted autologous fat was monitored and measured for volume changes at various time intervals after surgery using MRI. Precise determination cannot be done if fat cells are injected into existing fat tissue owing to a lack of discrimination that projects a signal of equally high intensity. MRI investigations were undertaken with a magnetic induction of 1.5 T. Parallel, serial slices with a thickness of 5 mm and a gap of 10% (0.5 mm) were used. Preoperatively, the intended areas of augmentation were subjected to a planning scan (Fig. 6). MRI studies were performed at day 6, 3, 6, and 9 months, and 1 year after fat injection (Fig. 7). No one exactly corresponding slice could be detected from the six individual investigations. Volume evaluation, however, was not based on the individual depicted slices but instead on the sum of the slice area (a1), the slice thickness (t), and the gap (g). It should also be mentioned that the colleague performing the calculations was unaware

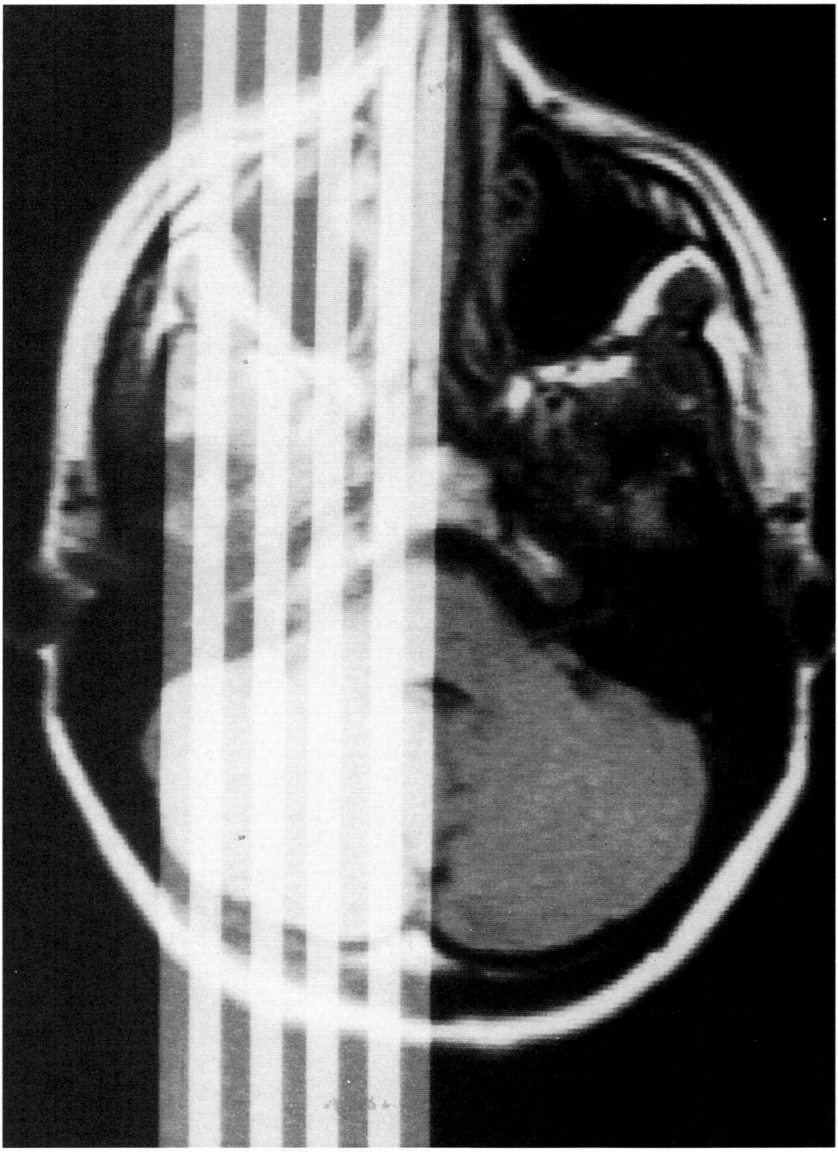

Figure 6 MRI scan of the head region. (Courtesy of Lippincott.)

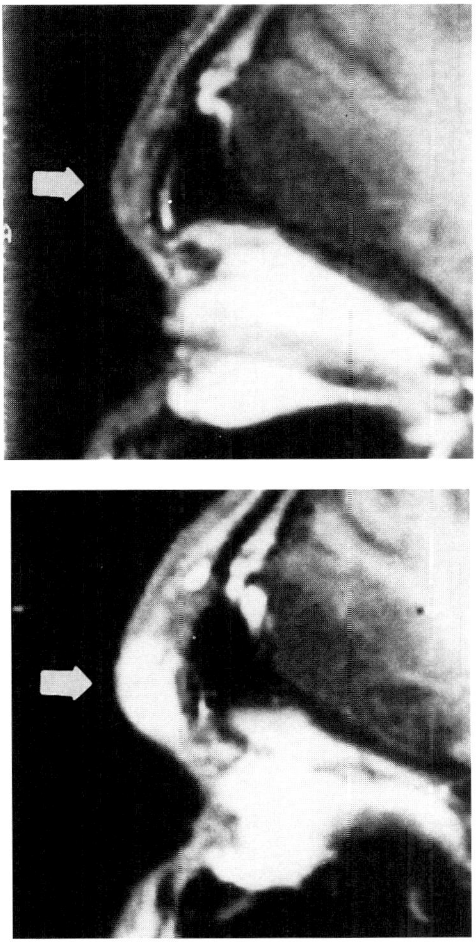

a

b

Figure 7 Follow-up magnetic resonance imaging. The arrow indicates the augmentated area. (**a**) The preoperative view. (**b**) Six days after transplantation.

of the original fat tissue volumes injected. Comparisons made between volume evaluations and the original amount of fat injected yielded an average discrepancy of less than 5%.

Computer-assisted calculations of the volumes were performed integrally by using the following formula: $v = (t + g) \cdot \Sigma(a_i)$ (Fig. 8). The results demon-

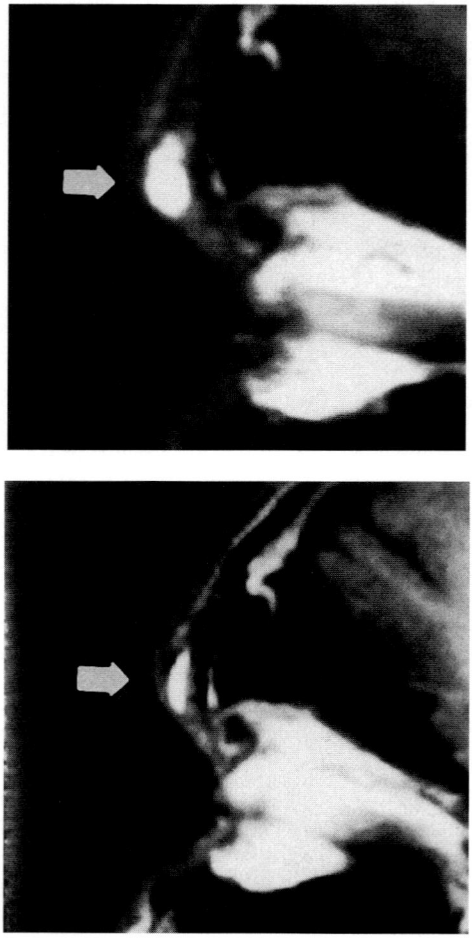

c

d

Figure 7 (**c**) At 3 months. (**d**) At 6 months.

strated volume losses up to 49.3% (range 37–61%; SD ±8.1) within the 3 months. During the interval between 3 and 6 months, the average percentage of loss increased another 6% to approximately 55.3% (range 45–68%, ±6.4%). No further decreases in volume were noted from the 7th (55.4 ±6.3%; range 46–69%) through the 12th (55.6 ± 6.3%; range 47–68%) months after injection (Fig. 9). The high–average 49% volume loss occurring between the first and third

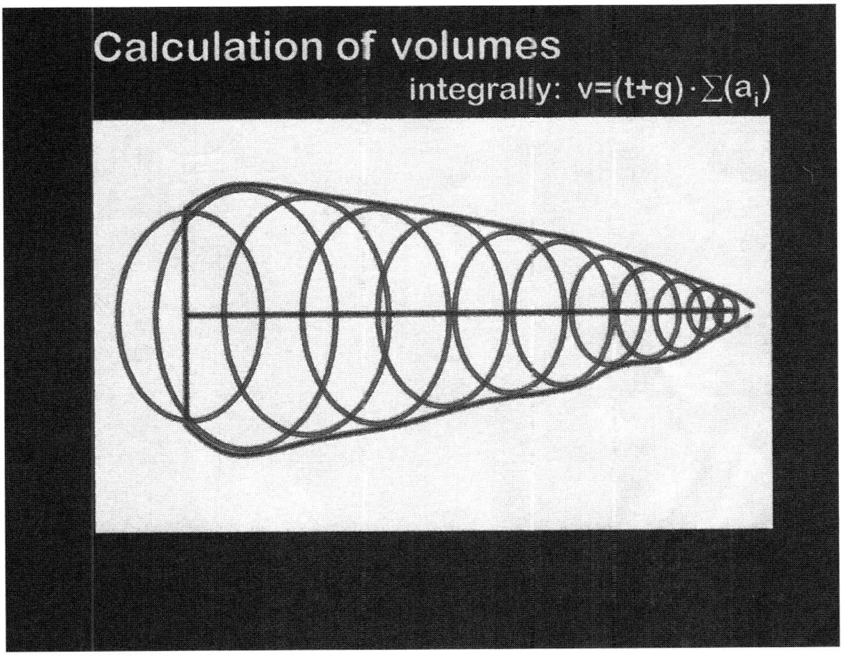

Figure 8 Computer assisted calculations of volumes using the formula $v = (t + g) \cdot \Sigma(a_i)$.

months was attributed to the breakdown of nonusable material within the first 6 months. No further volume loss could be detected during the final 6 months of the study period.

III. DISCUSSION

The authors use the tumescent technique involving the injection of hyaluronidase directly into the donor site. Our own histological investigations have yielded results indicating that through the use of hyaluronidase, the number of nontraumatized cells increased from 40 to 52%.

The percentage of intact cells is less than optimal with both techniques, and is surely responsible for the high reabsorption rate. This, however, is in sharp contrast to the results of Illouz,[4] who reported a figure of more than 70% "viable" cells by using the wet technique and more than 90% viable cells with

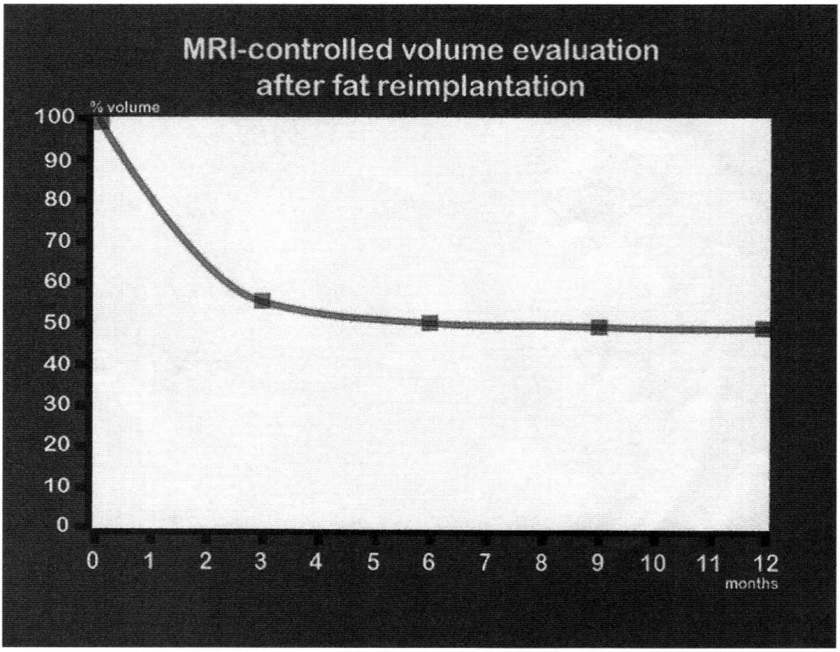

Figure 9 Volume of fat decreases until sixth month following which there is no further decrease.

the dry technique. Asken[8] alleges that by using a small-caliber cannula and lower pressures fewer damaged fat cells can be harvested. In our experience, the diameter of the blunt suction cannula ranging from 4 to 5 mm as well as various suction pressures between 0.8 and 1.0 atm cause less trauma than smaller cannulae and lower suction pressure.

The success of lipofilling is dependent on the method of procurement and also the preparation of the aspirated fat tissue. Some surgeons reinject the fat without any preparation, whereas others centrifuge the aspirate to separate healthy cells from dead ones and to eliminate oil and any blood constituents. We carefully wash the suctioned fat in Ringer's solution. The fat is then injected without any added medication. The antilipolytic effect of insulin[9–11] has not yet been proven experimentally or in clinical cases. The same can be said regarding the systemic administration of vitamin E.

An effort should always be made at overcorrection for defects that can be corrected in one session, because of the high rate of resorption. Larger defects require multiple treatments. In contrast to the pearl technique[3] where the fat is placed through a skin incision, the injection technique provides the needle with concealed subcutaneous access to the soft tissue defect and placement by withdrawing the needle. The liposuctioned fat achieves better contouring with this approach than with the pearl technique. Contouring is also dependent on the transplant site. A firm base such as the forehead achieves better results than a loose area like the cheeks.

In the literature, resorption of fat ranges between 20 and 80% and are based on clinical impressions. Until now, there has been a lack of objective volume evaluations documented in the literature. Even when reabsorption of the fat tissue appears to be total by clinical evaluation, MRI investigation can still detect approximately 40% of the transplanted tissue in most patients. MRI also optimized visualization of the fat tissue by using a T1 relaxation interval and allows the surgeon for research reasons to determine the extent of fat reabsorption over a certain period of time. It is also helpful to be able to differentiate between connective tissue and scarring, which projects a signal of high intensity, compared with fat tissue, which appears lighter. The connective tissue as well as scarring sends off a weaker signal that produces a darker image. MRI can demonstrate cellular viability by means of vascular depiction in patients with larger tissue volumes. This is, however, usually not the case with smaller tissue deposits. It is also very difficult to differentiate between healthy fat tissue and oil cysts, which project a similar signal. Nevertheless, the extent of reabsorption can be conclusively determined.

Among the listed advantages of fat tissue transplantation is the essential benefit of using autologous tissue. This is especially true when comparing treatments involving collagen injection. Reabsorption rates with collagen are considerably higher, and administration must be frequently repeated to maintain the augmenting effect. The injection of bovine collagen in some patients results in allergic reactions and treatment must, therefore, be discontinued.[11]

The technique of autologous fat transplantation must be critically judged. Despite the numerous advantages of augmentation with autologous tissue, the basic disadvantages of this method lie in the cumbersome nature of the procedure itself as well as the proved high rate of reabsorption within the first 6 months after treatment. A prerequisite for improving the method would be the transplantation of only intact viable fat cells. This would require a change in the technique of aspiration as well as the possibility of separating defective cells from healthy intact cells before reinjection.

REFERENCES

1. Hoerl HW, Feller A-M, Steinau HU, Biemer E. Autologe Fettgewebeinjektion nach Liposuction—keine Methode zur Brustaugmentation. Handchirurgie 2:59, 1989
2. Chaychir A, Benzaquen J. Liposuction fat grafts in face wrinkles and hemi-facial atrophy. Aesthetic Plast Surg 10:115, 1986
3. Ellenbogen R. Free autogenous pearl fat grafts in the face—a preliminary report of a rediscovered technique. Ann Plast Surg 16:179, 1986
4. Illouz Y-G. Present results of fat injection. Aesthetic Plast Surg 12:175, 1988
5. Peer LA. Loss of weight an volume in human fat grafts. Plast Reconstr Surg 5:217, 1950
6. Bauer R, van de Flierdt E, Busch U, et al. Quantifizierung der Herzfunktion mit der Kernspintomographie. Nuklearmedizin 3:203, 1988
7. Bauer R, Bauer U. Nuclear magnetic resonance (NMR) imaging. In Kriegel, ed. Handbook of Nuclear Medicine. Vol 1/1. Stuttgart: Fischer, 1985:229–255
8. Asken SA. Liposuction Surgery and Autologous Fat Transplantation. Norwalk, CT: Appleton & Lange, 1988:24–56
9. Sidmann RL. The direct effect of insulin in organ cultures of brown fat. Anat Rec 124:723, 1956
10. Smith U, Böstrom S, Johansson R, Nyberg G. Human adipose tissue in culture. V. Studies on the metabolic effects of insulin. Diabetologia 12:137, 1976
11. Solomon SS, Turpin BP, Duckworth WC. Comparative studies of the antilipolytic effect of insulin and adenosine in the perifused fat cell. Horm Metab Res 12:601, 1980

5

History of Autologous Fat Transplant Survival

Melvin A. Shiffman
Private Practice, Cosmetic and Reconstructive Surgery, Tustin, California

The survival of free fat used as an autograft is operator dependent and requires delicate handling of the graft tissue, careful washing of the fat to minimize extraneous blood cells, and installation into a site with adequate vascularity.

There is evidence that fat cells will survive and that filling of defects is not from the residual collagen following cell destruction. There is some loss of fat after transplant, and most surgeons tend to overfill the recipient site.

I. HISTORICAL REVIEW

Verderame (1909)[1] reported that autogenous fat grafts in ocular surgery became reduced in size and advised the use of a larger transplant than seemed necessary to fill the defect. Lexer (1911)[2] claimed that manipulation and tearing of the graft at the time of transfer would cause a greater degree of graft shrinkage. Kanavel (1916)[3] felt that graft survival was improved by not using sutures to secure the graft, careful hemostasis, and aseptic technique. He transplanted sheets of fat varying from 0.25 to 1.0 in in thickness to prevent adhesions and contractures and lessen deformity about tendons, nerves, blood vessels, and joints. He felt that fat can be transplanted into any ordinary field with assurance that it will not act as a foreign body. Clinically it appears to live, become a part of the structure in which it is placed, and persists for many months and probably years. Davis (1917)[4] concluded that omentum, transplanted freely beneath the skin in a mass 1 in in diameter maintains the greater part of its bulk. Lexer (1919)[5] reported excellent clinical results with very large fat grafts, but stated that up to 66% of the fat autografts were absorbed and significant overcorrection should be used. He also stated that multiple small grafts would turn to scar, whereas larger grafts would remain fatty tissue. Mann (1921)[6] performed free transplant of omentum fat and stated that it remained seemingly viable for as long as 1 year and retained small percentage of its fat.

Neuhof (1923)[7] examined available experimental and clinical evidence and concluded:

1. Transplanted autologous fat undergoes practically the somesame changes as transplanted bone.
2. The transplant dies and is replaced either by fibrous tissue or by newly formed fat.
3. Newly formed fat occurs through the activity of a large wandering histocyte-like cell which takes on fat and becomes a fat cell.

Guerney (1938)[8] noted that autogenous fat grafts should be transplanted in larger bulk than required, since 25–50% of the graft survives 1 year after transplantation. He studied transplanted 1.7-mm^3 (average size) fat grafts over a period of 12 months in rats and concluded:

1. Liberation of fat by contiguous cell probably gives rise to fatty cysts.
2. Phagocytosis of liberated fat was assisted by polymorphonuclear leukocytes.
3. The percentage of normal fat in any surviving graft gradually increased throughout the year.
4. A certain portion of the transplanted tissue gained an adequate blood supply early and continued to survive, whereas the remainder of the graft degenerated and was gradually eliminated from the site of the implant without evidence of gross scar.
5. Crushed grafts eventually disappeared, attesting to the devastating effect of trauma on the vitality of a graft.
6. Single pieces of fat remain viable for at least 1 year, whereas grafts of a similar size cut into smaller pieces may last as long as 6 months, but the majority disappear by the third month.
7. Absolute hemostasis is essential, since even a slight hemorrhage jeopardizes the viability of the graft.
8. Although slight infection results in only a small loss of tissue, gross infection leads to a loss of the whole graft.
9. Phagocytes do not use their fat to form new fat cells during the first year after transplantation.

Hilse (1928)[9] showed histologically that free fat transplants regenerate fatty tissue without any exception. He referred to the histocyte filled with fat as a "lipoblast."

Green (1947)[10] used fat and fat-fascia autografts in the treatment of osseous defects secondary to osteomyelitis. He presumed that transplanted fat would become connective tissue and then bone, closing the defect. Wertheimer and Shapiro (1948)[11] studied fat physiology and determined that fat develops from primitive adipose cells, the structure of which is like that of the fibroblasts of connective tissue.

Peer (1950)[12] implanted autogenous fat (a single piece compared to a piece cut into 20 segments) into the rectus muscle. Grafts were removed at intervals from 3 to 14 months. Grossly all grafts were surrounded by a connective tissue capsule, and upon sectioning, the bulk of the graft contained fatty tissue. Single grafts (the size of a walnut) lost 45% of their weight, whereas multigrafts lost 79% of their weight. He concluded that the fat grafts appeared like normal fat tissue 1 year or more after transplantation.

Bames (1953)[13] noted that circulation in grafts is established in about 4 days after transplantation by anastomosis between host and graft blood vessels. Traumatized fat grafts lose much more weight and volume than gently handled transplants (50% loss after 1 year). Normal-appearing adipose cells were present in all the transplants. Dermal fat grafts provide a readily available transplantation material for establishing normal contour in small breasts instead of foreign implants.

Hansberger (1955)[14] proposed that histocytes phagocytose the lipid and do not replace graft fat. After the graft of mature autotransplanted fat goes through initial ischemia, fat cells either necrose or dedifferentiate into immature cells. Under suitable conditions, the immature fat cell revert to mature adipocytes.

Schorcher (1957)[15] reported using autogenous free fat transplantation to treat hypomastia. He noted that the connective elements remained intact with fat shrinkage to 25% of the original size by 6–9 months. He believed that if the graft was in several pieces, it would receive better nourishment from the recipient site.

In rats, Van and Roncari (1978,[16] 1982)[17] demonstrated conversion of adipocyte precursors into adult adipocytes both in vitro and in vivo. Saunders et al. (1981)[18] studied fat autograft survival and observed initial adipose tissue breakdown followed by revascularization. There is early breakdown of fat cells with formation of cyst-like lipid deposits and infiltration by host histocytes.

Illouz (1985)[19] opined that the human body is an excellent culture medium and the fat cells apparently survive by intercellular lipolysis and osmosis until they are revascularized. The area to be augmented should be overcorrected by

30%, because approximately 30% necrosis of fat cells results when using the wet technique.

Illouz (1986)[20] reported that fat transplantation in one patient biopsied 9 months and 16 months later showed normal fat cells.

Asken (1987)[21] found that 90% of fat extracted by liposuction appears viable, assuming it is not traumatized either by handling or by high suction pressure. Damage incurred by the adipocytes is inversely related to the diameter of the instrument used for harvesting and injection.

Campbell et al. (1987)[22] noted, both morphologically and biochemically, that adipocyte integrity and metabolism remain intact when subjected to liposuction. Johnson (1987)[23] examined liposuctioned fat and noted that 90% or more of the fat cells remained viable. He found that there was 75–85% of the original fat present after transplantation 3 months previously. Agris (1987)[24] claimed that trauma and desiccation injured transplanted fat cells. Bircoll (1988)[25] stated that the American Society of Plastic and Reconstructive Surgery (ASPRS) report[26] of 30% survival and Peer's report[27] of 50% survival of autologous fat transplantation are based on the older technique of bulk fat transfer. Biopsies show 80% survival of fat after 1 year and additional bulk by 10–20% of fibrous tissue. Fat transplants must be placed into the fatty subcutaneous tissue.

Billings and May (1989)[28] analyzed the histology of free fat grafts and noted the following:

Time (days)	Histology
First 4 days	Cellular infiltrate: polymorphonuclear cells, plasma cells, lymphocytes, eosinophils With vessels of graft: red blood cells were clumped together, white blood cells were in process of diapedesis (passage of blood cells through intact vessel walls) No degeneration of graft endothelial cells and fibroblasts of the stroma
Fourth day	Engorgement and dilatation of smaller stromal vessels with abundant and red blood cells and diapedetic white blood cells (anastomoses between smaller graft vessels and host red blood supply) Increased number of eosinophils in cellular infiltrate Foreign body–type giant cells often seen
10 days	Areas of necrotic adipose tissue Regenerative proliferation of original fat cells mostly at periphery of lobules—includes proliferating adipose

Time (days)	Histology
	cells of the graft and host round "histocyte-like" cells that took up lipid and enlarged
14–21 days	Further adipose cell breakdown
	Increasing number of large host histocytes which appear to be picking up lipid with formation of droplets within their cytoplasm
30–60 days	Increasing numbers of large histocytes which peaks at 2 months
	Coalescing of fat globules in the cytoplasms

Markman (1989)[29] has suggested that the number of fat cells may increase, through differentiation of existing preadipocytes, when fat cell reach a "critical size."

Illouz (1990)[30] reported that fibroblast-like precursor cells are able to multiply and give rise to fibroblasts or cells that resemble fibroblasts. When these cells are stimulated to absorb fat vacuoles with insulin or dexamethasone, they do not because adipocytes. He noted that adipocytes are very fragile and have a short life span outside the body. The cells live longer if mixed with normal saline and kept at a moderate temperature. They do not tolerate excessive manipulation, refrigeration, or major trauma such as grinding.

Hudson et al. (1990)[31] demonstrated a greater cell size and lipogenic activity (using measurement of the activity of lipogenic enzyme adipose tissue lipoprotein lipase [ATLPL]) in the gluteal-femoral area compared to the abdomen. Facial fat was found to have small cells with almost no ATLPL activity. This may have implications for donor site suitability.

Nguyen et al. (1990)[32] compared suctioned fat, aspirated fat, and excised fat 9 months after implantation. Suctioned fat was obtained by using 1 atm negative pressure, and on microscopy found only 10% of the fat cells with intact cell membrane. In all the grafts, fat was replaced with fibrosis and only a small number of surviving adipocytes were still present.

Kononas et al. (1990)[33] compared the loss of fat following transplant between surgically excised fat cut into small pieces and suctioned fat which was centrifuged. Weight loss was 59% for excised fat and 67% for suctioned fat. Ersek (1991)[34] used a wire wisk to agitate harvested fat and then strained it. He reported disappointing results even with repeated injection and concluded that little, if any, autologous fat survives in its new site.

Courtiss et al. (1992)[35] reported marginal success in fat grafting of two patients with postliposuction depressions. Asaadi (1993)[36] reported 5-year successful retention of fat transplanted to a right trochanteric posttraumatic depressed scar.

Samdal et al. (1992)[37] measured blood flow and the amount of surviving fat following needle abrasion of the recipient site in rats. Abrasion was performed by a criss-cross pattern with 20 strokes using an 18-gauge needle in the subcutaneous tissue prior to transplant and compared this to controls without abrasion. They found that the mean weight of the fat transplant had shrunk to 44.6% of the original weight in the abraded group and 33.5% in the control group. The mean blood flow in fat was 0.165 mL/min/g in normal fat, 0.120 mL/min/g in the controls, and 0.187 mL/min/g in the abraded group. Microscopic examination of the transplanted fat varied from oil cysts, connective tissue, and inflammatory cells in some specimens and completely normal fatty tissue in others. Fat survival varied for 0 to 90%. They concluded that fat transplant survival was unpredictable.

Eppley et al. (1992)[38] reported that the addition of basic fibroblast growth factor delivered by dextran beads to fat grafts resulted in a larger weight maintenance of fat at 1 year than controls.

Group A Fat alone
Group B Fat with dextran beads
Group C Fat with dextran beads soaked with cytochrome *c* (nonmitogenic control solution)
Group D Fat with dextran beads soaked with basic fibroblast growth factor

Group	Weight retention after 12 months (%)
A	48.8
B	79.6
C	75.2
D	93.8

Histologically, Eppley et al. noted extensive interlacing collagen formation between the adipocytes which provides support for the known effects of basic fibroblast growth factor on mesenchymal cell lines. There was an increased uniformity in adipocyte size seen in 1-year grafts compared to 1-month grafts, which may indicate a possible maturation of these more "immature" cells. Whether this represents repair of damaged adipocytes, preadipocyte differentiation, conversion of infiltrating macrophages or fibroblasts, or entrapped lipid material is speculative.

Carpaneda and Ribeiro (1993)[39] examined fat 2 months after transplantation and noted viable tissue only in the peripheral zone of 3.5 mm diameter cylindrical grafts. There was 60% loss of grafted tissue which occurred closer to the center. In 1994, they reported that graft viability depends on the thickness and geometrical shape and is inversely proportional to the graft diameter if the diameter is greater than 3 mm.[40] The maximum percentage of viability is 40% when the graft is no greater than 3 mm thick.

Niechajev and Sevchuk (1994)[41] reported 50% fat survival over 3.5 years after single fat transplantation with 50% overcorrection. They found that fat obtained under maximum negative pressure (–0.95 atm) results in partial breakage and vaporization of the fatty tissue. About two thirds of the fat withstood the trauma of aspiration. Low pressure (–0.5 atm) resulted in smaller cell size (29% smaller than with aspiration at –0.95 atm), and they assumed that high pressure causes mechanical distention of the adipocytes, which increases the risk of and sometimes causes cell breakage.

Courtiss (1994)[42] stated that fat grafting remains controversial and poorly understood and that, "some surgeons have some impressive results, but most of us have many disappointing results." Fagrell et al. (1996)[43] examined fat 6 months after implantation in the ears of rabbits. The fat implanted was obtained by:

1. Fat cylinder retrieved with 4.5-mm internal diameter syringe pushed into the fat and pulling the piston back
2. Excised fat, 1 mg in weight
3. Aspirated fat using 2-mm (14 gauge) cannula and syringe

The tissue was examined by light microscopy and computer-assisted image analysis. There was no difference between the weight of the 6-month excised specimen (no weight loss) between the fat cylinder and the excised fat, but there was a 59% loss of weight of the aspirated fat. The conclusion was that fat aspiration is traumatic and breaks up the cells. However, there was histological evidence of viable fat cells in all transplants.

Jones and Lyles (1994)[44] harvested fat with a 60-mL syringe and a 3.0-mm pyramid cannula and locked the plunger at 35 mL. The harvested fat was washed three times with normal saline and gently agitated. Cell cultures were prepared and maintained for 1 day to 2 months. Microscopy disclosed maintenance of mature adipose cells without dedifferentiation into a precursor phenotype. There was very little evidence of cellular damage or debris.

Using photographs over a 6-year period of time, Coleman (1995)[45] demonstrated long-term survival of liposuctioned fat transplanted into the nasolabial fold. He stated that fat can migrate as the pressure of excess tissue forces the transplanted fat to shift and that fat can die from inadequate nutrition and oxygen from competition with other transplanted parcels of fatty tissue.

Placement of fat into multiple tunnels allows closer location to nutrition. He concluded that fat survival is technique dependent and the primary reason for failure of long-term correction of the nasolabial fold is initial inadequate correction.

Sattler and Sommer (1997)[46] found that autologous fat dried over sterile swabs and frozen at –20°C (lower temperatures down to –70°C are preferable) up to 2 years and then thawed at room temperature contain only fat cells and no fibrous debris.

Ullmann et al. (1997)[47] added Cariel, a modified serum-free cell culture medium (MCDB 153), to aspirated human fat prior to reinjection into mice. Cariel contains essential and nonessential amino acids, vitamins, inorganic salts, trace elements, buffers, thyroxine, growth hormone, insulin, and sodium selenite. There was 46% of the weight of the fat remaining after 15 weeks in the group with Cariel compared to 29% in the control group without Cariel. They concluded that the addition of nutrients enriched with anabolic hormones enabled the survival and take of more adipose cell in the graft. US Patent (Lindenbaum) Composition and methods for enhancing wound healing. Patent No. 5461030. Date of patent: October 24, 1995.

REFERENCES

1. Verderame P. Ueber fettransplantation bei adharenten knochennarben am orbitalran. Klin Monatsbl f Augenheilkd 1909;7:433
2. Lexer E. Ueber freie fettransplantation. Klin Ther Wochenschr 1911;18:53
3. Kanavel AR. The transplantation of free flaps of fat. Surg Gynecol Obstet 1916;23:163–176
4. Davis CB. Free transplantation of the omentum, subcutaneously and within the abdomen. JAMA 1917;68:705–706
5. Lexer E. Fatty tissue transplantation. In Die Transplantation. Part I. Stuttgart: Ferdinand Enke, 1919:265–302
6. Mann FC. The transplantation of fat in the peritoneal cavity. Surg Clin N Am 1921;1:1465–1471
7. Neuhof H. The Transplantation of Tissues. New York: Appleton, 1923:74
8. Guerney CE. Experimental study of the behavior of free fat transplants. Surgery 1938;3:679–692
9. Hilse A. Histologische ergebuisse der experimentellen freien fettgewebstronsplantation. Beitr 2 Path Anal U Z Allg Path 1928;79:592–624
10. Green JR. Repairing bone defects in cranium and tibia. South Med J 1947;40:289
11. Wertheimer E, Shapiro B. The physiology of adipose tissue. Physiol Rev 1948; 28:451

12. Peer LA. Loss of weight and volume in human fat grafts: with postulation of a "cell survival theory." Plast Reconstr Surg 1950;5:217–230
13. Bames HO. Augmentation mammoplasty by lipotransplant. Plast Reconstr Surg 1953;11:404
14. Hansberger FX. Quantitative studies on the development of autotransplants of immature adipose tissue of rats. Anat Rec 1995;122:507
15. Schorcher F. Fettgewebsver pflanzung bei zu kneiner. Brust Munchen Med Wochenschr 1957;99(14):489
16. Van RLR, Roncari DAK. Complete differentiation of adipocyte precursors: a culture system for studying the cellular nature of adipose tissue. Cell Tissue Res 1978;195:317–322
17. Van RLR, Roncari DAK. Complete differentiation in vivo of implanted cultured adipocyte precursors from adult rats. Cell Tissue Res 1982.225:557–561
18. Saunders MC, Keller JT, Dunsker SB, Mayfield FH. Survival of autologous fat grafts in humans and mice. Connect Tissue Res 1981;8:85
19. Illouz Y-G. New applications of liposuction. In: YG Illouz, ed. Liposuction: The Franco-American Experience. Beverly Hills, CA: Medical Aesthetics, 1985:365–414
20. Illouz Y-G. The fat cell "graft": a new technique to fill depressions. Plast Reconstr Surg 1986;78:122–123
21. Asken S. Autologous fat transplantation: Micro and Macro techniques. Am J Cosm Surg 1987;4:111–121
22. Campbell GLM, Laudenslager N, Newman J. The effect of mechanical stress on adipocyte morphology and metabolism. Am J Cosm Surg 1987;4:89–94
23. Johnson GW. Body contouring by macroinjection of autogenous fat. Am J Cosm Surg 1987;4(2):103–109
24. Agris J. Autologous fat transplantation: a 3-year study. Am J Cosm Surg 1987; 4(2):95–102
25. Bircoll M. Autologous fat transplantation: an evaluation of microcalcification and fat cell survivability following (AFT) cosmetic breast augmentation. Am J Cosm Surg 1988;5(4)283–288
26. ASPRS Ad-Hoc Committee on New Procedures: Report on Autologous Fat Transplantation. September 30, 1987
27. Peer LA. The neglected free fat graft. Plast Reconstr Surg 1956;18:233–250
28. Billings E Jr, May JW. Historical review and present status of free fat graft autotransplantation in plastic and reconstructive surgery. Plast Reconstr Surg 1989; 83(2):368–381
29. Markman B. Anatomy and physiology of adipose tissue. Clin Plast Surg 1989; 16:235
30. Illouz Y-G. Fat injection: a four year clinical trial. In: GP Hetter, ed. Lipoplasty: The Theory and Practice of Blunt Suction Lipectomy. 2nd ed., Boston: Little, Brown, 1990:239–246
31. Hudson DA, Lambert EV, Block CE. Site selection for fat autotransplantation: some observations. Aesthetic Plast Surg 1990;14:195–197

32. Nguyen A, Pasyk KA, Bouvier TN, et al. Comparative study of survival of autologous adipose tissue taken and transplanted by different techniques. Plast Reconstr Surg 1990;85:378–386
33. Kononas TC, Bucky LP, Hurley C, May JW Jr. The fate of suctioned and surgically removed fat after reimplantation for soft-tissue augmentation. A volume and histologic study in the rabbit. Plast Reconstr Surg 1993;91:763–768
34. Ersek RA. Transplantation of purified autologous fat: a 3-year follow-up is disappointing. Plast Reconstr Surg 1991;87:219–227
35. Courtiss EH, Choucair RJ, Donela, MB. Large-volume suction lipectomy: an analysis of 108 patients. Plast Reconstr Surg 1992;89:1068–1079
36. Asaadi M. Successful autologous fat injection at 5-year follow-up. Plast Reconstr Surg 1993;91:755–756
37. Samdal F, Skolleborg KC, Berthelsen N. The effect of preoperative needle abrasion of the recipient on survival of autologous free fat grafts in rats. Scand J Reconstr Hand Surg 1992;26:33–36
38. Eppley BL, Sidner RA, Plastis JM, Sadove AM. Bioactivation of free-fat transfers: A potential new approach to improving graft survival. Plast Reconstr Surg 1992; 90:1022–1030
39. Carpaneda CA, Ribeiro MT. Study of the histologic alterations and viability of the adipose graft in humans. Aesthetic Plast Surg, 1993;17:43–47
40. Carpaneda CA, Ribeir MT. Percentage of graft viability versus injected volume in adipose autotransplants. Aesthetic Plast Surg, 1994;18(1):17–19
41. Niechajev I, Sevchuk O. Long-term results of fat transplantation: clinical and histologic studies. Plast Reconstr Surg, 1994;94:496–506
42. Courtiss EH. Surgical correction of postliposuction contour irregularities. Plast Reconstr Surg 1994;94:137–138
43. Fagrell D, Enerstrom S, Berggren A, Kniola B. Fat cylinder transplantation: An experimental comparative study of three different kinds of fat transplants, Plast Reconstr Surg 1996;98(1):90–96
44. Jones JK, Lyles ME. The viability of human adipocytes after closed-syringe liposuction harvest. Am J Cosm Surg 1997;14:275–279
45. Coleman S. Long-term survival of fat transplants: Controlled demonstrations. Aesthetic Plast Surg 1995;19:421–425
46. Sattler G, Sommer B. Liporecycling: Immediate and delayed. Am J Cosm Surg 1997;14:311–316
47. Ullmann Y, Hyams M, Ramon Y, Peled IJ, Linderbaum ES. Enhancing the survival of aspirated human fat injected into mice. Plast Reconstr Surg 1998;101: 1940–1944

6

Complications of Fat Augmentation

Melvin A. Shiffman
Private Practice, Cosmetic and Reconstructive Surgery, Tustin, California

There are very few serious complications of autologous fat transplantation. Since it is the patient's own tissue, there is no rejection phenomenon and there should not be a scar capsule surrounding the fat or autoimmune reaction.

The harvesting of large amounts of fat using liposuction techniques is prone to any of the complications of liposuction surgery. If small amounts of fat (under 50 mL) are retrieved, then one may expect the possibility of bruising or infection in the donor site.

The reinjection of fat may be associated with[1–10]:

1. Loss of fat volume
2. Need for repeat injection
3. Bruising, hematoma
4. Swelling: especially with overinjection
5. Prolonged erythema
6. Tenderness, pain
7. Fibrous capsule
8. Infection: rare
9. Mass: usually fat cyst(s) from too much fat in one place
10. Microcalcifications: in fat transfer to the breast area
11. Loss of sight from retinal artery occlusion with injection into glabella area
12. Central nervous system damage: with injection into glabella area

Unilateral blindness has been reported in four cases and severe central nervous system damage in one case following fat injection to the glabellar area.[7–9]

Loss of fat volume is the most common problem, and prior to fat transplant the patient should be informed of this and the possible need for reinjection.

Autologous fat augmentation of the breast is controversial because of the risk of microcalcifications. Necrotic fat tends to form small calcifications, which are usually round but may be irregular in shape. Fat injection to the breast, as advocated by Fulton (1992),[11] is performed by inserting the fat into the retropectoral area, pectoral muscle, and retromammary area. The fat is not injected into the breast parenchyma and, therefore, should not be confused with calcifications possibly associated with breast malignancy. Bircoll[2,12] claimed that there is a 1.4% incidence of microcalcifications following fat augmentation of the breast. The calcifications are periparenchymal and are not multidensity, rod-like, punctate, or branching spicules.

REFERENCES

1. Bircoll M. Autologous fat transplantation to the breast. Plast Reconstr Surg 1988;82:361–362
2. Bircoll M. Autologous fat transplantation: An evaluation of microcalcification and fat cell survivability following (AFT) cosmetic breast augmentation. Am J Cosm Surg 1988;5(4):283–288
3. Skouge JW. Autologous fat transplantation in facial surgery. In: WP Coleman III, CW Hanke, TH Alt, S Asken, eds. Cosmetic Surgery of the Skin: Principles and Techniques. Philadelphia: Decker, 1991:239–249
4. Grazer FM. Special reconstruction and therapeutic procedures. In: FM Grazer, ed. Atlas of Suction Assisted Lipoectomy in Body Contouring. New York: Churchill Livingstone, 1992:319–329
5. Vizcaino JM, Montilla PB, Ruiz JB. Complications of autografting fat obtained by liposuction, Plast Reconstr Surg 1990;85(4):638–639
6. Chiu DTW, Edgerton BW. Repair and grafting of dermis, fat and fascia. In: JG McCarthy, ed. Plastic Surgery, Philadelphia: Saunders, 1989:508–526
7. Coiffman F. Lipoinjection complications. In: U Hinderer, ed. Plastic Surgery 1992. Vol II. Amsterdam: Excerpta Medica, 1992:759–760
8. Niechajev F, Sevcuk O. Long-term results of fat transplantation: clinical and histologic studies. Plast Reconstr Surg 1994;94:496–506
9. Teimourian B. Blindness following fat injections. Plast Reconstr Surg 1988;82:361
10. Illouz Y-G. Fat injection: hope or reality? In: Y-G Illouz, YT DeVillers, eds. Body Sculpting by Lipoplasty. Edinburgh: Churchill Livingstone, 1989:389
11. Fulton JE Jr. Breast contouring by autologous fat transfer. Am J Cosm Surg 1992;19:273–279
12. Bircoll M. Autologous fat transplantation to the breast (Letter). Plast Reconstr Surg 1988;82:361–362

7

Fat Transfer with Rice Grain–Size Fat Parcels

Giorgio Fischer
Fischer Institute, Rome, Italy

Augmentation and restoration of various areas of the body has been practiced for over a century and is an important part of the practice of cosmetic surgeons. Various autologous, alloplastic, allogeneic, and xenograft materials have been used with differing reports on success. The materials most commonly used have been autologous fat, dermal grafts, silicone, Goretex, Alloderm, bovine collagen, and bovine cartilage.[1]

In 1926, Willi[1] (Fig. 1) described the use of autologous fat delivered with a syringe (lipoinjection) for the correction of facial lines and the loss of tissue (Fig. 2). This technique employed by Willi represents the method that is most similar to our current approach. Reports using liposuctioned fat vary widely with respect to harvest technique, processing, refinement of tissue, and method of replacement. The data are inconsistent with respect to survival of the transplanted tissue with survival rates ranging from 30 to 80%.[2] The author has noted 80–90% graft survival in all cases using rice grain–size fat parcels.

I. PREOPERATIVE EVALUATION

The initial consultation should elicit the patient's chief complaint and reason for seeking surgical intervention. This is paramount in all anticipated esthetic procedures. A thorough medical and surgical history should be obtained. Particular attention and consideration should be given to prior esthetic procedures.

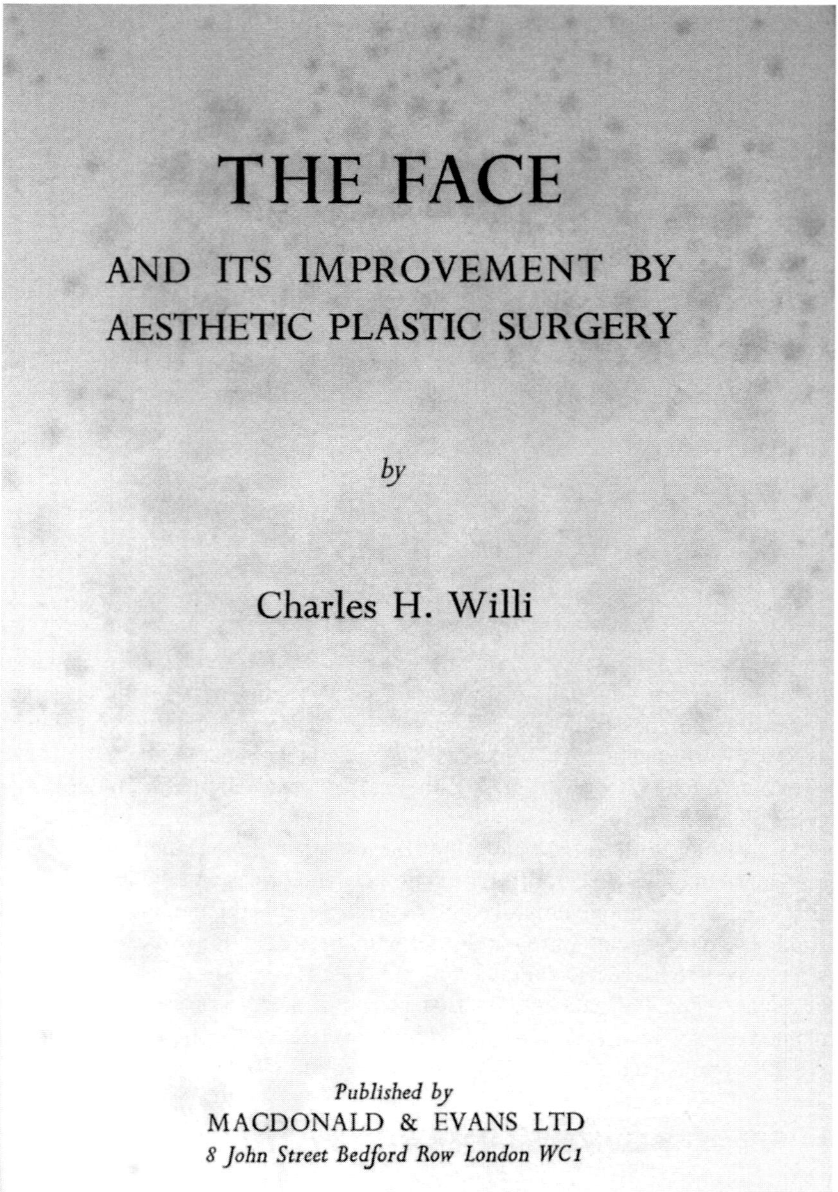

THE FACE

AND ITS IMPROVEMENT BY
AESTHETIC PLASTIC SURGERY

by

Charles H. Willi

Published by
MACDONALD & EVANS LTD
8 John Street Bedford Row London WC1

Figure 1 Publication by Dr. Charles Willi in 1926.

THE AUTHOR

Figure 1 *continued*

IMPLANTATION

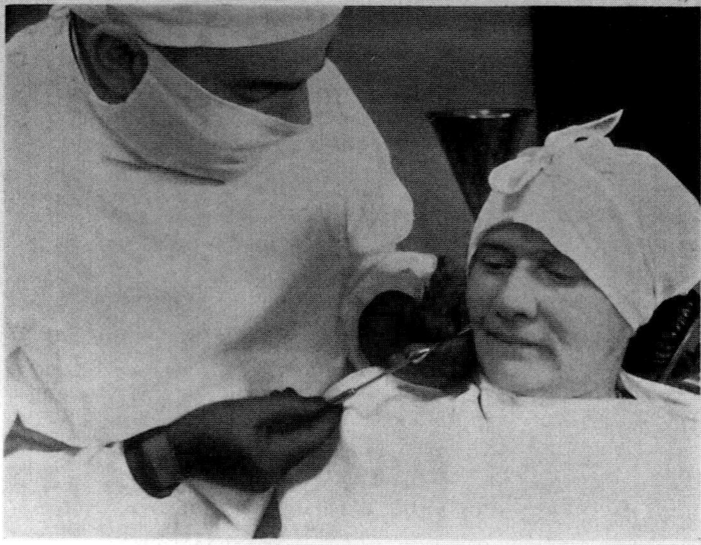

For the correction of facial lines, and the loss of tissue

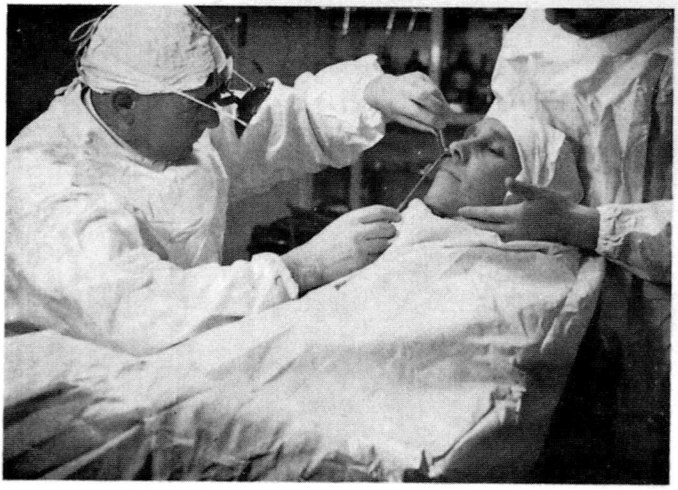

The technique involves the use of fat either taken from the patient's own body or prepared from Bovin cartilage, and transferred to the required place by a newly designed syringe as shown on this photograph.

Figure 2 Dr. Charles Willi describes the use of autologous fat transfer with syringe and needle to correct facial wrinkles and loss of tissue.

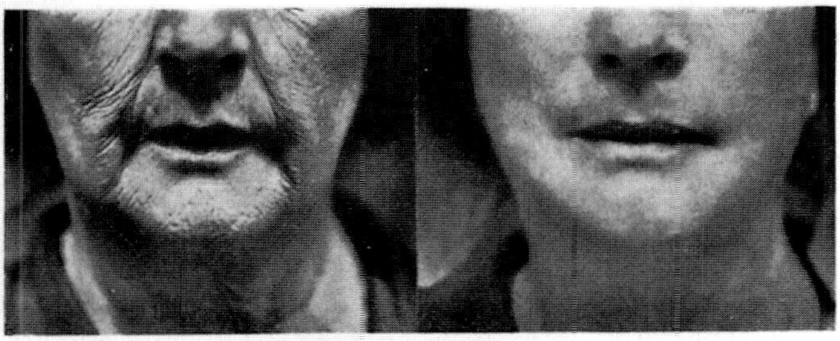

Restoration of contour and peeling process and fat implantation round the mouth.

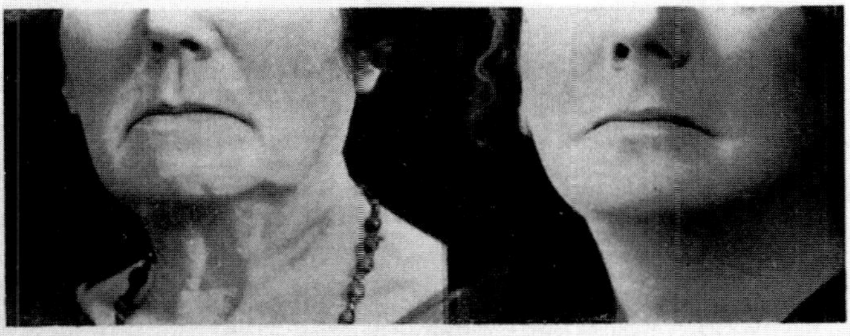

Restoration of chin, neck and mouth.

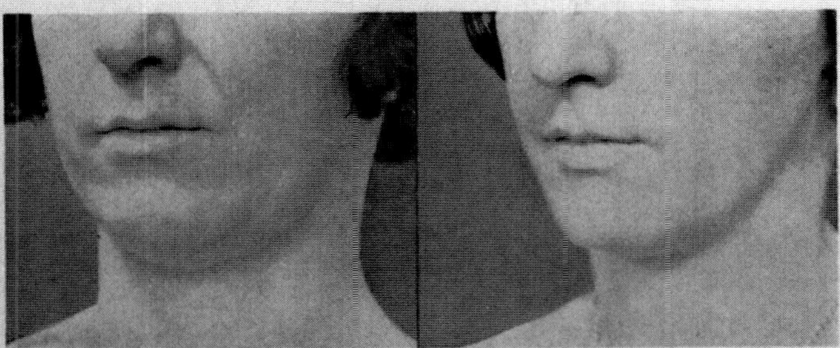

Though still young, this lady suffered from sagging skin and its bad effects: heavy contour, deep nose-to-mouth lines, and double chin. A face-lift and removal of the double chin restored the youthful contour.

Figure 2 *continued*

Preoperative photography is essential to study harmony or disharmony and anthropometrics and to allow the surgeon systematically to evaluate and plan the areas to be grafted. Autologous fat transfer can be used as an adjunct to other cosmetic procedures.

II. HARVESTING

All harvesting procedures are performed using intravenous sedation and modified tumescent anesthesia for harvest. The donor site is selected based on available fat and the anatomical area in need of contour improvement. Donor sites include but are not limited to the cervicofacial areas, lateral and medial thigh, abdomen, flank, and medial aspect of the knees, with the middle plane of fat being harvested from these anatomical sites. The donor site is prepared and draped in the usual sterile fashion for liposuction surgery. Local anesthesia with 25 ml lidocaine and 1:100,000 epinephrine is adequate if small amounts of fat are to be harvested. The modified Klein tumescent technique of anesthesia for liposuction surgery is employed if large quantities of fat will be harvested. Approximately 25% of the normal amount administered for large-volume liposculpture is utilized, which allows for a more concentrated aspirate upon harvesting. A curved single-holed cannula with a cutting end is introduced either with a sharp trocar or a stab incision made with a #11 blade. Alternatively, an 18-gauge needle can also be used for harvest of fat. The use of a blunt cannula for harvesting will yield primarily liquified fat. A sharp harvest cannula should be used which allows the harvesting of fat in a spaghetti strand–like configuration of rice grain–shaped tissue. The cannula is attached to a 1 ml Luer-Lock syringe, so that it is not necessary to transfer the fat from harvest syringe to the syringe used for transfer. This limits manipulation of the graft and decreases the chance of fat cell damage. The use of a small-volume syringe also aids in the transfer of more exact amounts of fat to each esthetic unit. Slow, constant, and gradual negative pressure is applied during the harvest procedure. The harvest cannula is directed in a spokewheel fashion to prevent depression in the skin overlying the donor site. Aspiration from the donor site should be slow and gradual to maintain the integrity of the rice grain–sized fatty parcels to be grafted. The fat remains in the syringe used for harvest and is set aside in an upright position.

III. PROCESSING AND PURIFICATION OF FAT

The harvested materials will contain lidocaine, lactated Ringer's, and intact viable fat cells. The fat can be processed in three ways. (1) Centrifuge the fat at 2000–3000 rpm for 2–3 min. Centrifuge allows the graft to be isolated and separated quickly, but it has the risk of damaging viable fat cells. (2) Wash the aspirate with normal saline and retain the adipose tissue. (3) Allow the aspirated material to settle into two distinct layers over a period of 5–6 min in the syringe used for harvesting (Fig. 3). The bottom and most dense layer contains lidocaine and lactated Ringer's solution, and the top layer comprises solely viable fat cells.

The third technique is the one I use, and it is used exclusively to avoid unnecessary manipulation of the graft. With the atraumatic technique of harvesting described, no oil layer is encountered, which indicates little or no trauma to the delicate fat cells upon harvesting. An excess oil layer is an indication that excessive pressure was used during harvesting which resulted in rupture of the fat cells. The bottom layer is decanted from the syringe and discarded, and the top layer, which consists of usable fat, is saved and set aside for grafting. The syringes can be used for reaspiration if they are not filled to capacity at the end of harvest. The fat remains in the syringe used for harvesting, and the cannulae used for transfer are attached.

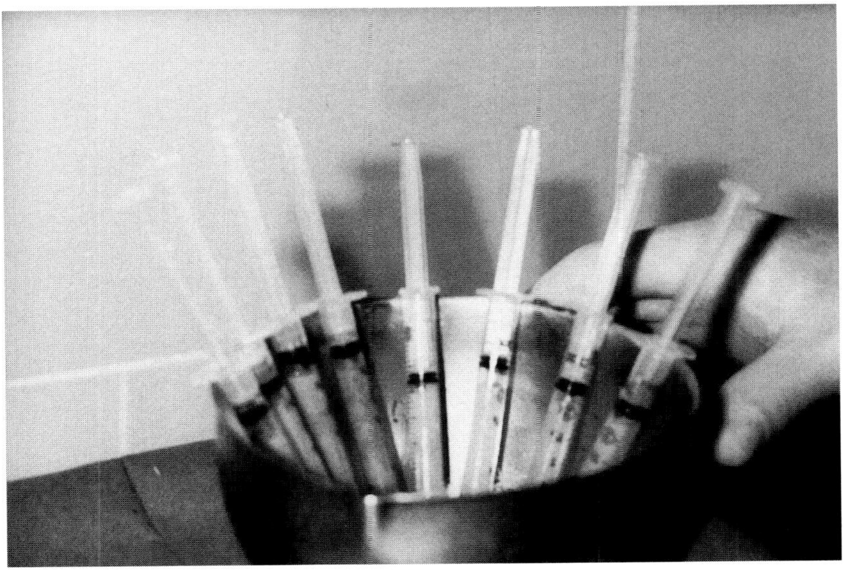

Figure 3 Fat is harvested with 1cc syringes and placed in an upright position to allow separation.

IV. TRANSFER AND CONTOURING

The recipient site is marked and the area to be augmented is prepared in sterile fashion. The entry site for the cannula is anesthetized with a minimal amount of 2% lidocaine and 1:100,000 epinephrine. Nerve blocks are useful in the face to prevent distortion of the soft tissues at the donor site. A small stab incision is made with a #11 blade. The transfer cannula (Fig. 4) is attached to the 1 ml Luer-Lock syringe used for harvesting. The fat is slowly deposited in a continuous fashion maintaining positive pressure in the areas to be augmented. Small rice grain–sized parcels of viable fat cells are transferred on insertion and withdrawal of the cannula. The cannula is directed in a spokewheel fashion over the recipient site. If resistance is met, the blunt cannula is withdrawn 0.5–1.0 cm to avoid damage to the local microvasculature and prevent hematoma formation. It is extremely important not to overfill, which decreases the chance of long-term graft viability and increases the chance of early graft death and loss. It is common practice to repeat the procedure if necessary rather than initially overfilling the area to further augment or refine the grafted areas. Rice grain–size parcels are utilized so that neovascularization to the graft can be more readily attained. Parcels of fat in this

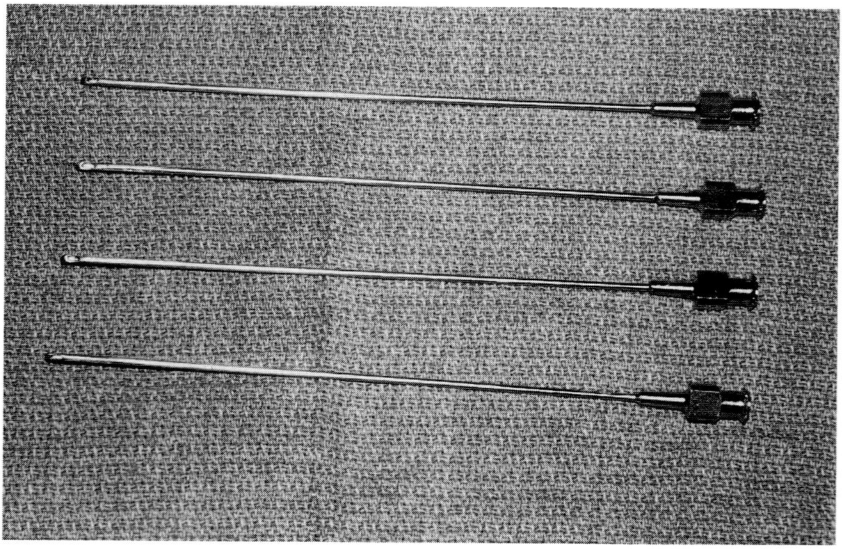

Figure 4 Cannulae used for injecting the recipient site.

size and configuration allow a greater surface area being available for the local microvasculature to reestablish the blood supply to the free graft more easily and in less time. The graft is placed in the subdermal tissue plane to give tension to the lax overlying tissue. Graft is also placed intramuscularly or adjacent to muscle to restore lost muscle mass associated with age-related muscular atrophy.

V. POSTOPERATIVE CARE

Sterile dressings are applied to the donor site and Reston (3M Company) is placed over the treated area. The patient is instructed to remove the Reston dressing in 3 days and return for postoperative evaluation in 7 days. Postoperative photographs are taken at 7 days and 3 months after the initial procedure. The patient is placed on oral antibiotics for 7 days postoperatively. After 3 months, the restored area is reevaluated and additional augmentation is performed shortly thereafter if needed.

VI. DISCUSSION

Autologous fat transfer is a versatile procedure that is minimally invasive and provides long-lasting results. Its applications are wide, and it can be used to treat age-related deformities by restoring the atrophied subcutaneous tissues, to enhance existing facial elements, and to treat posttraumatic deformities. The technique uses sound biological and anatomical principles to achieve optimal results. It avoids unnecessary manipulation of the graft during harvesting and transfer and processing is eliminated. Recontouring using autologous rice grain–size fat parcels involves precise layering of the graft to ensure a natural appearance and to ensure maximal graft viability.

REFERENCES

1. Willi CH. The Face and Its Improvement by Aesthetic Plastic Surgery. London: MacDonald & Evans, 1926:15–41
2. Coleman SR. Long term survival of fat transplants: Controlled demonstrations. Aesthetic Plast Surg 1995;19:421–425

8
Preoperative Consultation

Melvin A. Shiffman
Private Practice, Cosmetic and Reconstructive Surgery, Tustin, California

The patient being seen for autologous fat transplantation should have routine history taken. Specifically, the patient should be questioned about a history of easy bruising, bleeding tendencies, or bleeding problems following any prior surgery. Family history is evaluated as to relatives with malignant hyperthermia, coagulopathy, coronary artery disease, and diabetes mellitus.

Any patient who has had prior treatment with antiobesity medication should be carefully questioned about any symptoms such as edema, heart murmur, shortness of breath, and chest pain. If symptoms are present, then cardiac consultation with an echocardiogram should be obtained. If there are no symptoms, then the surgeon has a choice as to whether or not to have the patient seen by a cardiologist. The patient must be informed of the possibility of sudden hypotension if general anesthesia is going to be used, and neosynephrine must be on hand for treatment of hypotension. Antiobesity medications of concern are in Table 1.

With small amounts of fat removal (under 500 mL), there is little risk of problems with lidocaine in normal saline solution. However, when the tumescent technique with 500 mg lidocaine in each 1000 mL of normal saline is used for contouring liposuction, the patient who has been on cytochrome P450 inhibitors preoperatively may be subject to lidocaine toxicity.[1] To prevent toxicity, the amount of lidocaine should be drastically reduced or the patient kept off the offending drugs for 3–14 days, depending upon the half-life of the drug. Cytochrome P450 inhibitors are in Table 2, although not all these medications may affect the lidocaine level.

Physical examination should consist of at least the heart and lungs as well as the areas to have fat harvested and the areas to have fat injected. Photographs

Table 1 Patient Questionnaire on Antiobesity Medications

HAVE YOU EVER TAKEN ANTIOBESITY MEDICATIONS?	______
Indicate which medication:	
Aminorex fumarate	______
Clobenzorex	______
Dexfenfluramine (Redux)	______
Diethylpropion	______
Fenfluramine (Pondomin)	______
Fenproporex	______
Phenmetrazine	______
Phentermine (Tonamin, Adipex-P, Fastin)	______
Zantryl (Oby-Cap, Oby-Trim)	______
Patient Signature:__________________	
Date:________________	

must be taken of the defects to be filled or augmented, and postoperative photographs should be taken at a reasonable length of time (6–12 months) postoperatively.

The patient who is a smoker has increased risk for loss of fat transplant (vascular compromise), probably delayed healing, and possibly infection. Smoking must be stopped at least 2 weeks prior to surgery. The patient who refuses to stop smoking has significant risks, and therefore the physician should consider canceling any planned surgery.

The surgeon must discuss with the patient the procedure of autologous fat transplantation, any material risks and complications, and any viable alternatives as well as their risks and complications. The patient should be informed about the complete surgical fees (surgeon, facility, anesthesia).

If fat transplantation is to be performed in association with liposuction for contouring, then the patient should be informed of the material risks of liposuction and viable alternatives and their risks and complications.

I. RECORDS

The medical record is the physician's best defense. Included in the record should be the history and physical examination with significantly negative findings as well as positive findings. The presenting complaint, diagnosis, and proposed treatment must be clearly indicated. The discussion of surgery, alternatives, and risks and complication is an important part of the record.

Table 2 Drugs Which Inhibit Cytochrome P450

Drug	Plasma half-life
Acebutolol	—
Amiodarone (Cordarone)	30–40 min at 10%
Atenolol	
Carbamazepine (Tegretol, Atretol)	25–65 h
Cimetidine (Tagamet)	—
Chloramphenicol (Chloromycetin)	68–99% excretion in 72 h
Clarithromycin (Biaxin)	5–7 h
Cyclosporin (Neoral, Sandimmune)	10–27 h
Danazol (Danocrine)	—
Dexamethasone (Decadron)	—
Diltiazam (Cardizem)	3.0–4.5 h
Erythromycin	—
Esmobolol (Brevibloc)	—
Flucanazole (Diflucan)	20–50 h
Fluoxetine (Prozac)	1–3 days after acute administration; 4–6 days after chronic administration; norfluoxetine (active metabolite) 4–16 days
Flurazepam (Dalmane)	47–100 h
Isoniazid (Nydrazid, Rifanate, Rifater)	Excreted with 24 h
Itracanazole (Sporanox)	—
Ketoconazole (Nizoral)	Biphasic 2 h after first 10 h then 8 h
Labetalol (Normodyne, Trandate)	6–8 h thereafter
Methadone (Dolophine)	—
Methylprednisolone (Medrol)	—
Metoprolol (Lopressor)	3–7 h
Metronidazole (Flagyl)	8 h
Miconazole (Monistat)	—
Midazolam (Versed)	—
Nadolol (Corgard, Corzide)	—
Nicardipine (Cardene)	Average 8.6 h
Nifedipine (Procardia, Adalat)	2 h (extended release in 8 h)
Pentoxifylline (Trental)	1.0–1.6 h
Pindolol (Visken)	—
Propranolol (Inderal)	4 h
Propofol (Diprivan)	1–3 days
Quinidine	12 h
Sertraline (Zoloft)	Average 26 h
SSRI antidepressants	—
Tetracycline	—
Terfenadine (Seldane)	Mean 6 h
Thyroxine	—
Timolol (Timolide, Timoptic)	—
Triazolam (Halcion)	1.5–5.5 h
Valproic acid (Depakene)	6–16 h
Verapamil (Calan, Isoptin, Verelan)	—

The record must be either written in readable script or typewritten. There is a general tendency for physicians to include minimal information and write as quickly as possible to save time and effort. Unreadable records are a detriment to the physician if litigation should occur.

II. LABORATORY STUDIES

If general anesthesia is contemplated or if liposuction is estimated probably to be over 1000 mL, then preoperative hemoglobin (Hgb) and hematocrit (Hct) are essential. Occasionally, liposuction, even with tumescence, can result in significant blood loss.

Other laboratory studies such as prothrombin time (PT), partial thromboplastin time (PTT), platelet count, and human immunodeficiency virus (HIV) can be performed as necessary depending upon the patient's history.

III. INSTRUCTIONS

Both preoperative and postoperative instructions are given to the patient. The avoidance of medications such as aspirin and ibuprofen for 10–14 days before surgery should be explained. Postoperative activities and wound care must be described.

REFERENCE

1. Klein JA, Kassardjian N. Lidocaine toxicity with tumescent liposuction: a case report of probable drug interactions. Dermatol Surg 1997;23:1169–1174

9

Fat Transplant to the Glabella and Forehead

Felix Rüdiger G. Giebler
Vincemus Klinik, Eider, Germany

Fat is a biogenetic material, but it is not an ideal soft tissue filler, especially for the forehead region. Injectable autologous fat appears to be compatible, but the stress of the adipocytes during transplantation seems to limit fat as an injectable soft tissue filler. Anatomical and histological factors should be respected when injecting fat into the facial regions.

It has been the efforts of many cosmetic surgeons to compensate for the defects of the soft tissue with a filler.[1] Different fillers have been injected in the past, including paraffin, Teflon, petrolatum, oil of fruits, lanolin, beeswax, autologous skin, silicone, Goretex, collagen implants, and fat. To be successful filler, an injectable implant must have specific qualities:

1. No side effects
2. Replaceable
3. No allergic potential
4. Easy handling
5. Wide spectrum of usage
6. Availability
7. Inexpensive
8. Normal to the touch
9. No functional changes
10. Long lasting

Most of the alloplastic materials partially dissolve within the aggressive milieu of the body. They lead to a prolonged foreign body reaction, and this "bland infection" is the space filler in the profile defect of the soft tissue. Biogenetic implants feel more natural, the filling of the space is caused by the material itself, but there is a regression of the material and foreign body reaction as well. Hyaluronic acid, collagen (acting as a biological sponge),[2] and fat are the natural materials found within the skin. There is the choice of using homologous material or a xenograft.

Autologous fat transplantation invokes questions of its lasting effect. The vulnerability of fat cells has not been solved. Mechanical stress, prolonged contact of the fat cells with different surfaces in the syringes, and the cannulae may be some of the reasons. Another problem is how the transplanted adipocytes come into contact with the arteries and veins. Fat cells should be transplanted from a less vascular to a more vascular area of the body, and the fat for the transplantation should be obtained from areas of the body that are persistent deposits (alpha 2 receptors that do not respond to diet).[3]

Before filling up the soft tissue defect, the physician must make an exact anatomical diagnosis of the profile defect. It should be determined whether the area of the anatomical defect is inflatable or not, and each defect has to be corrected with an appropriate filler such as:

1. Skin: collagen, hyaluronic acid
2. Subcutaneous: fat
3. Muscle: plastic surgery/fat
4. Bone: alloplastic implant/reconstructive surgery

Fat is not an adequate filler for either the dermis or the epidermis but is adequate for the subcutaneous layer. Fat cannot be injected into the skin; it can only be placed into the prepared space.

I. GLABELLAR AND FOREHEAD

The folds on the front of the forehead are functional folds caused by the muscularis frontalis, the muscularis corrugator, and the muscularis procerus. In these upper face folds, "bands" are attached from the skin to the fascia. Apart from this, the soft tissue is stretched flatly over a bony prominence, so that persistent pressure is easily obtained on the transplant. Fat is not the ideal material for forehead or glabellar augmentation, since a space must be prepared for the fat injection, a special stab incision must be made, and the fat should be placed in different layers by blunt preparation.

Indications for fat transplantation to the face include (modified Skouge)[4]:

1. Central cheek depression
2. Diffuse age-related lipoatrophy
3. Facial hemiatrophy

4. Flattened chin
5. Flattened upper and lower lips
6. Nasolabial grooves
7. Supra eyebrow grooves
8. Glabellar depression

II. EQUIPMENT

The equipment used is modified from Fournier.[5] A 13- or 14-gauge blunt needle is attached to a 10 mL syringe with a Luer-Lock. The syringes are previously washed with an isotonic saline solution leaving a residual of 0.5 mL in the syringe to enable better suction and to prevent contact of the fat with the air. After harvesting, the syringes are decanted (5–15 min), and only yellow fat is used. Specimens streaked with blood should be discarded. Negative pressure is applied to the syringe by retraction of the syringe plunger mechanically steered with thumb and fingers. The same syringe is used for harvesting and injecting. There is minimum handling of the graft with a diminished risk of possible contamination.

III. ANESTHETIC

For local anesthetic, modified Klein's[6] solution is used for the donor side (hip or trocanteric area) consisting of 1000 mL saline solution, 50 mL Xylonest (prilocaine hydrochloride) with 1:200,000 epinephrine (adrenaline), 10 mL natriumhydrogenbicarbonate 8.4%, and 10 mg triamcinolone. For the recipient site, 0.5% prilocaine without epinephrine is used.[7]

Further equipment utilized includes a spinal needle, #11 blade (for stab incision), blunt stab preparator (for subcutaneous space), blunt transplantation needle with one hole, top-opened blunt transplantation needle (14 gauge), and blunt stab dissector.

The injection site of the glabella is infiltrated with a spinal needle containing 0.5% Xylonest without epinephrine so that vascularization in the recipient area is not diminished. Access to the skin is reached by an extra stab incision, cranial to the groove, in order to prevent injury to the vascularization and the innervation on the saddle of the nose.

IV. INJECTION TECHNIQUE

A multilayered fanning technique should be used with fat placement while retracting the needle. Overcorrection up to 30% would be wise. After implantation into the furrow, the fat is pushed into place. It is important that the smoothing finger tip is creamed to slide easily on the skin, otherwise the fat can be pushed back out of the stab incision. Ice packs are used to prevent bruising and no bandages are applied. The patient is encouraged to remain quiet with the head in an upright position to prevent ecchymosis and displacement of the graft.

V. RESULTS

Loss of volume is often noticed within the first 14 days but becomes stable for the next few months (Fig. 1). A touch-up is planned 1–3 months after the first injection using fat which is stored in the freezer at –30°C.

VI. EXPECTATIONS OF SURVIVAL

Various reports claim a wide range of survival statistics of fat cells based upon clinical impressions. Peer[8] and Wetmore[9] claimed a survival rate of 40–60% after transplantation. The different aspects of these survival rates were discussed

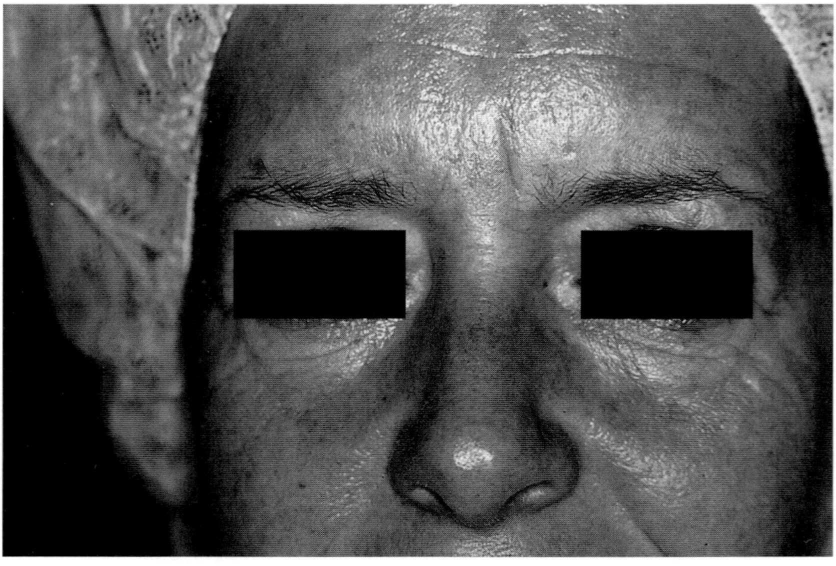

Figure 1a 62-year-old female patient with a hollow appearance. **(b)** Immediately after injection of 1.8 mL of fat into the glabellar region with adjunctive lipoaugmentation of the malar region (3 mL each side). **(c)** One month after fat injection.

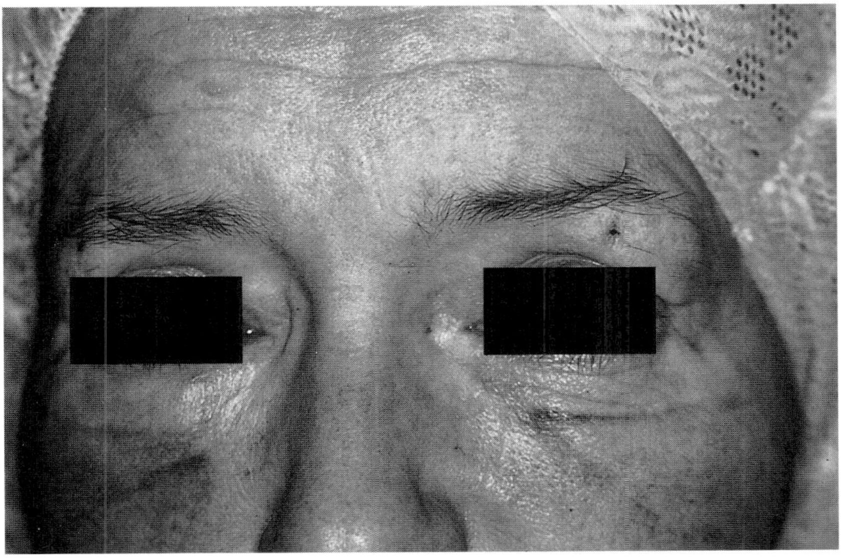

Figure 1b *continued*

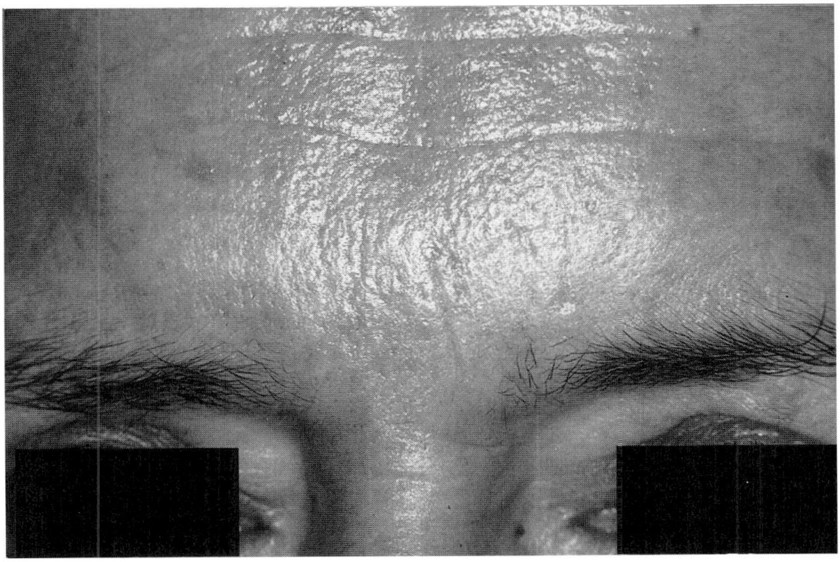

Figure 1c *continued*

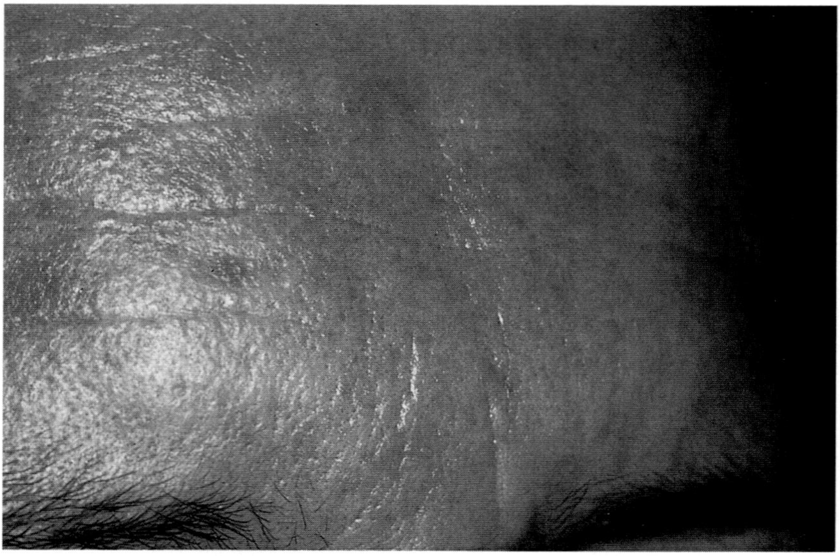

Figure 1d Three months after lipocontouring of the glabellar fold.

by Shiffman,[10] who came to the conclusion that there could be consistently good results. In the author's series, a survival rate of up to 2 years was found, but most of the patients had a remarkable loss of volume after 3 months, so that there was a need for touch-ups to enable a "stable" result.

VII. COMPLICATIONS

1. *Fat embolism:* Injecting into a vessel with a loss of unilateral vision after the fat injection is the most serious complication reported.[11] The glabellar area is a highly vascularized area and injections of collagen-related embolism have also been reported.[12]
2. *Hematoma and ecchymosis*
   a. May be caused by blood remaining in the unwashed sample.
   b. May occur from disrupting the vessels in the angioarchitecture of the glabellar area.
   c. Hematoma may be caused by the extra incision.
3. *Overcorrection:* Overcorrection should be done up to 30%. There is no problem when the patient is adequately informed. If there is a "true"

lasting overcorrection, it can be corrected by pressure, the use of ultrasonic pressure application, or local syringe liposuction.

4. *Sliding of the graft:* Sliding of the graft into the adjacent nasal portion of the upper lids, an infrequent complication, can be corrected by suction removal.
5. *Infection:* Infections seem to be very rare. Perhaps this is due to the influence of the local anesthetic which acts as an antibacterial.[13]
6. *Calcifications*
7. *Oil cysts*
8. *Short-lasting effect*

The expectation of survival of the adipocytes depends upon a number of points.

1. Contact of the adipocyte with foreign materials
2. Pressure within the syringes
3. Temperature
4. Angioarchitecture of the donor and recipient sides
5. Diameter of the harvest and donor cannula
6. Duration of the transplantation operation
7. Pressure in the recipient area
8. Amount of the concomitant hematoma
9. Fat storage of the cells

VIII. DISCUSSION

The glabellar frown lines are not the ideal place in which to put fat. These functional folds are skin folds caused by muscle traction. The best way to treat these grooves is to weaken the muscle (e.g., by botulinum toxin [Botox] or excision) and then lifting up the remaining skin fold with a skin filler such as collagen or hyaluronic acid. Fat should be placed into the subcutaneous layer. The long-lasting effect, longer than 4–6 months, is only achieved when the muscle is excised. There is no perfect material for the augmentation of the glabella, but with implanting fat there is no allergic or foreign body reaction potential. Fat is used when there is a conspicuous allergic history and when there is hypersensitivity to collagen (bovine collagen).

In the aging face there is generally a loss of soft tissue so that augmentation is the answer. Nowadays, there are a number of fillers for the skin and the appendages of the skin. For correction of the defect, it is wise to use fat in the fat layer and collagen or hyaluronic acid in the skin. Longevity is only guaranteed by a permanent foreign body reaction around an alloplastic material. Within the face, with its complicated mimic structure, these foreign body reactions may interfere with the swinging lightness of the facial lines. The fluent interaction of the muscles should not be impaired by a persistent lump of a foreign body.

REFERENCES

1. Giebler FRG. Injizierbare Implantante Aest Chirurgie. Einhorn-Presse Verlag Reinbek 1996;387–390
2. Giebler FRG. Vor- und Nachteile des Biogenen Implantates Collagen Plast Und Wiederherstellungschir. Einhorn-Presse Verlag Reinbek 1996;319–323
3. Hiragun A, Sato M, Mitsui H. Establishment of a clonal cell line that differentiated into adipose cells in vitro. In Vitro 1980;16:685
4. Skouge JW. Autologous fat transplantation. In: WP Coleman, ed. Cosmetic Surgery of the Skin: Principles and Techniques. Philadelphia: Decker, 1991;206–216
5. Fournier PF. Liposculpture: The Syringe Technique. Paris: Arnette, 1991
6. Klein JA. The tumescent technique for liposuction surgery. Am J Cosm Surg 1987;4:263–267
7. Sattler G, Rapprich S, Hagedorn M. Tumescent Localanethesie Untersuchung zur Pharmakokinetik von Prilocain. Z Hautkr 1997;522–525
8. Peer LA. The neglected "free fat graft": its behavior and clinical use. Am J Surg 1956;92:40
9. Wetmore SJ. Injection of fat for soft tissue augmentation. Laryngoscope 1980;99:50
10. Shiffman MA. A complication of liposuction surgery. Am J Cosm Surg 1997;14:349–350
11. Teimourian B. Blindness following fat injections. Plast Reconstr Surg 1996;82:361
12. Zyderm collagen implant. Physician Instruction for use.
13. Thompson KD, Welykyj S. Antibacterial activity of lidocaine in combination with a bicarbonate buffer. J Dermatol Surg Oncol 1993;19:216–220

10
Multilayered Malar and Mental Lipotransplant Surgery

Julius Newman
Graduate Hospital, Philadelphia, Pennsylvania

Autologous fat grafts have been used for many years to fill soft tissue or bony defects successfully. Fat was originally harvested in large portions through surgical incisions with the advent of liposuction surgery, with direct manual aspiration being allowed for nonsurgical excision of an abundant supply of viable fat. The author began using fat transfer in 1984 utilizing a special automatic fat injection instrument but ultimately switching to a 3-mL syringe with a special long injection needle. Frozen fat was tried but not continued. The use of centrifugation is now used to compact the fat, and there is no need for insulin in washing the fat.[1–3]

I. PATIENT CONSULTATION

The patient must be evaluated to determine if he or she is a suitable candidate for lipotransplant surgery. This involves assessing whether the deformity is subdermal and, therefore, correctable by fat transplantation. Availability of sufficient donor fat must also be confirmed. Sites that are utilized are the abdomen and medial and lateral thighs but not the neck area. Although many surgeons have preferences for specific donor sites, no study has demonstrated superiority of one site over another. This fact gives greater latitude for the patient to participate in the selection of the intended donor site. The author uses the same donor site for harvesting for repeat lipotransfer. The reason is that with the previous trauma of liposuction for fat harvesting, there is increased collagen formation in the healing process which will increase the staying power of the fat. Full explanation of the treatment and expected results are discussed in detail with the patient. This includes the fact that the amount of fat reabsorbed is entirely unpredictable. Any intended overcorrection must be discussed at the time to avoid initial postoperative disappointments. The possible need for additional touch-up corrections must also be emphasized.

In the past, the author stored frozen fat for future lipofilling procedures but found that after thawing there was excessive rupture of the cell walls with large quantities of free fat. The frozen fat technique has been discarded, and harvesting fat for each separate session has been adopted. The abdominal area is used for harvesting, but if no fat is available, the thighs are utilized. Originally, the fat was reinjected superficially, but by the mid 1990s, it was realized that the take was not good, and therefore a technique was developed for multilayering the fat. The injection is started deep, almost to the bone, and gradually worked upward and ending superficially. The multilayering technique has an increased blood supply and thus a better take of the injected fat.

It is explained to the patient prior to scheduling that it is not known how much fat will be retained and that it varies from individual to individual. Men do not retain as much fat as women. The patient should know that the first lipofilling is overcorrected by 30–40% and that he or she will be reevaluated in 3 months, with 40–50% of patients not needing a second session. At the end of 3 months, the patient is evaluated and a second fat harvesting injection is carried out if necessary. The patient is again evaluated at the end of a 6-month period, and usually about 25% of the patients need a third session. With the second and third lipotransfer sessions, there is a better take because of the laying down of collagen from previous fat harvesting.

II. TECHNIQUE

There are two scenarios for fat transfer: (1) is when the patient only requires fat transfer, and the syringe technique is used for this harvesting (Fig. 1). (2) If the patient undergoes body liposuction, the reduced-pressure technique is used, which allows for decrease in the ruptured fat cells. Midway in the suction tubing a trap is inserted to collect the fat during liposuction. When the trap is full, it is emptied and the trap is reinserted (Fig. 2). Fat is then placed in 10 mL syringes (Fig. 3) and centrifuged, which spins down the fat and separates debris, fluid, and bloody material (Fig. 4). The fat is then placed in 3 mL syringes (Fig. 5).

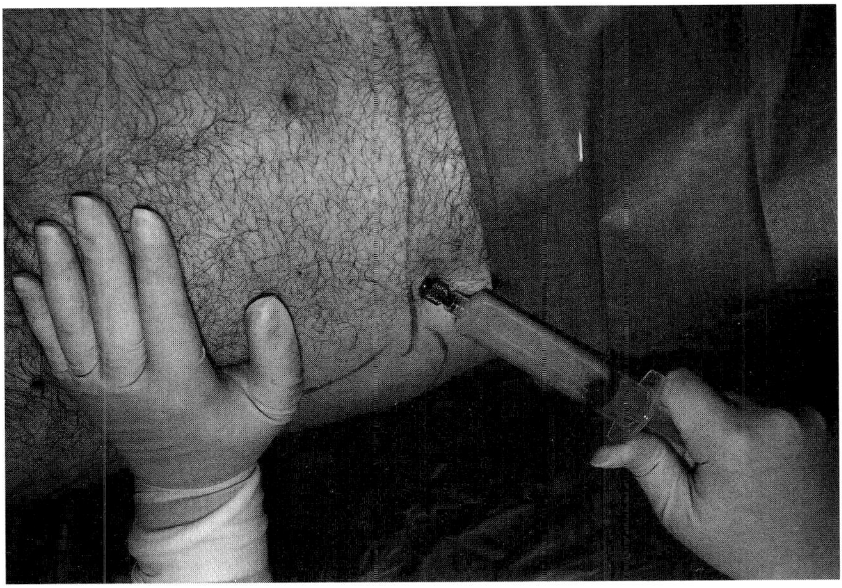

Figure 1 Harvesting fat with the syringe technique.

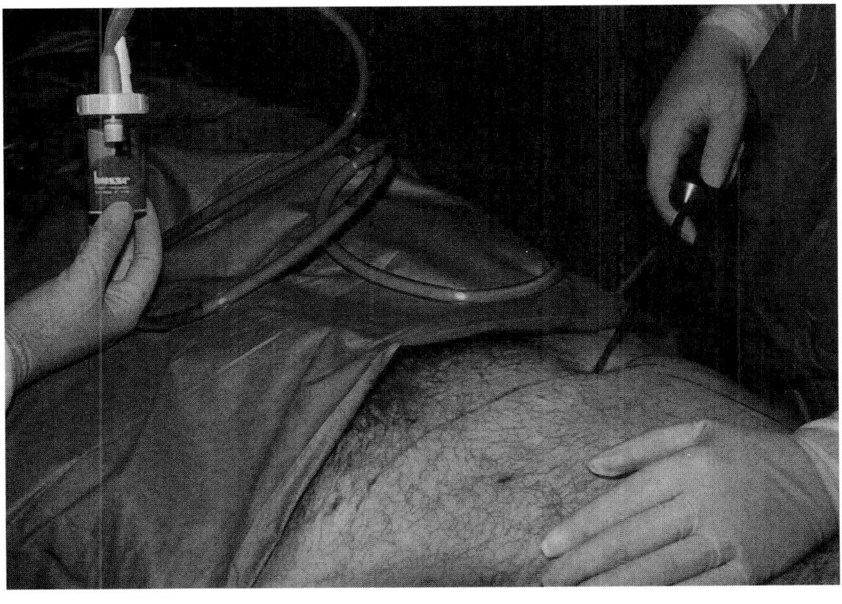

Figure 2 Harvesting fat with machine liposuction. Note the fat trap.

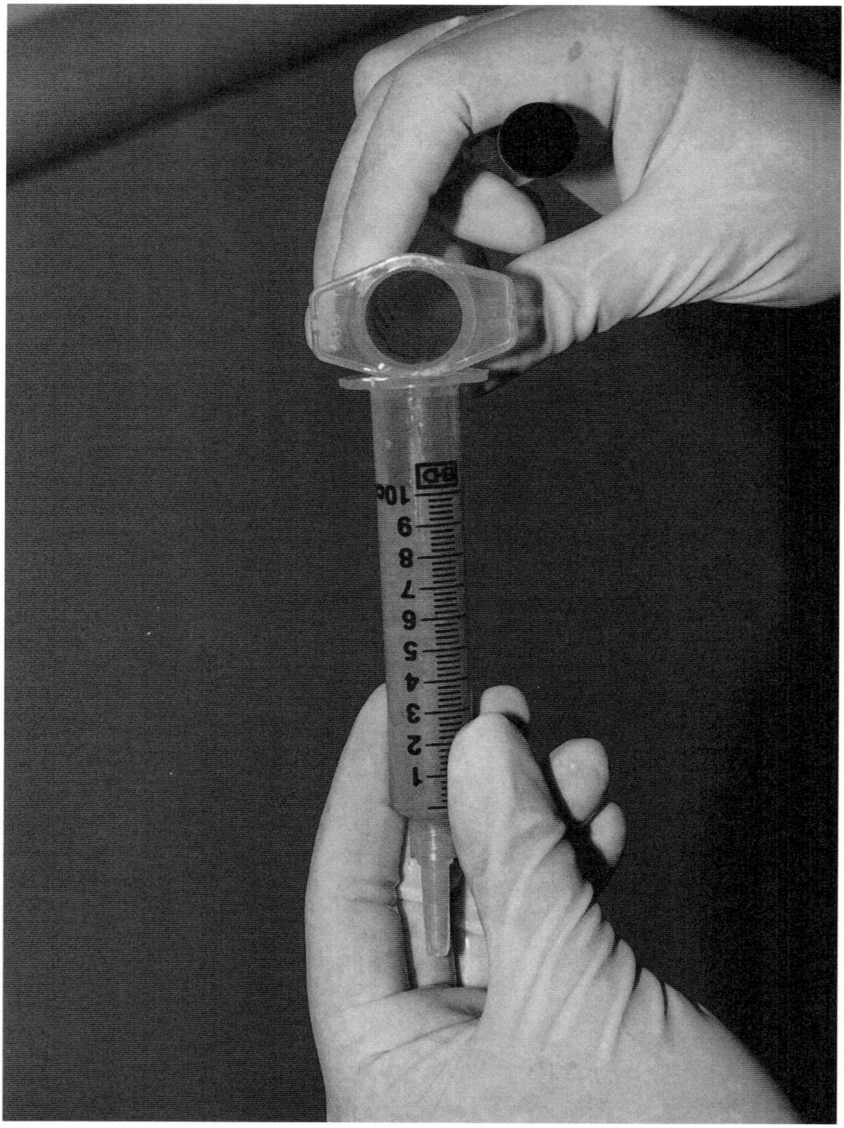

Figure 3 Fat is placed in 10 mL syringes.

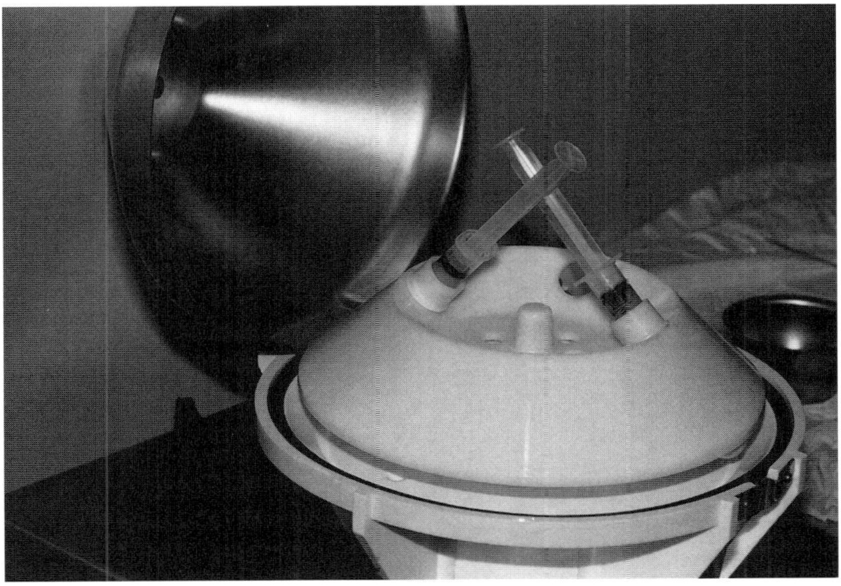

Figure 4 Centrifugation of fat.

Prior to malar transplants, bilateral infraorbital and front of the tragus blocks are performed where the injection needle will be inserted. For mental fat transplants, the mental nerve bilaterally and along the mandibular rim are blocked for the injection site entrance.

Using an 18-gauge needle, the injection site is pierced which allows the injection needle to slide in easily (Fig. 6). The lipofilling injection needle, which is blunt ended and open at the tip, is attached to the 3 mL syringe containing the fat (Fig. 7). The injection site for the malar lipofilling is just in front of the tragus (Fig. 8). The injection site for the chin is along the mandibular rim.

Because of the inability for large quantities of fat to develop a blood supply and survive, the multidirectional and multilayered technique was devised. The fat is injected very deeply, almost to the bone, and then layered while proceeding superficially. The injection needle is then directed in a multidirectional crisscrossing pattern, which allows the fat to be evenly redistributed. This concept allows for a better blood supply, which is the key to maintaining the viability of the fat.

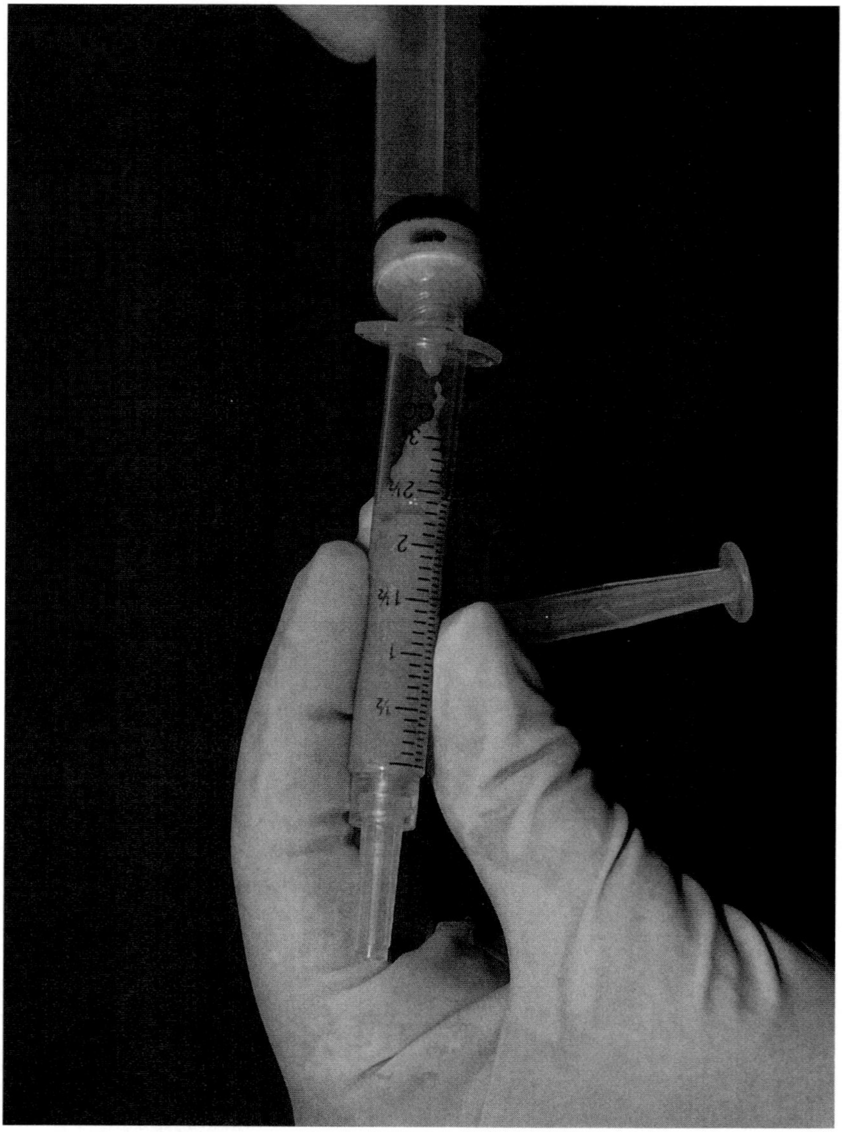

Figure 5 Fat is placed in 3 mL syringes after centrifuging.

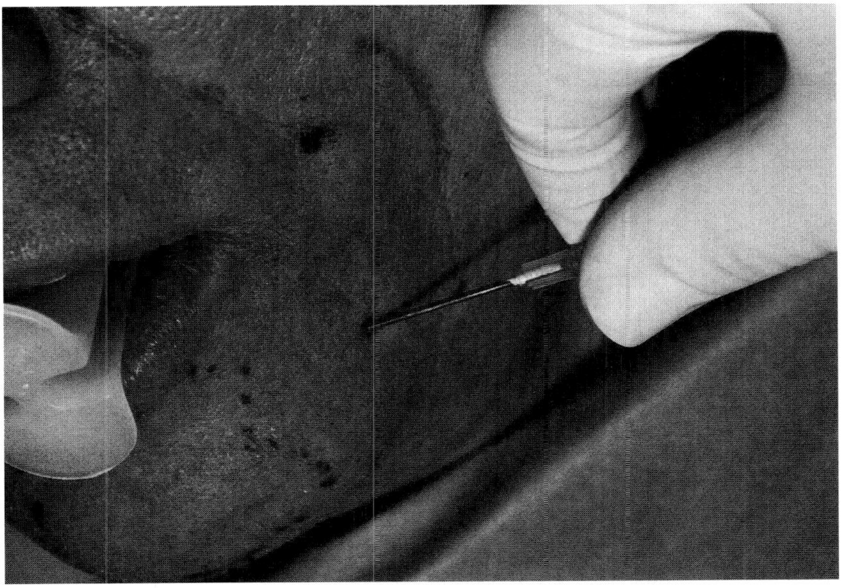

Figure 6 Piercing skin with 18-gauge needle before insertion of injection needle.

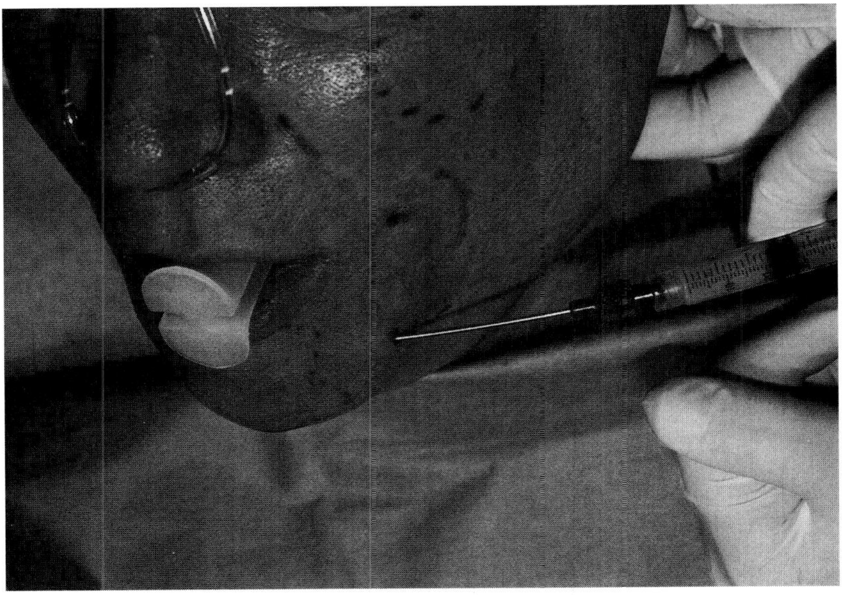

Figure 7 Blunt-tipped injection needle attached to 3 mL syringe inserted through 18-gauge needle puncture site.

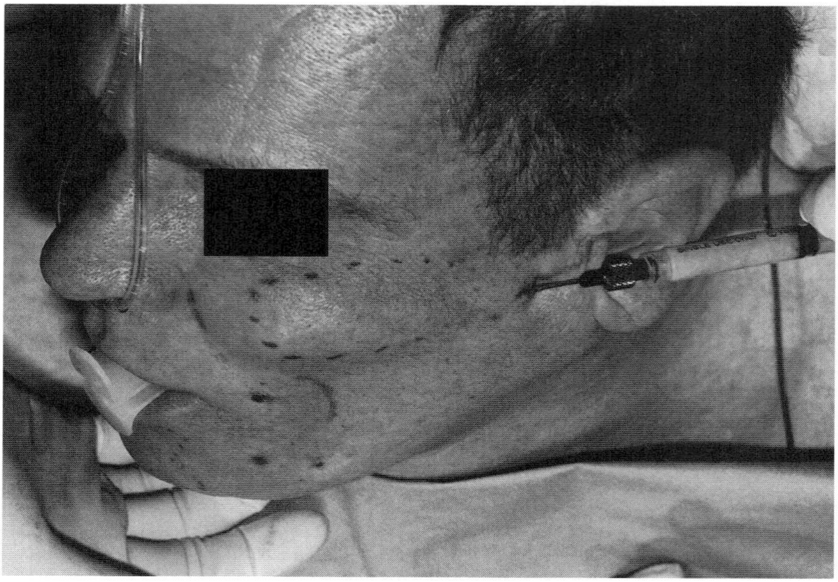

Figure 8 Injection site for malar lipofilling is just in front of the tragus.

Afro-American patients have a specific problem related to their chin and lower lip. Many have very prominent everted lower lips which requires a lower lip reduction. For patients only needing a small-volume augmentation combined with lower lip reduction, the author prefers to use lipofilling rather than chin implant (Fig. 9). For all chin augmentations, lipofilling is preferred, which alleviates the submental incision and possible hypertrophic scarring. In patients who have irregularities, deep creases, or grooves in the chin area, the chin is recontoured with fat transplants (Fig. 10). With malar augmentation, it is most important not to inject the fat superficially but to multilayer the fat deep and with crisscrossing motions that gradually end up superficially (Fig. 11).

III. POSTOPERATIVE CARE

There are no dressings, no antibiotics, and no instructions other than the fact that the areas will be somewhat distorted because of the overcorrection. This must be explained to the patient so that they do not panic. They may return to work the following day.

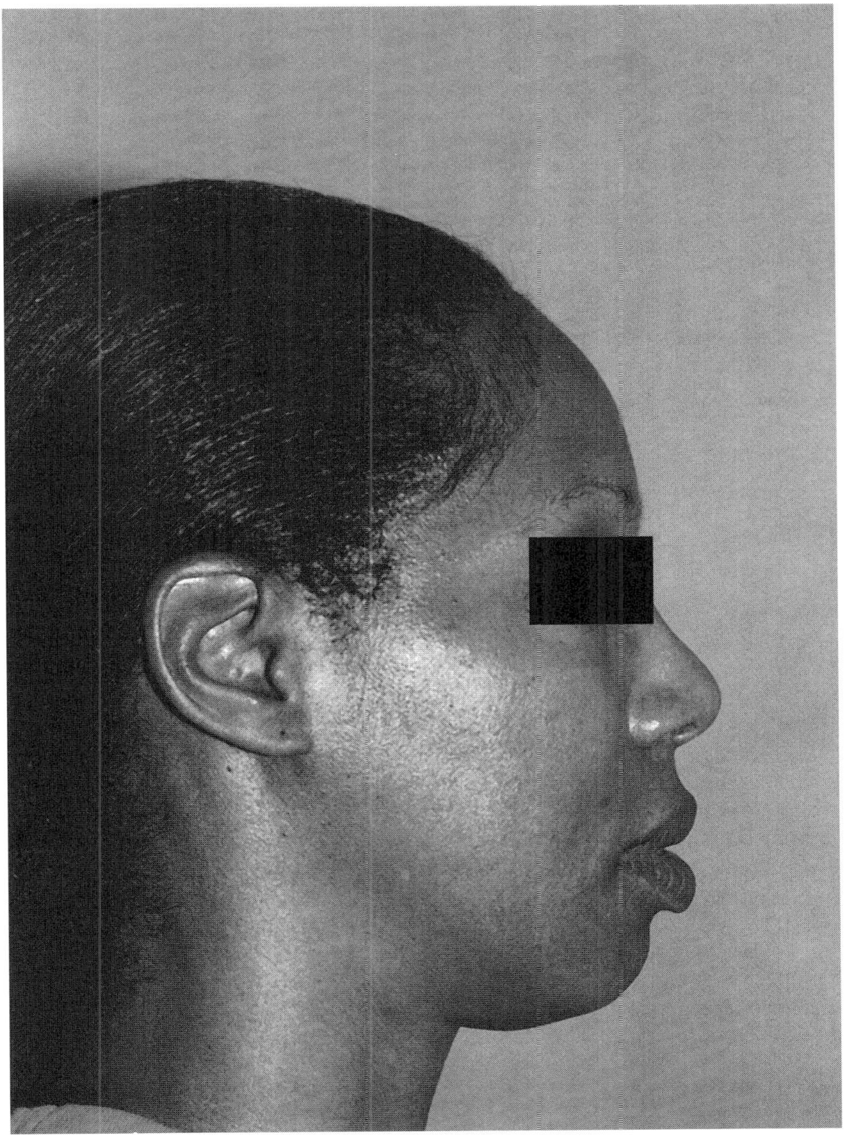

Figure 9a Small-volume chin augmentation with fat combined with lower lip reduction. Preoperative.

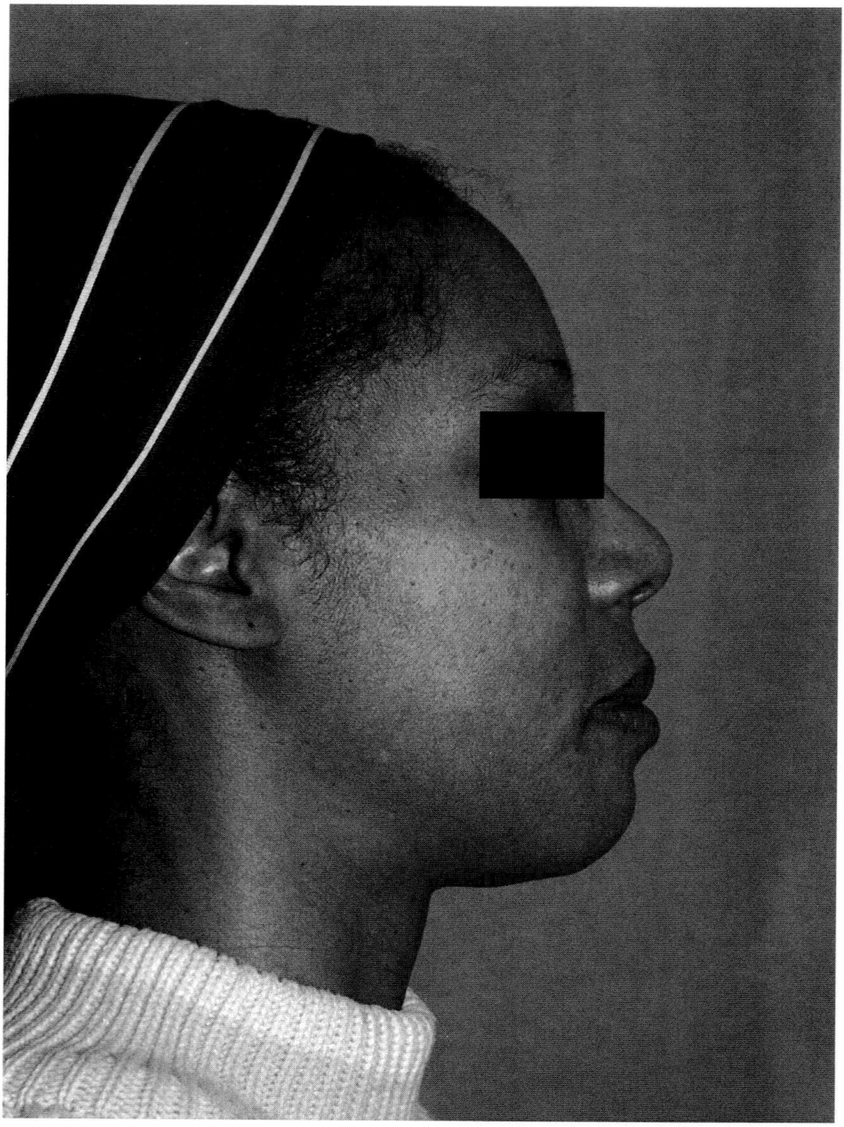

Figure 9b Sixteen months postoperative.

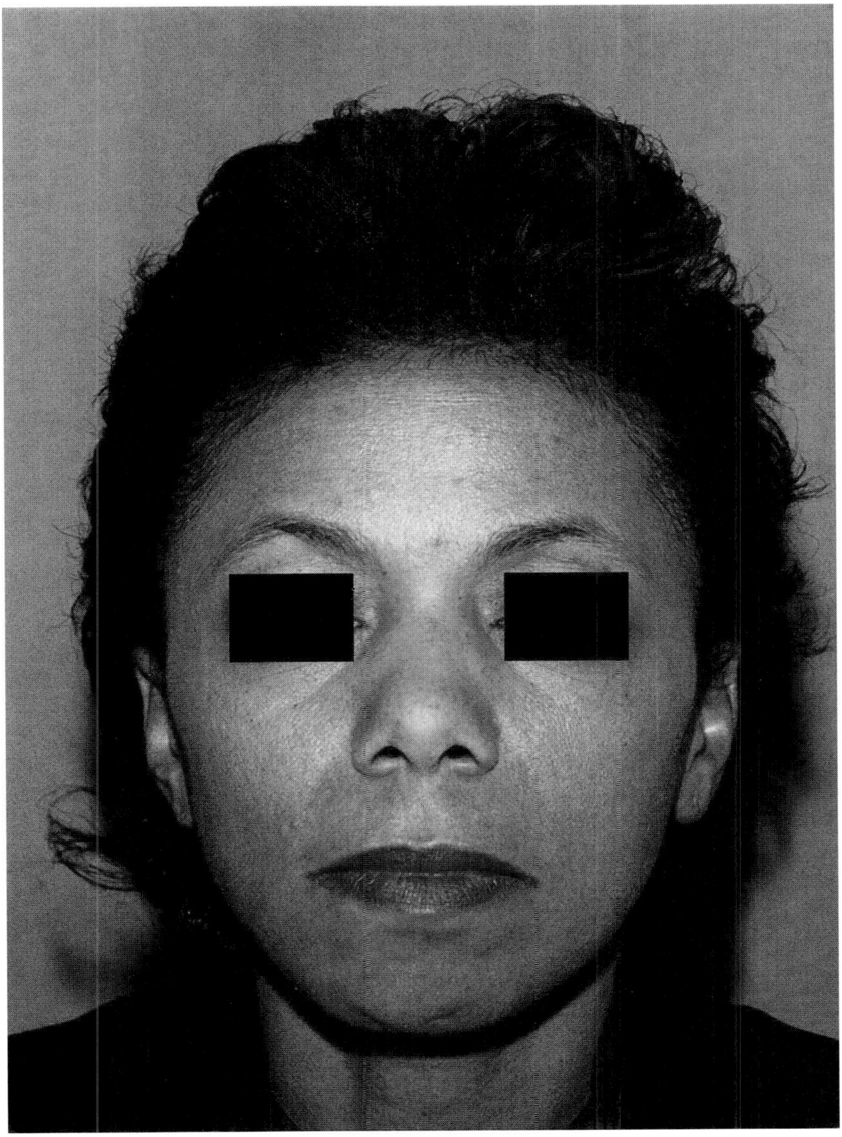

Figure 10a Contouring of irregularities of chin with lipofilling. Preoperative.

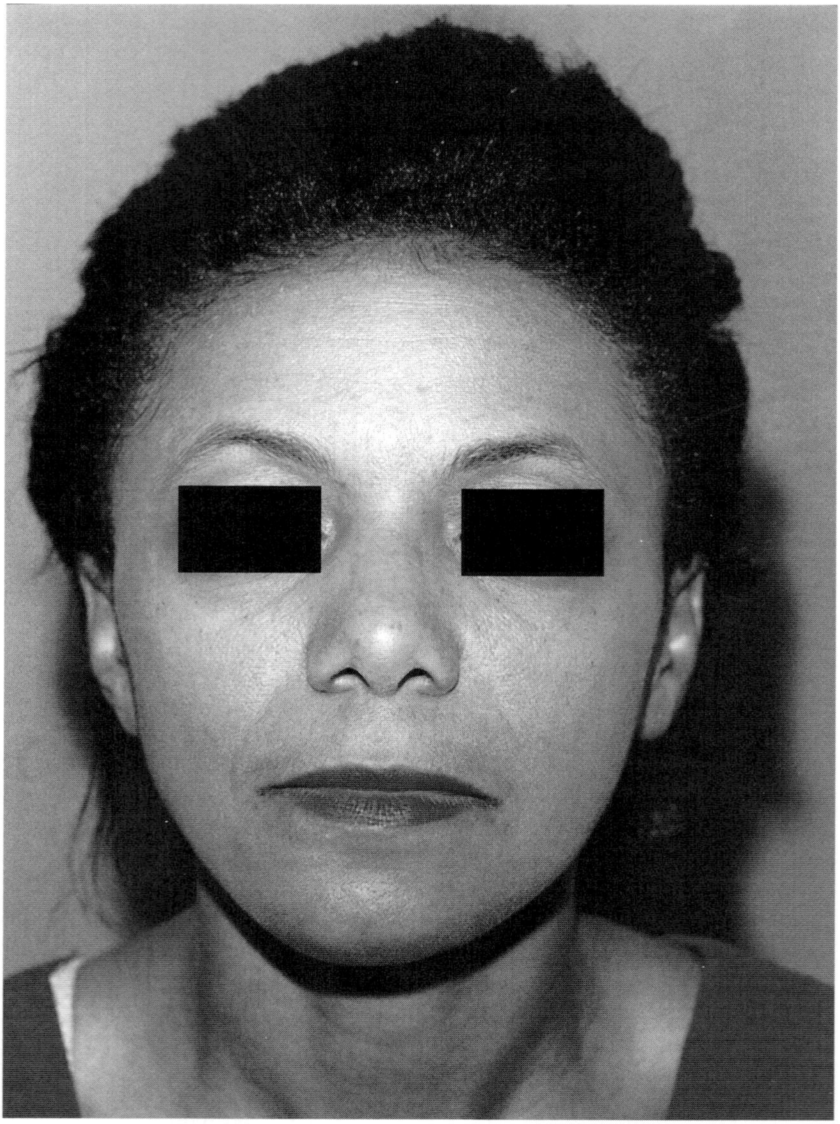

Figure 10b Postoperative.

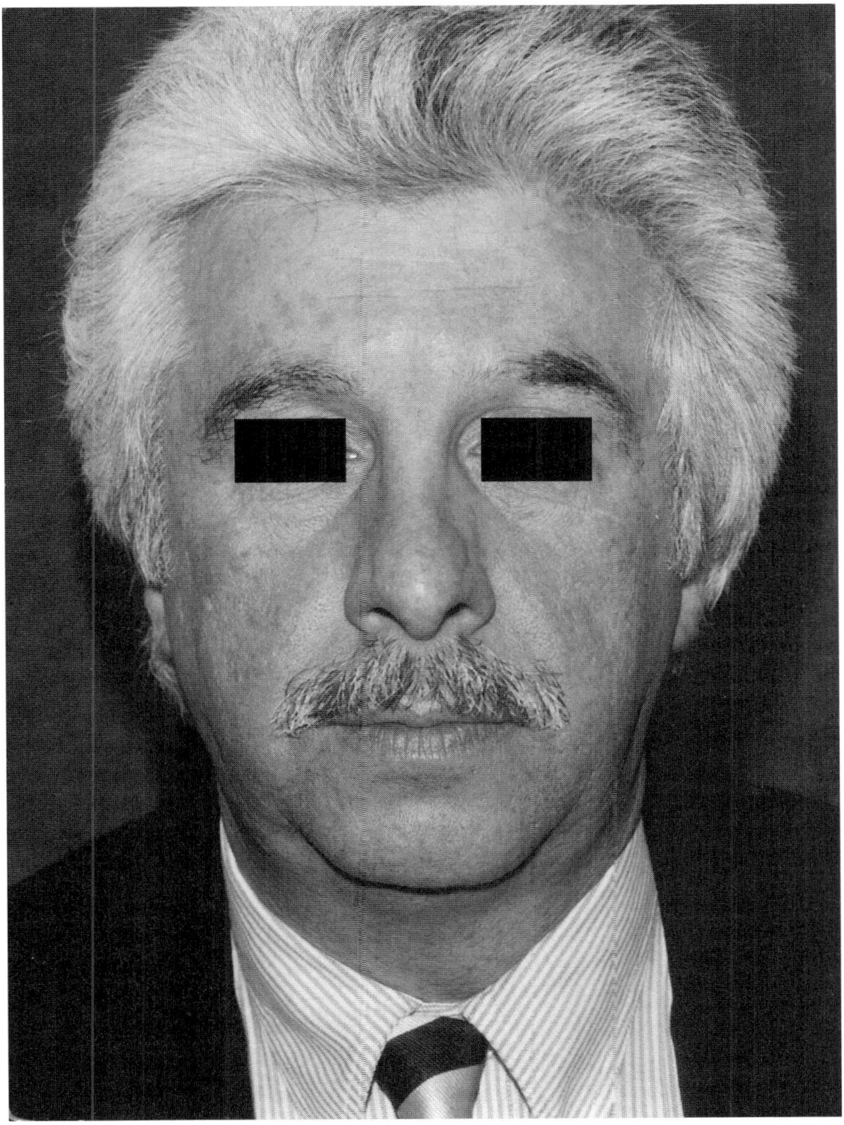

Figure 11a(1) Malar augmentation with fat combined with acrylic chin implant. Preoperative frontal.

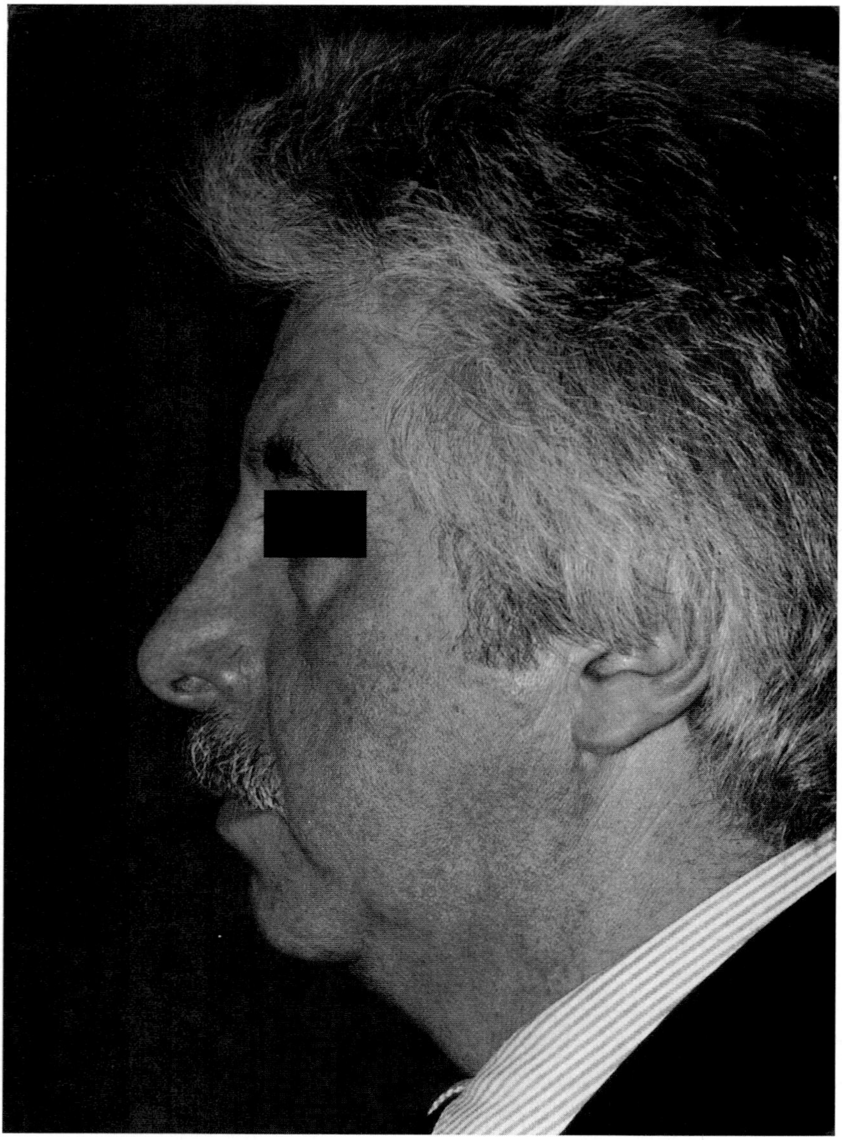

Figure 11a(2) Preoperative lateral.

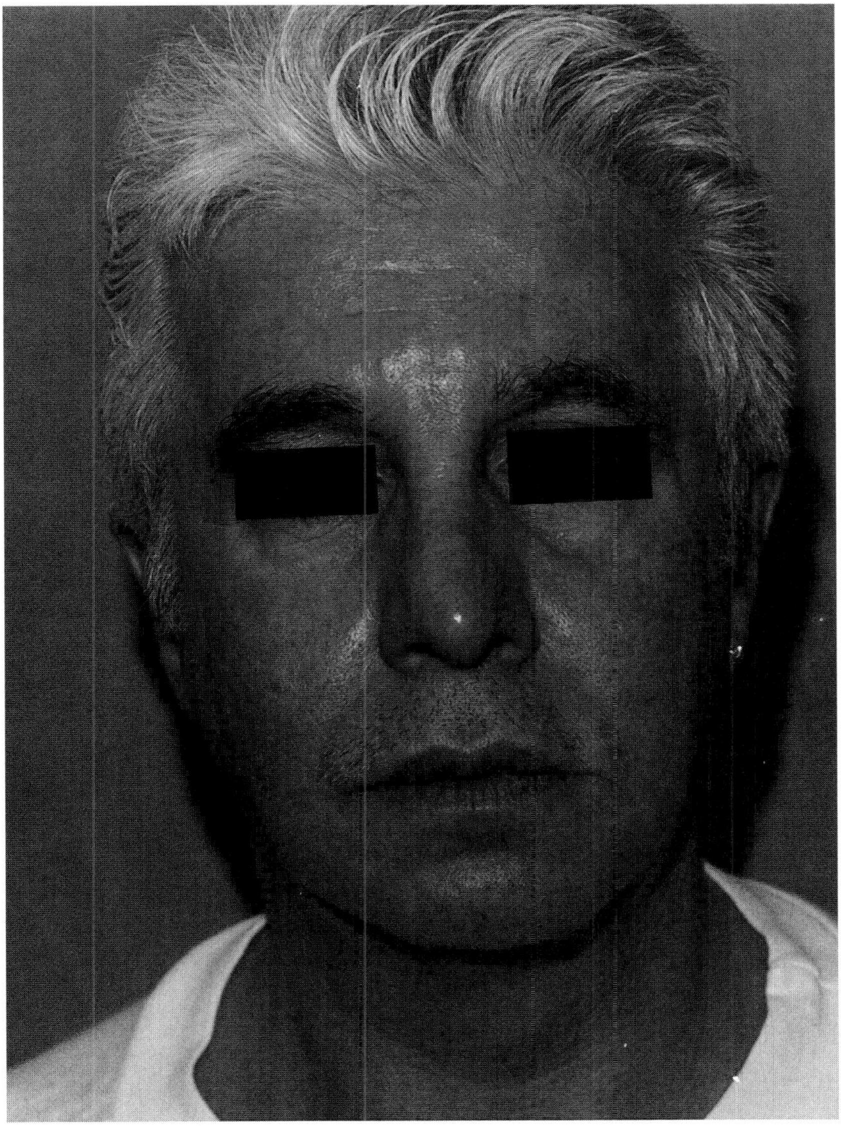

Figure 11b(1) Postoperative frontal using fat combined with a chin implant.

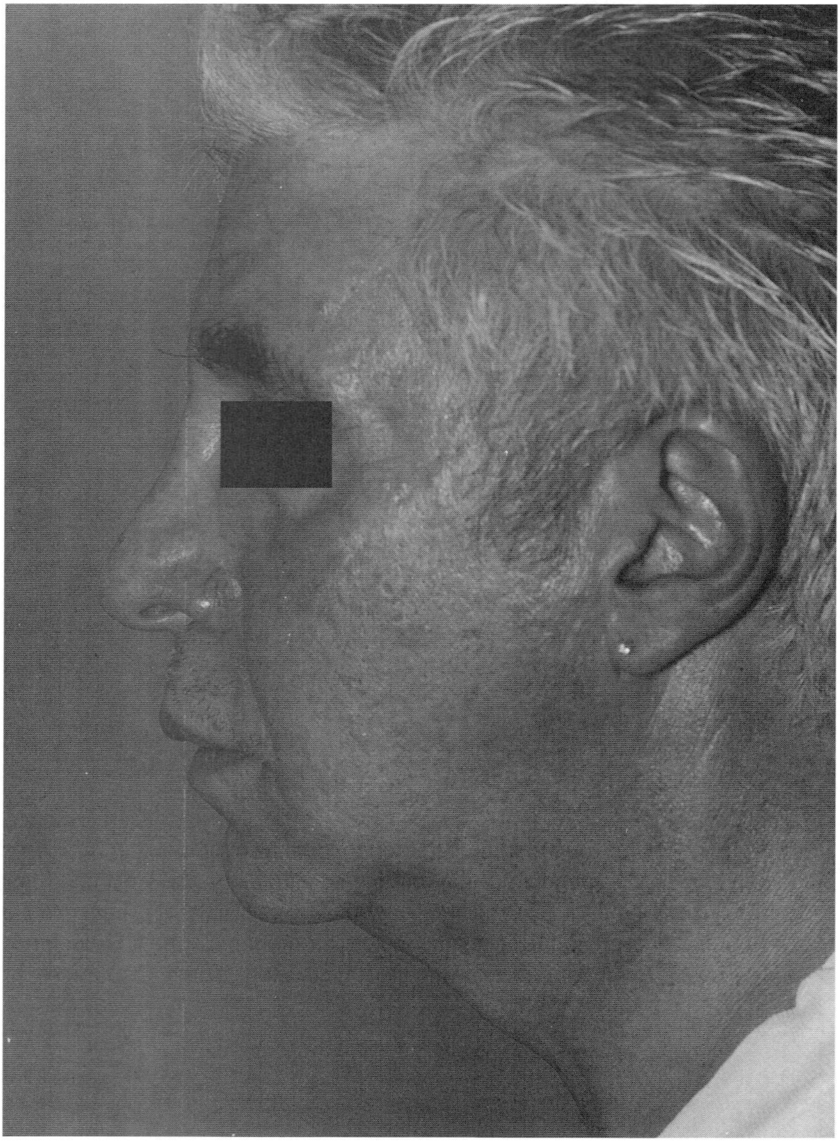

Figure 11b(2) Postoperative lateral. Fat transfer to cheeks and acrylic chin implant.

IV. COMPLICATIONS

At times, there are irregular areas due to the technique or irregular absorption of fat. There may be a lack of symmetry between the two sides. Infection is rare. There have been some reports of surgeons using the approach through the oral cavity. The author thinks this is a poor decision, because it will cause bacterial contamination. The external injection technique leaves no scars. In males, over-correction is done to a greater degree than with females, because they lose much more of their injected fat.

V. DISCUSSION

With the advancement of lipofilling techniques, the author has considerably cut down on the number of implants for malar and chin augmentation. Approximately 95% of cheek augmentation is done by fat transfer. Malar implants may be associated with a high incidence of infection, slippage, nerve damage, and irregularities. With fat transfer the cheek complex can be remodeled. In patients requiring a large chin implant, fat transfer is not done, because there is not enough blood supply to handle a large quantity of fat. At times, slight lipocontouring may be used along with the chin implant if there is some irregularity or divots of the chin. By filling in these irregularities, a smoother and more attractive chin contour can be achieved. Smaller chin augmentation is done with fat transfer.

REFERENCES

1. Newman J. Preliminary report on "fat recycling" lipo-suction fat transfer implants for facial defects. Am Cosm Surg 1986;3(2):67–69
2. Campbell GLM, Laudenslager N, Newman J. The effect of mechanical stress on adipocyte morphology and metabolism. Am Cosm Surg 1987;4(2):89–94
3. Brandow K, Newman J. Facial multilayered micro lipo-augmentation. Int Aesthetic Restor Surg, 1996;4(2):95–110

11

Lipotransfer to Nasolabial Folds and Marionette Lines

David S. Alkek
University of Texas, Southwestern Medical School at Dallas, Dallas, Texas

Autologous fat transfer has rapidly developed since the introduction of liposuction. Fat can be harvested in small amounts with a syringe and then reinjected into facial rhytids.[1,2] Guidelines have been developed for fat grafting including the use of fat grafts 3 mm in diameter or less, washing of fat with physiological solutions, reinjecting into multiple planes, and even using frozen fat for repeat injections.[3–7]

Instruments have been developed for microlipoextraction and injection. A ratcheted pistol-like device can be used to deliver precise measurement amounts of fat.

I. FACIAL LINES AMENABLE TO LIPOINJECTION

Wrinkles and creases are linear or multidirectional lines that involve only the epidermis and dermis. Facial folds are full dermis and subcutis elevations and depressions. Folds such as the nasolabial folds are amenable to correction by lipoinjection, since the subcutaneous fat is involved. The more superficial lines are best treated by intradermal injections with other substances such as collagen. Remember that "fat goes where fat is." If it is injected into the subcutis under a dermal crease, it will only elevate the crease along a mound of fat.

II. PATIENT CONSULTATION

When counseling a patent for facial rejuvenation, the surgeon should ask the patient what he or she wishes to improve. The patient should be given a hand mirror to show the surgeon the areas to be addressed. The doctor should be very familiar with the procedures that will best improve the patient's problem areas related the nasolabial and commissure areas. The patient may have heavy jowls and need a facelift rather than more fat injected. The patient may have swallow lines and lax skin and need laser resurfacing to tighten the collagen. The folds may be amenable to injection with less bulky materials such as collagen. Other permanent fillers may be used.

If the patient and physician are convinced that lipoinjection is the desired course, then the points that are unique to this procedure should be discussed. There will be an overcorrection of 50–100% initially because a half or more of the injected material will not persist. Two or three reinjections may be required to attain the desired result. There may be bruising, redness, infection, and lumping. These can be minimized with proper technique and postoperative care.

III. PROCEDURE

The donor site may be fat from the buttocks, abdomen, medial and lateral thighs, and almost any other site where fat is adequate. The outer thigh is preferred by Hudson[8], Pinski[9], and Fulton[10]. This fat is resistant to weight reduction and is easily reached from the inferior buttock crease, which leaves minimal scarring. The patient is told to shower and scrub the donor area with Hibiclens the night before and morning of the procedure.

Equipment for microfat transplantation is rather simple. Microcannulae for extraction and equipment required are shown in Figure 1 and listed in Table 1. For tumescent anesthesia, we have found that Field's (personal communications) variation of Klein's formula is better suited for the small amount needed in microliposuction (Table 2).

The patient is placed on his or her side with the back to the surgeon (Fig. 2). After cleansing the donor site and injecting the incision site with lidocaine with epinephrine, a small stab incision is made with a #15 blade in the lateral end of the inferior buttock crease. The long spinal needle is used to infiltrate 60–80 mL of tumescent fluid into the donor area. After waiting approximately 10–15 mins, the microcannula attached to a 10-mL syringe is inserted and fat is harvested. The syringe is held with the right (dominant) hand with the plunger suction maintained with the thumb and forefinger (Fig. 3). The left hand is placed on top of the donor area to steady and guide the cannula. The syringe with cannula is moved in and out maintaining suction with the hand only. Care should be used not to withdraw

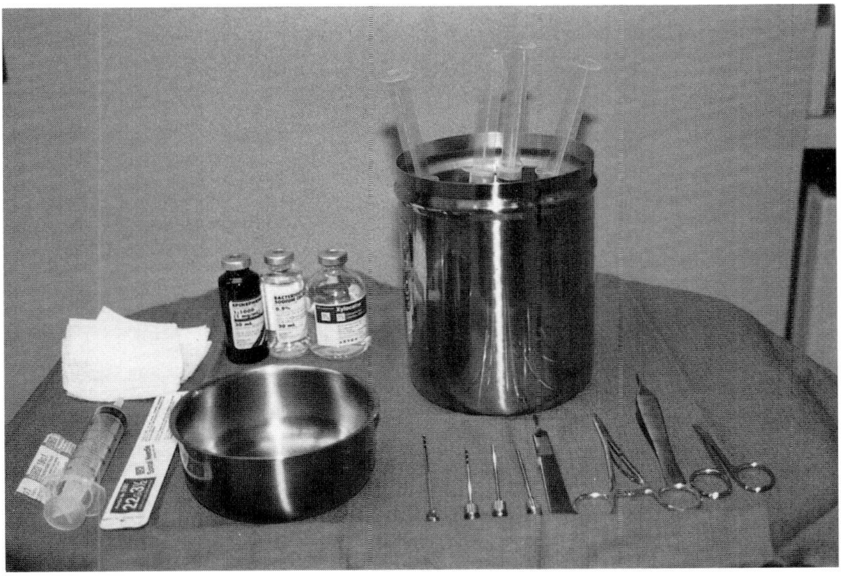

Figure 1a Set-up for microtransfer.

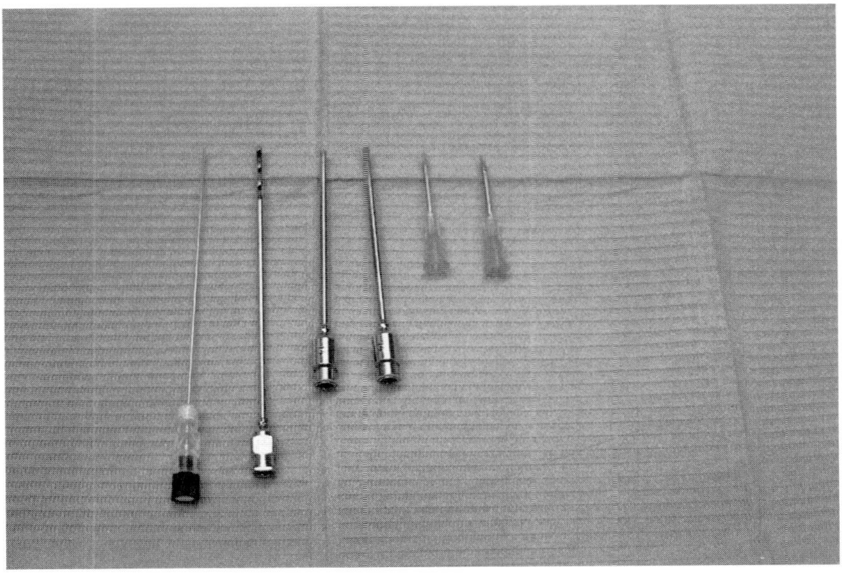

Figure 1b Microcannulae and 14- and 16-gauge needles.

Table 1 Equipment for Microlipotransfer

Equipment
Micro liposuction cannula
Five 10-mL syringes for tumescent anesthesia
Six 10-mL syringes for fat extraction
3-mL syringe for local anesthesia
Spinal needle
14- or 16-gauge needle for lipoinjection
Rack to hold syringe upright
Tumescent anesthetic solution
Lidocaine with epinephrine for local anesthesia
Saline or Ringer's solution

the cannula out too fast and suck air into the syringe. Four of five 10-mL syringes are filled with a mixture of fat and tumescent mixture in a short time. The first syringe may contain a fair amount of clear fluid, whereas the later ones may be blood tinged. Do not attempt to withdraw a full 10 mL of fat and fluid in each syringe. It becomes physically difficult to manipulate the syringe with one hand after 6–7 mL, and this will also leave room the syringe for washing the fat.

The cannula is removed from the syringe after it is adequately filled. The syringes are capped and placed upright in a rack or container with the tips down (Fig. 4). The fat will rise, whereas the mixture of tumescent fluid and blood will settle (Fig. 5). Allow several minutes before separating the fat. Usually the first syringes are ready by the time the last fat is extracted. The syringe is maintained with the tip down while the fluid and serum mixture is carefully ejected. Sterile saline or Ringer's lactate is immediately drawn into the syringe and the fat is washed in the same manner. There may be a darker yellow fluid layer above the fat. This is lipid from disrupted cells, which should not be reinjected. It can be discarded by placing the syringe tip up for awhile or simply leaving it behind in the syringe after injecting the fat into the recipient areas. The author does not transfer the fat to 3 mL syringes or centrifuge it before injecting, since the less the fat is manipulated, the better its chance for survival. Only clean yellow fat should be used for injection (Fig. 6).

The donor site incision is not sutured but covered with bacitracin and mutliple gauze 4 x 4s. It will drain for a few days. While the fat is setting, the patients

Table 2 Tumescent for Microliposuction

Component	Amount
1% Lidocaine	50 mL
Normal saline	30 mL
Epinephrine	0.8 mL

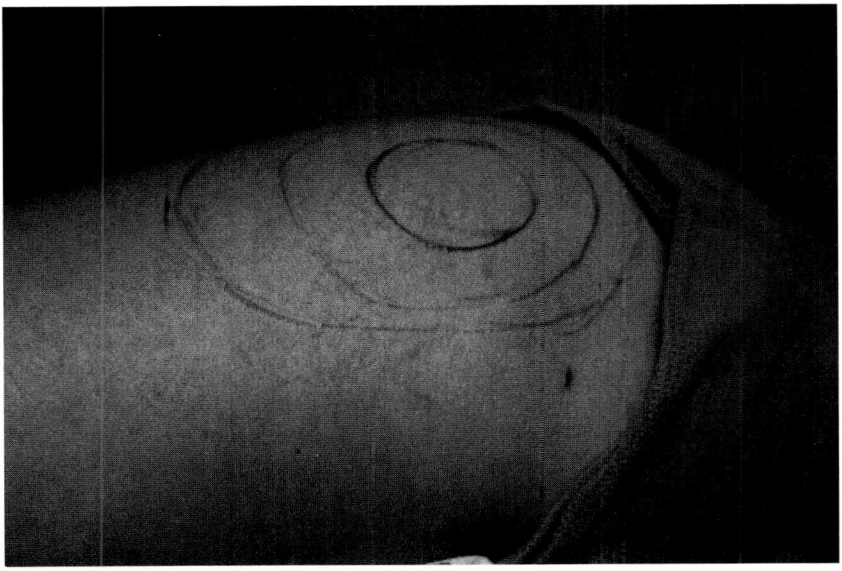

Figure 2 Patient in position for microliposuction with back to surgeon and lateral thigh exposed.

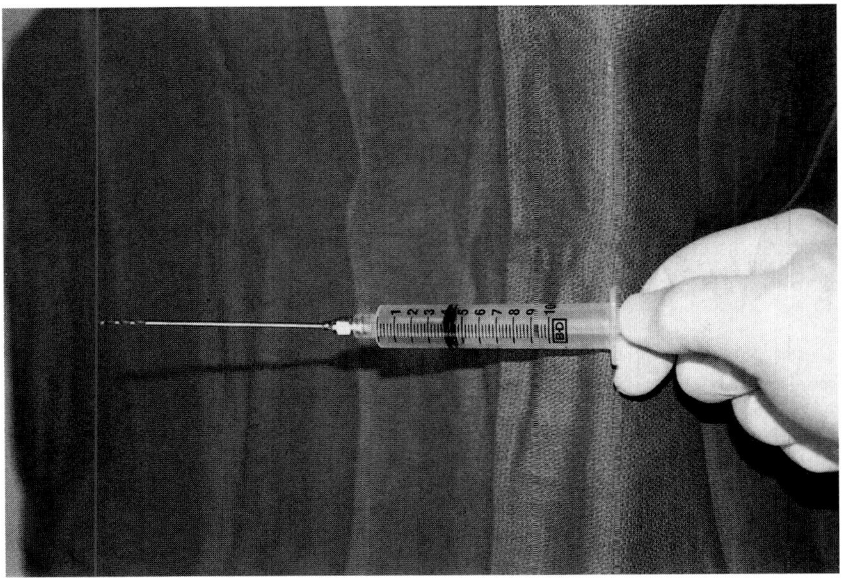

Figure 3 Maintaining suction on syringe with one hand.

Figure 4 Fat settling in syringes in rack.

is turned on his or her back and the areas to be injected cleansed with povidone-iodine (Betadine). Local anesthesia or nerve block with lidocaine with epinephrine is done before the fat is washed to allow the lidocaine swelling to subside and for vasoconstriction. A sharp 16-gauge needle is attached to the syringe for injection. This causes less trauma to the tissue than a blunt or large syringe.

For the nasolabial folds, the surgeon should stand above the head of the supine patient and insert the needle in between the ala nasae and nasolabial fold. The small puncture wound is best hidden here even if sutured. The needle

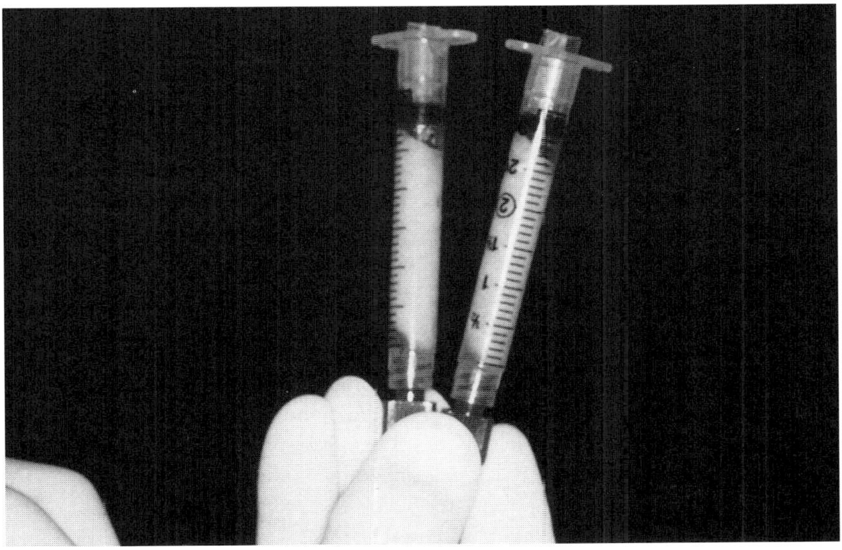

Figure 5 Syringes with fat above and anesthetic mixture below.

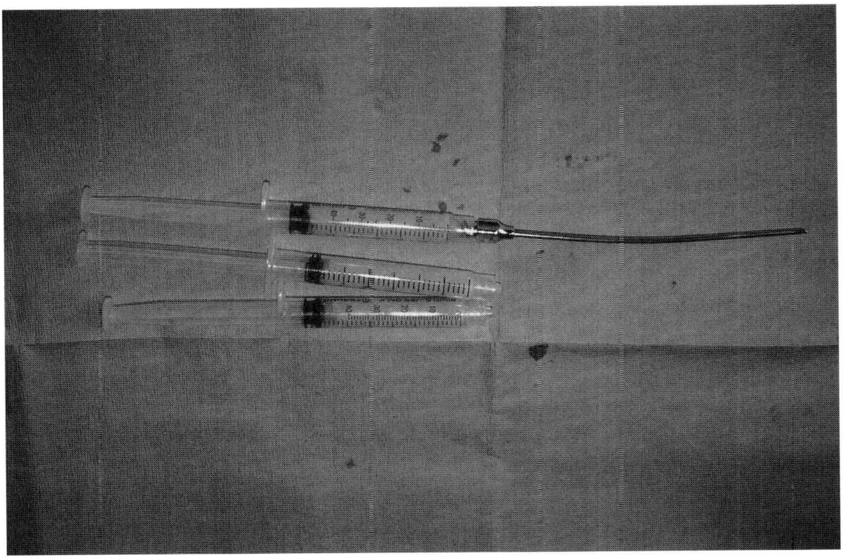

Figure 6 Clean fat ready for injection.

should be advanced the entire length of the fold and fat injected on withdrawal. An attempt should be made to inject the fat in a small, smooth stream along several lines and planes to prevent lumping. It is important to keep the needle in the subcutaneous layer. For the marionette lines, insert the needle at the lateral commissure of the lips and inject in several planes as withdrawing. A 50–100% overinjection should be made, which will correct nonviable fat cells and fluid. The deposited fat may be manipulated into the desired place, but pressure should be placed over the injection site to prevent fat expulsion. The injection sites may be closed with 5–0 fat absorbing gut.

Postoperatively the patient is given 1 mL betamethasone (Celestone) intramuscular (IM) or 20 mg of prednisone orally daily for 5 days along with cephalexin 500 mg or minocycline 50 mg bid for 7 days. The patient is sent home with ice packs to the face and told to maintain them until bedtime. Spot bandages with bacitracin to the injection sites of the face and a pressure bandage to the donor area are applied. The bandages and bacitracin are reapplied daily by the patient. The patient may lightly shower or sponge bathe but not soak in a tub for 4 days. The patient should be seen in 1 week and approximately monthly. The patient should return immediately if there is redness, swelling, pain, or bleeding.

IV. RETENTION OF INJECTED FAT AND TOUCH-UP

There is an expected decrease in the volume of the injected correction due to resorption of nonviable fat cells, serum, and tumescent solution. Fulton's study showed that there was an actual increase in the augmentation for 15–30 days. This is probably due to local swelling and inflammatory response. As the inflammatory cell response mobilizes and removes the nonviable material, the injected volume decreases for about 60 days. After that time, the graft volume remains stable. The areas where fat injection is best retained are the cheeks and chin and least retained in the lips and nasolabial folds. This is also true with collagen fillers, and this is probably because these areas of facial expression rapidly "wear out" the implants. That is why these areas "age" and need augmentation to begin with.

The patient should be reevaluated at 2 months and reinjected, "touched-up", if necessary. The original syringes with cleaned fat are carefully capped and frozen in a refrigerator. Approximately one half to one third of the original extracted fat should be saved. Before the reinjection, the fat-filled syringes are slowly defrosted to room temperature. The donor areas are prepared and injected in the same manner as the original injection. The overinjection should only approximate 50% this time. The patient should be warned that additional injection may be necessary for optimal correction.

V. COMPLICATIONS

Complications with microlipotransfer are few and usually can be prevented. Decreased survival of the injected fat is the biggest problem, although not truly a complication, and should be adequately discussed with the patient prior to the procedure. Bruising or slight erythema of the donor and recipient sites is to be expected but should not persist more than a week or so. Intense swelling accompanied by pain and purplish discoloration signifies a hematoma, which should be immediately evacuated to save the graft cells. Infection is a potential problem in areas treated around the mouth, and these patients must always be treated prophylactically with antibiotics.

Overfilling may be problem initially due to overinjection or underresorption of injected material. It may also be a problem later if the patient gains weight and the injected fat also bulges. These problems may be corrected with dilute triamcinolone (5 or 10 mg/mL) injection or microlipoextraction. Table 3 summarizes the technique for microlipotransfer.

VI. ALTERNATIVE FILLERS FOR FACIAL FOLDS

A number of autologous and nonautologous fillers have been developed successfully over the past few years (Table 4). Some of these are best suited for fine lines and dermal wrinkles, whereas others are amenable for deeper nasolabial or marionette folds. The fillers that best augment deep folds must be placed below the dermis; otherwise there is only a raised line in a depressed trough. These include Alloderm (LifeCell Corp., the Woodlands, Tex.), Goretex (W.L. Gore & Assoc. Inc. Flagstaff, Az), and SoftForm (Collagen Corp., Palo Alto, CA) that are FDA (US Food and Drug Administration) approved at this time.

Table 3 Summary for Microlipotransfer

1. No aspirin, nonsteroidal anti-inflammatory drugs (NSAIDs), or alcohol 3 days before procedure.
2. Tumesce the donor area.
3. Collect several syringes of fat and stand point down in a rack.
4. Wash the fat gently with saline or Ringer's solution and let the fat settle by gravity.
5. Inject the fat evenly in small amounts as the needle is withdrawn.
6. Carefully mold the fat into the desired place.
7. Pressure dressing over the donor area.
8. Steroids, antibiotics, and ice pack postoperatively.
9. Freeze unused fat in the same syringes.
10. Wait at least 2 months between injection touch-ups.

Table 4 Soft-Tissue Fillers

- I. Tissue derived
  - A. Human
    1. Isolagen
    2. Autolagen
    3. Dermalogen
    4. Alloderm
    5. Autologous fat
    6. Autologous dermis
  - B. Animal
    1. Zyderm/Zyplast
    2. Hylaform Gel
    3. Restylane
- II. Synthetic
  - A. Goretex
  - B. SoftForm
  - C. Artecoll
  - D. Liquid silicone

Artecoll (RoFil Medical Breda, The Netherlands) and liquid silicone are not available in this country.

Alloderm is decellurized, freeze-dried dermal tissue from human tissue banks. It is available to the surgeon as a solid sheet which is trimmed and inserted. It may be used to augment large areas, such as the submalar areas, or it may be rolled and inserted along the nasolabial fold, marionette line, or lips.

Goretex, expanded polytetrafluoroethylene (ePTFE), is produced as sheets, single or multiple threads, or hollow tubes. These can be trimmed before use to the needed size. A forceps is tunneled under the fold, grabs the tube or multiple threads, and pulls them through the tunnel. The Collagen Corporation calls this SoftForm, and introduced it as a kit with a tube of ePTFE fitted on a sharp trochar. They come in two widths and two lengths at this time. In the case of Goretex and Softform, the material is threaded in the upper part of the subcutaneous layer under the dermis.

Artecoll is a suspension of microsized Plexiglas beads in a biodegradable matrix. This is a foreign body which is encapsulated by the patient's fibroblasts and is permanent. The technique is very user sensitive and is unforgiving, much as liquid silicone. Augmentation must be gradual with several injection sessions, because removal or adjustment is very difficult.

Liquid silicone as a soft tissue filler has been discussed at length[11–13]. This material used properly in experienced hands has less side effects and is more versatile than any other filler. It comes closest to the ideal filling substance.

Other dermal fillers may be used for more superficial wrinkle lines in the face. These include autologous collagen materials such as Autolagen (Collagenesis Inc., Beverly, Mass.), Isolagen (Isolagen Technologies Inc., Daramus, N.J.), processed bovine collagen Zyderm/Zyplast (Collagen Corp., Palo Alto, CA), human collagen Dermalogen (Collagenesis Inc., Beverly, Mass.), or hyaluronic acid Hylaform Gel (Biomatrix Inc., Ridgefield, N.J.), Restylane (Q-Med, Uppsala, Sweden).

Autologen[14] is the patient's suspended intact collagen fibers in 4 or 6% solution. The collagen must be processed from the patient's dermis, which is collected and shipped frozen. The main disadvantage is that 2 in^2 of dermis is required to produce 1 mL of material. This might work well for a patient who has a face lift or tummy-tuck and also needs soft tissue augmentation.

Isolagen[15–17] is living fibroblast and precollagen material that is derived from the patient. A 3-mm punch of skin is sent to the Isolagen laboratories where the fibroblasts are cultured. The fibroblasts and precollagen material are then returned at required intervals for injection. This delivers living fibroblasts which continue to produce collagen over a long period of time. Up to four or more treatments may be required, but results have persisted over 3 years.

Zyderm/Zyplast bovine collagen has been in use for many years and is quite familiar to most surgeon who do cosmetic work. Its main disadvantage is that it is not permanent. However, experienced users realize that repeated injections will produce a permanent effect. This may be due to the inducement of local tissue reaction and collagen formation.

Dermalogen,[18] like Alloderm, is processed from human tissue bank dermis. This available in suspension for injection. The processing ensures freedom from viral infection or allergenicity. However, it is recommended to do a skin test 72 hs before treatment to rule out reaction to the suspending material.

Hylaform Gel[19] is purified, animal-based hyaluronic acid, which does not illicit foreign body reaction. It is available in 5.5% concentration, and several injections are required to obtain optimal correction. Restylane is a stabilized hyaluronic acid derived from bacteria. At 1 year, patient and physician satisfaction was reported as 60%. Both of these products resorb over time and are only available in Europe at present.

VII. DISCUSSION

Microlipotransfer is a safe and effective treatment for depressed areas of the face such as nasolabial and marionette folds (Fig. 7 and 8). The areas to be treated

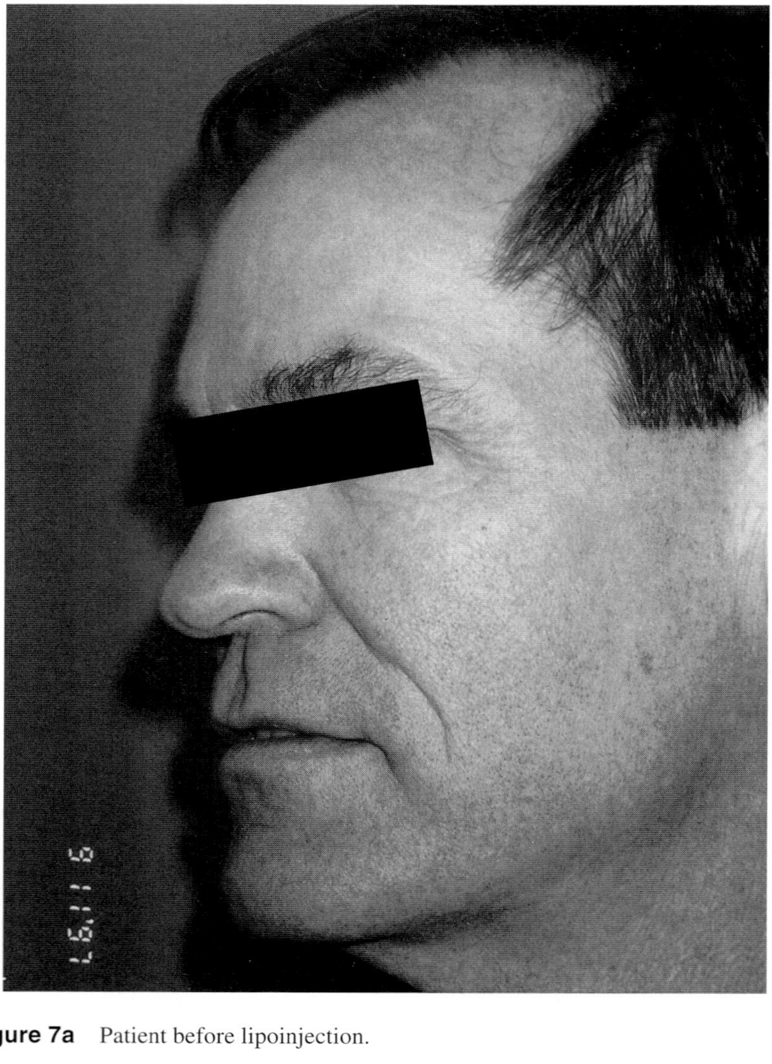

Figure 7a Patient before lipoinjection.

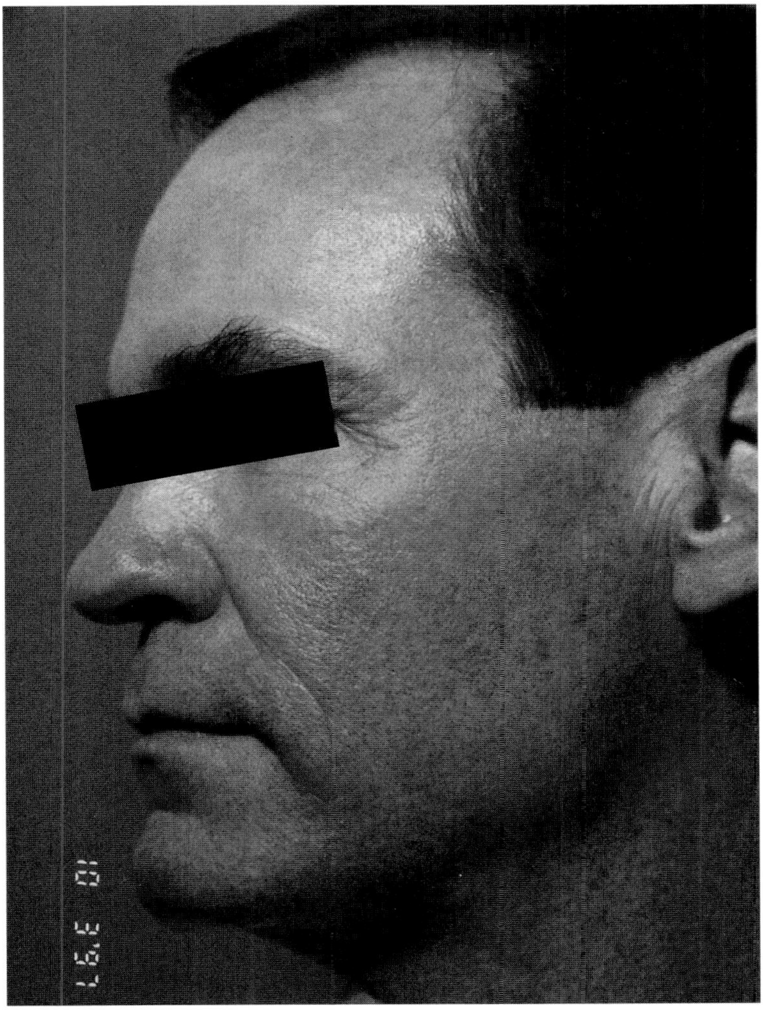

Figure 7b Patient after lipoinjection to nasolabial folds.

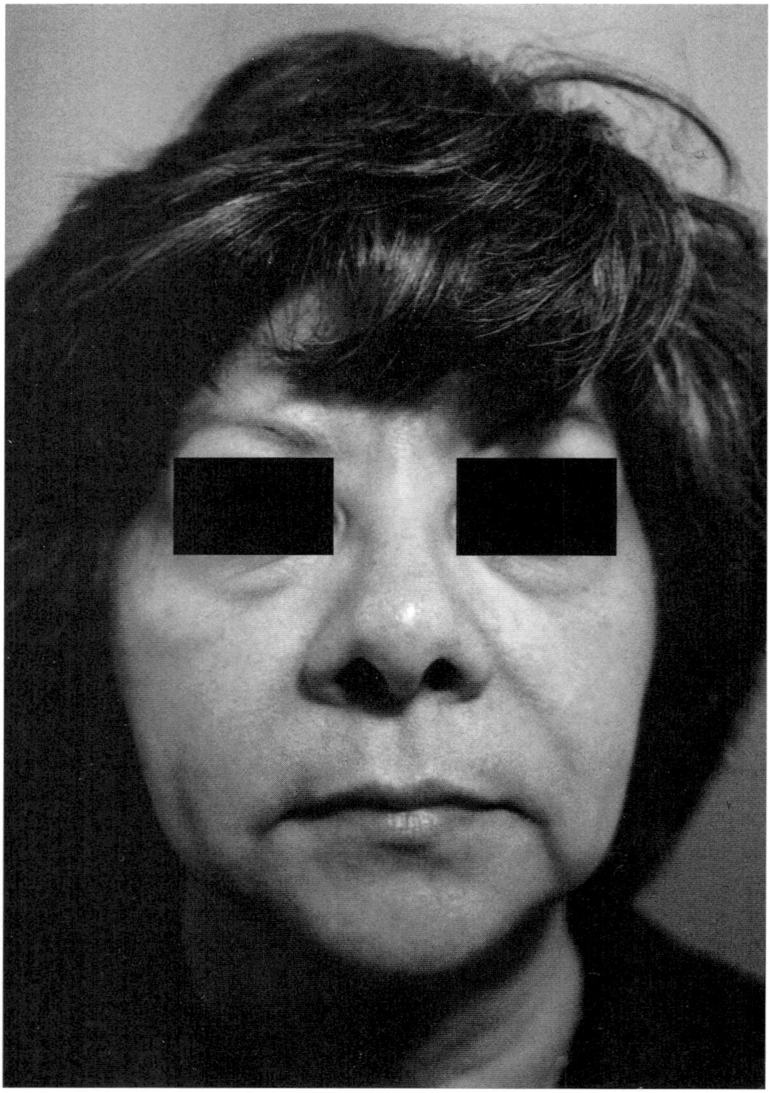

Figure 8a(1) Patient before lipoinjection.

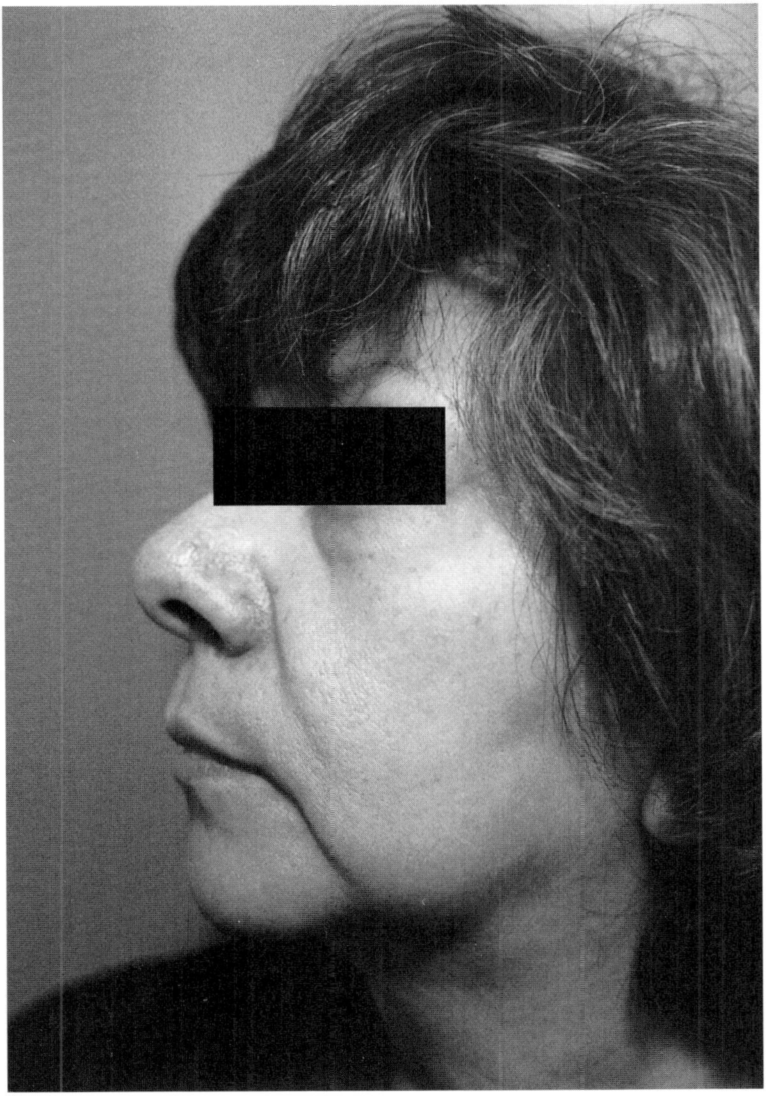

Figure 8a(2) *continued*

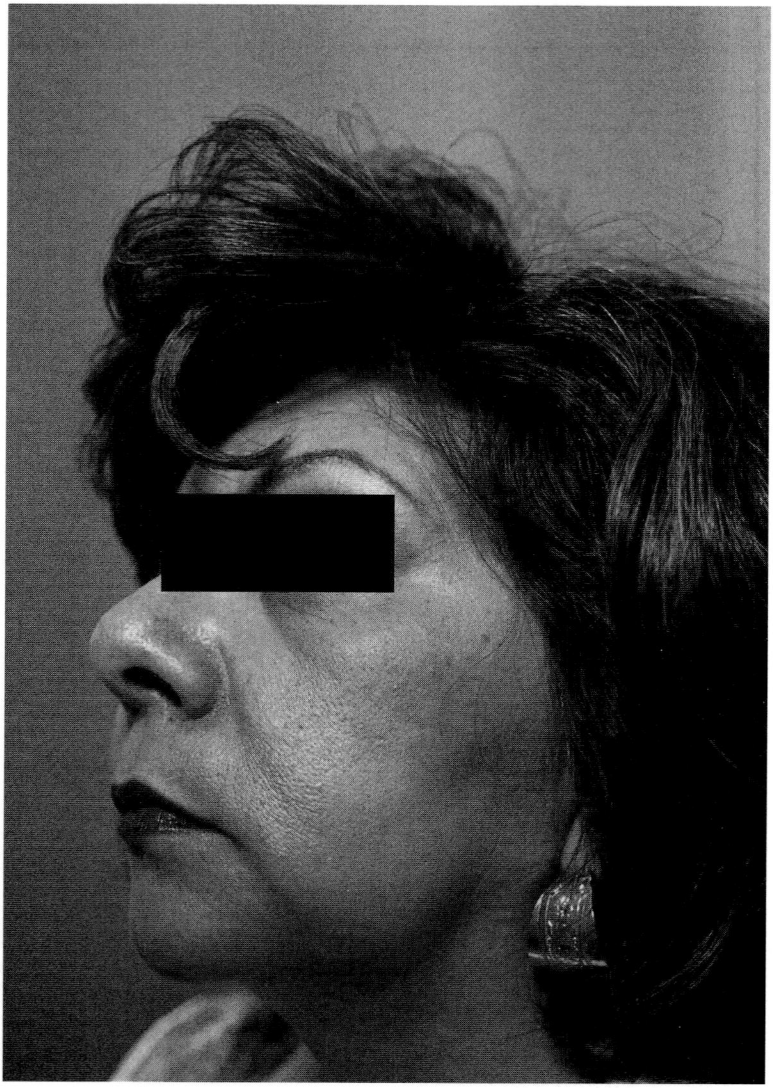

Figure 8b(1) Patient following lipoinjection to marionette lines and chin.

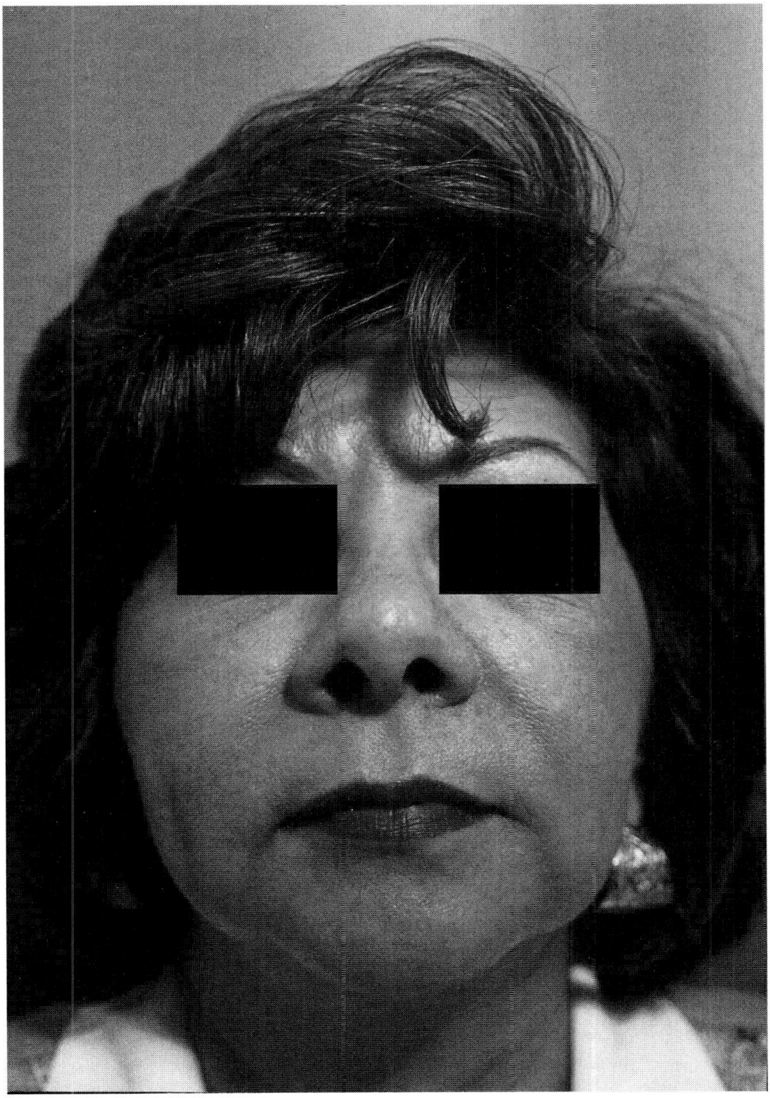

Figure 8b(2) *continued*

must involve the subcutaneous tissue, because that is where fat must be injected. There are allografts which may also be used to treat these types of defects. Each has its advantages and disadvantages. As will all treatment modalities, the surgeon must have proper hands-on training before attempting to use them. Some of the procedures are deceptively simple yet can produce disastrous results if used cavalierly.

REFERENCES

1. Fournier P. Facial recontouring with fat grafting. Dermatol Clin 1989;8:523
2. Chajchir A. Liposuction: fat grafts and wrinkles. Adv Plast Surg 1986;10:115–121
3. Newman J. Preliminary report on fat recycling. Am J Cosm Surg 1986;3:67–69
4. Asken S. Autologous fat transplantation: minor and macro techniques. Am J Cosm Surg 1987;4:111–115
5. Carpaneta C. Percentage of graft viability versus injection volume in adipose auto transplant. Adv Plast Surg 1994;94:496–506
6. Niechajev I, Sevchuk O. Long-term results of fat transplantation: clinical and histologic studies. Plast Reconstr Surg 1994;94:496–506
7. Jackson RF. Frozen fat - Does it work? Am J Cosm Surg 1997;14(3):339–343
8. Hudson DA, Lambert EV, Block C. Site selection for auto transplantation. Aesthetic Plast Surg 1990;14:194–199
9. Pinski JB, Roenick HH. Jr. Autologous fat transplantation: long term follow up. J Dermatol Surg Oncol 1992;18:179–184
10. Fulton J, Suarez M, Silverton K, Barnes T. Small volume fat transfer. Dermatol Surg 1998;24:857–865
11. Orentreich D, Orentreich N. Injectable Fluid Silicone. In: HH Roenigk, HH Roenigk Jr. New York: Marcel Dekker, 1989;1349–1395
12. Webster RC, Fuleihan NS, Hamdan US, Gaunt JM, Smith RC. Injectable silicone: report of 17,000 Facial treatments since 1962. Am J Cosm Surg 1986;3(2):41–48
13. Alkek DS. Lip augmentation with silicone. J Dermatol Surg Oncol 1991;17: 747–748
14. Alster T, West T. New options for soft tissue augmentation. Skin and Aging 1998; 6:32–36
15. West T, Aster T. Autologous human collagen and dermal fibroblasts for soft tissue augmentation. Dermatol Surg 1998;24:510–512
16. Alkek DS. Isolagen, a new autologous collagen. Cosm Dermatol 1998;11:30–32
17. Alkek DS. A new autologous collagen technique. Am J Cosm Surg 1998;15: 325–328
18. Elson M. Soft tissue augmentation with allogenic human tissue matrix. Cosm Dermatol 1998;11:24–28
19. Duranti F, Salti G, Bovani B, Calandra M, Rosati ML. Injectable hyaluronic acid gel for soft tissue augmentation. Dermatol Surg 1998;24:1317–1325

12
Autologous Fat Transfer to the Lips

Steven B. Hopping
George Washington University and Center for Cosmetic Surgery, Washington, D.C.

The lips have always been the focus of beauty and even eroticism. Every culture has shown an interest in enhancing lips. Lips have been elongated, pierced, painted, tattooed, made fuller, made thinner, and even caricaturized. The current enthusiasm for full voluptuous lips has resulted in a significant demand on the cosmetic surgeon to provide such results. This is true certainly in our female patients, but even male patients desire this more youthful lip fullness. Both male and female models are seen depicted with full voluptuous lips, and patients desire a similar look.

There have been various techniques of lip augmentation. Many different filler devices have been utilized to create full lips. The US Food and Drug Administration (FDA) has discouraged companies from marketing filler materials as suitable for lip augmentation, such as collagen and GoreTex (Flagstaff, AZ).

Microsilicone injection to the lips in the past was a common practice that effectively provided fuller lips. This technique is presently illegal in the United States, although it is still used elsewhere. The use of GoreTex, SoftForm, Alloderm (Branchberg, NJ), Collagen, Dermalagen (Beverly, MA), Autologen, and fascia are just a few materials that are used presently to augment lips. There are numerous others. Autologous fat is an obvious choice for lip filling owing to its accessibility, low cost, and ease of placement.

Autologous fat is the most frequently utilized material for lip augmentation in our practice.

I. PREOPERATIVE CONSIDERATIONS

In evaluating the patient preoperatively, it is important that the patient be appropriately counseled in terms of proper expectations from lip augmentation. Patients with thin lips, regardless of filling technique, are always going to have somewhat thin lips. It is difficult to change the size of the lip, and lip augmentation with any material merely enhancing the natural lip shape. There is a definite limit to how much augmentation can be performed in any given lip before the results look unnatural and distorted. Asymmetries of the lip, associated with smiling, can be improved but not resolved with lip augmentation. It is important to observe the patient when smiling and in repose. These limitations must be emphasized to the patient preoperatively.

In evaluating the patient's lips, a judgment should be made in terms of the length of the lower third of the face in relationship to the upper two thirds of the face (rule of thirds) (Fig. 1). The length from the columella to the vermillion border of the upper lip should be noted. In some patients, this length is excessive and would be ideally shortened with a lip lift. According to Juri[1], the ideal length of the upper lip from the base of the columella to the cupid's bow in a woman is 14 mm. The relationship of the lower lip to the upper lip should be evaluated. In general, the lower lip should have 25% more volume or fullness than the upper lip. The shape of the vermillion border and the cupid's bow of the upper lip should be observed. The desired shape of the upper lip should have fullness of the cupid's bow and should have two definite anatomical mounds at the highest point of the cupid's bow on each side of the midline. The lower lip should have more subtle mounds on either side of the midline. The normal vertical distance of the upper lip should be approximately 10 mm and of the lower lip 12–14 mm.

II. TECHNIQUE

Fat is best harvested from the lower extremities (lateral thigh, knee, or buttocks). Fat can be harvested from the abdomen, but occasionally abdominal fat can be somewhat fibrous and the harvesting process somewhat bloody. A 10- to 15-min massage of the abdomen postinfusion of tumescent anesthesia helps minimize this problem. Fat harvested from the neck is generally scant and too fibrous to be effectively utilized for fat transfer.

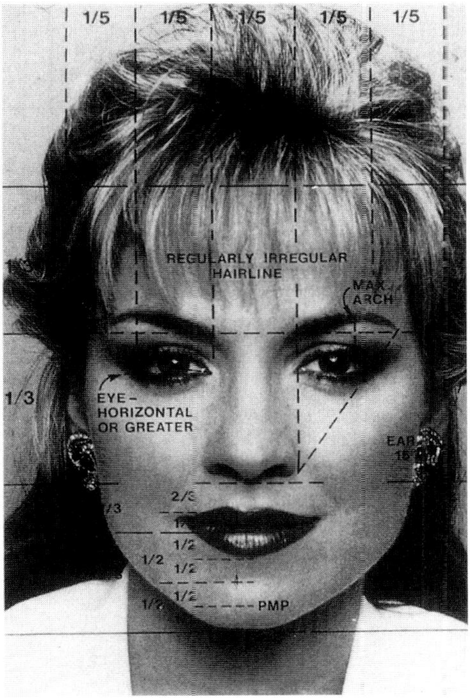

Figure 1 Ideal proportions of the face and lips.

III. TECHNIQUE

A. Harvesting of Fat

The donor area is injected with a modified Klein's tumescent solution (1000 mL of saline mixed with 50 mL of lidocaine 1% plain and 2 mL of epinephrine 1:1000). It is important to allow a 10- to 20-min interval to achieve maximum hemostasis and anesthesia prior to proceeding with fat harvesting. Gentle massage of the donor region for 10 min improves hemostasis, reduces bleeding, and effectively diffuses the tumescent solution more uniformly to the tissues. This is particularly helpful when the abdomen is selected as the donor area.

Harvesting is most easily accomplished with a 14-gauge blunt cannula with one or two small ports (Fig. 2). The cannula is attached to a 10-mL syringe,

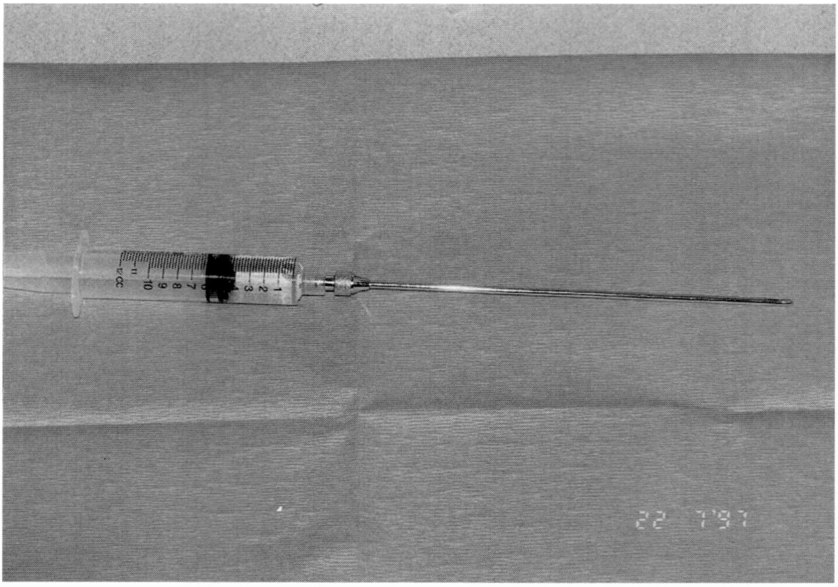

Figure 2 Harvesting system (10-mL syringe primed with tumescent solution and a 14-gauge cannula).

which provides the best system for collecting fat. The 10-mL syringe must be "primed" with a small amount of saline or tumescent solution to avoid an "air lock" and allow the creation of a continuous vacuum and liquid interface. The 10-mL syringe is easily utilized by pulling the plunger backwards and holding it in position with one thumb to achieve the desired vacuum effect. Larger syringes are often more difficult to use in this manner without a lock. The cannula is gently moved through the tissues to collect fat with a minimum of trauma, which is exceedingly important. An attempt is made to harvest between 30 and 50 mL of fat at each harvesting session. A minimum of at least three fat transfer sessions are planned for each patient and spaced approximately 4–6 weeks apart. Patients are informed that between 15 and 25% of the fat on the average will survive with each treatment; necessitating a series of treatments to achieve the desired results. There is tremendous individual variation in the amount of surviving fat.

The 10-mL syringes are placed vertically in a test tube carrier with the plunger at 12 o'clock. They are oriented vertically to allow fat to separate from the infranatant fluid and blood (Fig. 3). The fat, if bloody, can be washed with the same tumescent solution until all visible red cells are gone. This is easily accomplished by adding 2 mL of the same tumescent solution to the syringe and then turning the syringe back and forth to allow the solution to rinse through the fat. The syringe is then replaced in its vertical orientation to again allow separation of the cleaner fat from the infranatant fluid. The fluid is then decanted, allowing pure fat to remain in the syringe. The fat, once cleaned, floats to the top of the syringe and separates from the fluid, red cells, and debris. This allows a no-touch technique in cleaning and separating the fat from the infranatant, which is then discarded. The fat is then transferred, again using a no-touch technique, to 3-mL syringes with the syringe transfer adaptor or three way stopcock (Fig. 4). The 3-mL syringe is ideal for controlling fat injections through an 18-gauge sharp needle or blunt cannula (Fig. 5).

We have abandoned spinning the fat for a number of reasons. It does not appear to be necessary and two *Pseudomonas aeruginosa* infections occurred, which were traced back to the centrifuge device as the source of contamination. The less manipulation and handling of the fat, the more risk free the procedure.

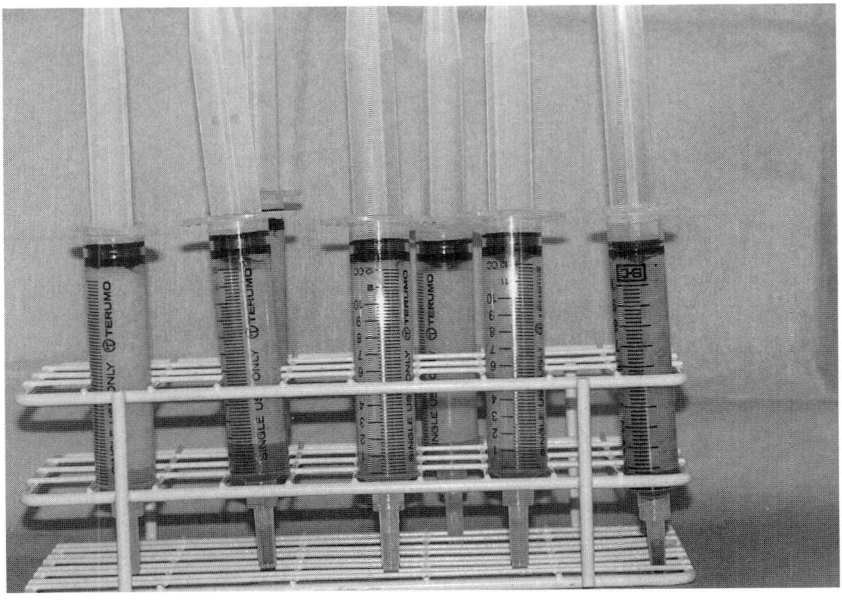

Figure 3 A test tube stand facilitates separation of fat and infranatant.

Figure 4 "No-touch" transfer from 10-mL "harvesting" syringe to 3-mL "injecting" syringe.

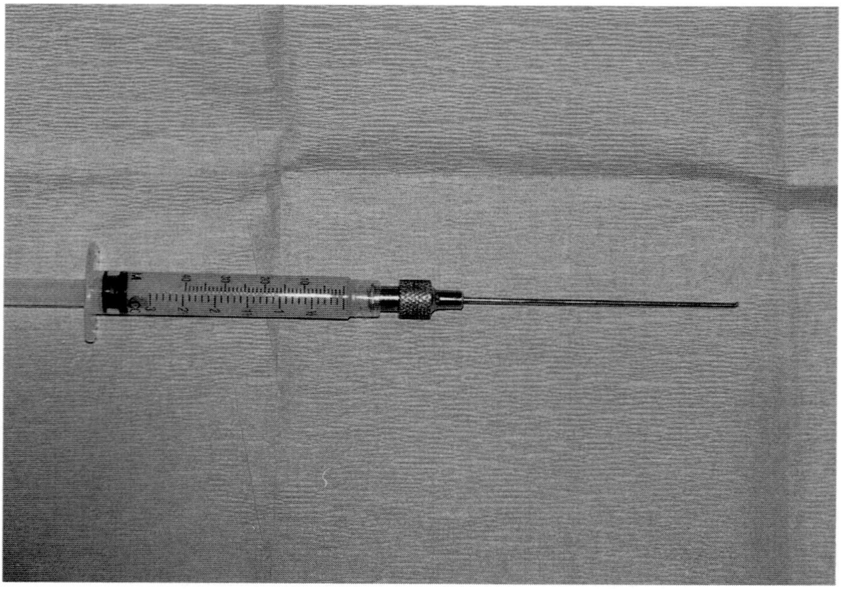

Figure 5 3-mL syringe and a blunt #18 cannula allows safe, controlled "microdroplet" technique of fat transfer.

B. Injection Technique

Utilizing an 18-gauge blunt cannula/3-mL syringe system allows for the safe placement of small aliquots of fat at literally any level anywhere on the face. Fat can be layered into the depths of the lip, into the muscle of the lip, and into the vermillion borders. Fat is also injected into the marionette depressions below the corners of the lips. It is important to inject very small quantities of fat, 0.25 to 0.5 mL at a time, in multiple areas. Conceptually, successful micro-lipotransfers create multiple, isolated droplets of fat, not puddles or lakes of fat. Keeping this concept in mind will help maximize the number of fat cells, achieving vascularity and therefore viability. On the average, 2–3 mL of fat is placed in the upper and lower lips. An 18-gauge needle is used to create a small stab incision at the lateral commissures, through which the 18-gauge blunt cannula can be easily placed. It is important in microlipoaugmentation of the lip that a blunt cannula be used to avoid laceration of the labial vessels and resultant hematoma. Moving the cannula and depositing microdroplets of fat at various planes along the lip will give the best results. Deposit the fat as the cannula is being withdrawn to prevent inadvertent injection into a vessel.

Injecting with the right hand, the upper lip is injected from the patient's right commissure and the lower lip is injected from the patient's left commissure. This allows the left hand to control the recipient lip and guide the cannula to the desired level. An 18-gauge NoKor (Becton Dickinson & Co., Franklin Lakes, NJ) needle is ideal for creating the entrance site for the 18-gauge blunt cannula (Fig. 6), but an 18-gauge needle works equally well. It is not necessary to suture the injection incision left with the Nokor needle or 18-gauge needle. The incision heals quite inconspicuously as compared to a #11 or #15 blade incision. Massage of the injected lip is very important to improve symmetry and reduce lumpiness.

IV. FAT STORAGE

The remaining fat not utilized at the first fat transfer session is placed in 3-mL syringes. Each syringe is individually labeled and then all of them placed in a labeled, sealed cellophane package. The patient's name, date of injection, and birth date are included. The sealed plastic bags are then stored alphabetically in a freezer solely designated for fat storage. It is ideal that the freezer have a backup power source to ensure against a defrosting disaster in the event of a power outage. The freezer should be checked every day by the staff, and a freezer log is kept to ensure the integrity of the system. Alarms and date meters are available for freezers and have been advocated by some practitioners.

We always plan a minimum of three microlipotransfers to the lip, scheduled 4–6 weeks apart, depending on the patient. We anticipate a 15–30% survival

Figure 6 An 18-gauge NoKor needle is ideal for creating an entrance for the 18-gauge blunt cannula. No suturing is required.

with each treatment. Therefore, three treatments should translate into approximately a 45–90% correction rate depending on the individual patient. Obviously, some patients have a better long-term fat survivability than others and may require fewer injections to achieve the desired results. The patient must be committed to repeat injections if the long-term results are to be realized.

When patients return to the office, the amount of fat required for each individual treatment is taken from the freezer and allowed to defrost slowly to room temperature. The areas to be infused are anesthetized with either 1% lidocaine (Xylocaine) or 1% lidocaine with 1:200,000 epinephrine. Once the fat is thawed, it is injected using an 18-gauge needle or blunt cannula. An effort is made to minimally overcorrect (approximately 10%) so as not to distort the patient's features and to maximize long-term fat survival.

V. COMPLICATIONS

Fortunately, the complications with fat compared to other alloplastic materials is quite low. The most common problem encountered with fat is when it does not give the desired long-term results. The second most common problem with fat is that the lips are sometimes distorted for days following the fat injection procedure. We have avoided this by injecting less rather than more at each session. This technique improves survivability and also results in less distortion in the immediate postinjection phase. It also results in fewer hysterical calls from patients or their family members on the evening of fat transfer. The majority of our patients have a fat treatment and return to their usual activities either that day or certainly the following day.

Bruising is another possibility, and again is best avoided by the use of a blunt cannula for injection and immediate pressure on the injection areas following treatment. Ice is definitely helpful.

Infections are a significant complication in fat injections, and can cause tissue loss, tissue scarring, and necrosis. We have experienced two infections, both involving culture-proven *Pseudomonas aeruginosa*. Both of the infections were traceable to a contaminated centrifuge. Since we no longer centrifuge our fat, further difficulties with infection of fat have not been encountered. Patients are not routinely given antibiotics or pain medication for this procedure.

Hematomas can certainly happen in the richly vascularized lip and can be minimized by using the blunt needle injection technique. Asymmetry is certainly not uncommon, but it is easily corrected with further fat injections.

Scarring is possible but has not been a problem utilizing the 18-gauge NoKor needle and 18-gauge blunt cannula. Hypesthesia or anesthesia has not been a problem utilizing fat as it is in some of the other augmentation materials.

Lipomatous formation after fat transfer has been seen in a number of cases.[2] This presents as cystic lumps of the lip that must be distinguished from a mucocele of the minor salivary glands. Treatment is incision and drainage. A cholesterol-like secretion is usually encountered, and no further treatment is generally necessary.

Mucocele or cyst formation can also occur and is related to blockage of one or more minor salivary glands. These are best treated by simple incision and drainage, and a yellow fluid is often expressed. No further treatment is generally necessary.

VI. RESULTS

Figures 7 and 8 show two examples of women desiring fuller lips after three fat transfer treatments spaced 6 weeks apart.

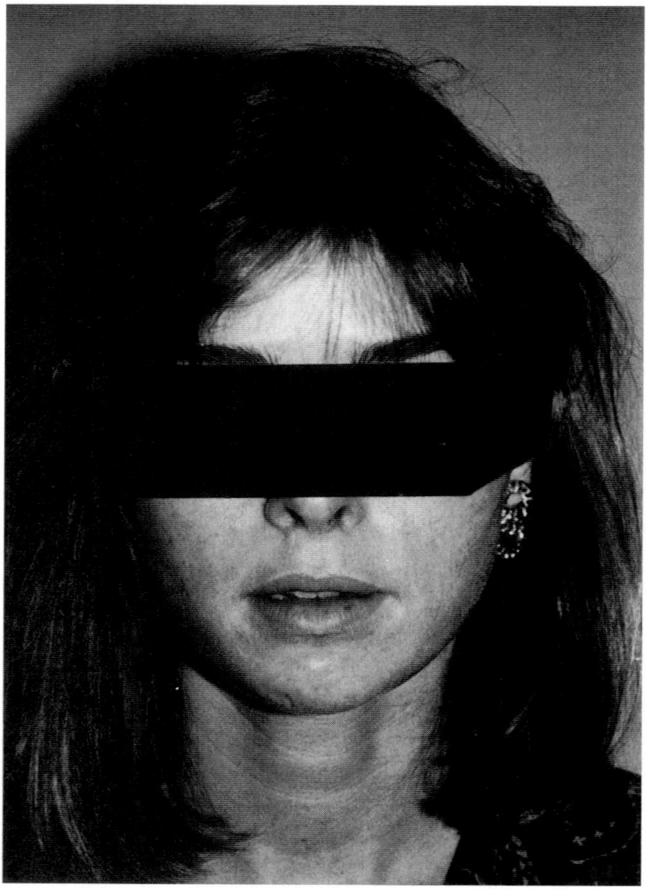

Figure 7a Patient with thin lips.

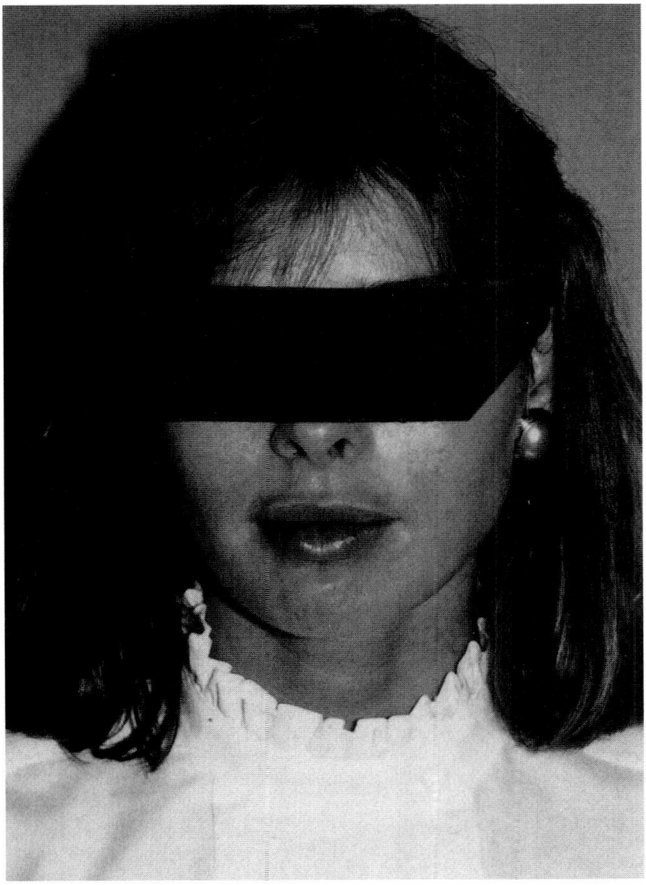

Figure 7b Results after three fat transfers spaced 6 weeks apart.

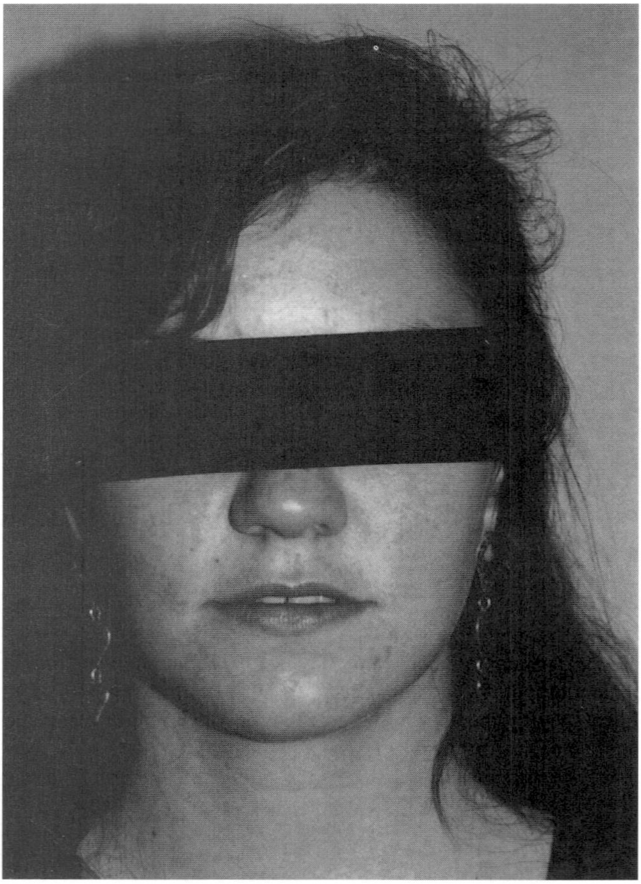

Figure 8a Before lip augmentation.

VII. DISCUSSION

Fat is an ideal material for lip augmentation, since it is readily accessible, is well tolerated in the lip, and has a natural feel. Voluptuous full lips are a desirable esthetic quality. There is a large demand on the part of patients to achieve such results. Fat is an excellent material to augment lips and is our lip augmentation treatment of choice. Patients must be made aware of the alternatives, of the

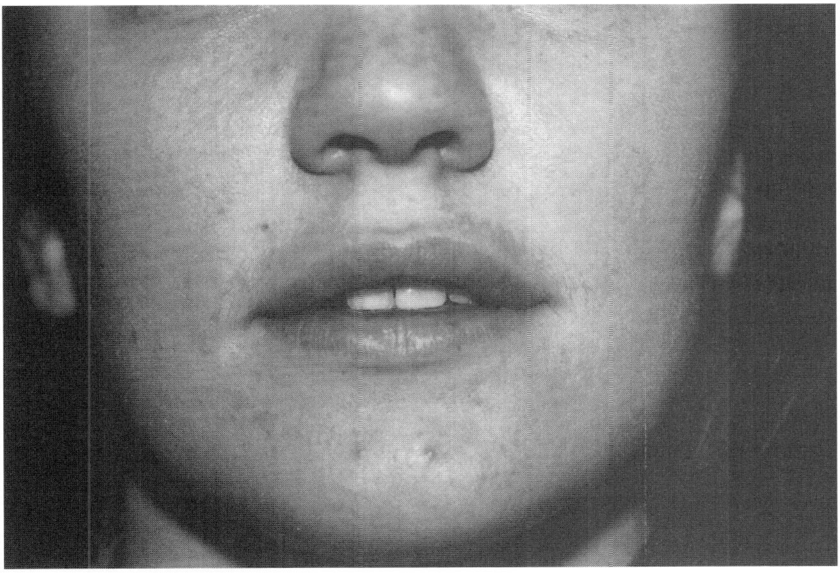

Figure 8b After lip augmentation.

limitations, and of the long-term sequelae of fat as well as other techniques. Fat absorption remains the biggest challenge with autologous fat lip augmentation.

In elective surgery, we must always weigh the risks against the advantages. Fat augmentation of the lips has the marked advantage of very few complications. It is always our first recommendation to patients expressing an interest in lip enhancement.

REFERENCES

1. Juri J. Personal communication, 10/98
2. Khawaha H, Hernandez-Perez E. Lipomatose formation after fat transfer. Int J Cosm Surg 1998;6(2):144–145

13
Eyebrow Lift with Fat Transfer

Giorgio Fischer
Fischer Institute, Rome, Italy

Traditional eyebrow lift is with surgical incisions involving the eyebrow area or coronal approach as well as endoscopic lift with smaller incisions. In 5–10 years, the surgery may have to be repeated because of continued skin relaxation and fat absorption or shifting with the aging process.

The presence of autologous fat, which can be easily obtained, sometimes in association with body contouring procedures, makes fat an ideal filler to lift the eyebrows. Aging of the face involves skin laxity and loosening of fat in the forehead and eyebrow region, which results in drooping of the eyebrows. Simply injecting fat in the proper areas will reverse the aged appearance (Figs. 1a, b). and can be reutilized as often as necessary.

I. TECHNIQUE

One month before fat transfer to the forehead, botulinum toxin (Botox) is injected into the mid to upper portion of the forehead. This is repeated 1 month prior to each of the follow-up reinjections.

Intravenous sedation with modified tumescent anesthesia is utilized for harvesting the fat. A limited amount of solution is used to allow a more concentrated aspirate. Donor sites include the cervicofacial area, lateral and medial thighs, abdomen, flank, and medial knees. A single-hole sharp cannula or 18-gauge needle is used for the harvest and the fat is retrieved in 1- or 3-mL Luer-Lock syringes. It is not necessary to transfer the fat prior to injection, so that there is limited fat

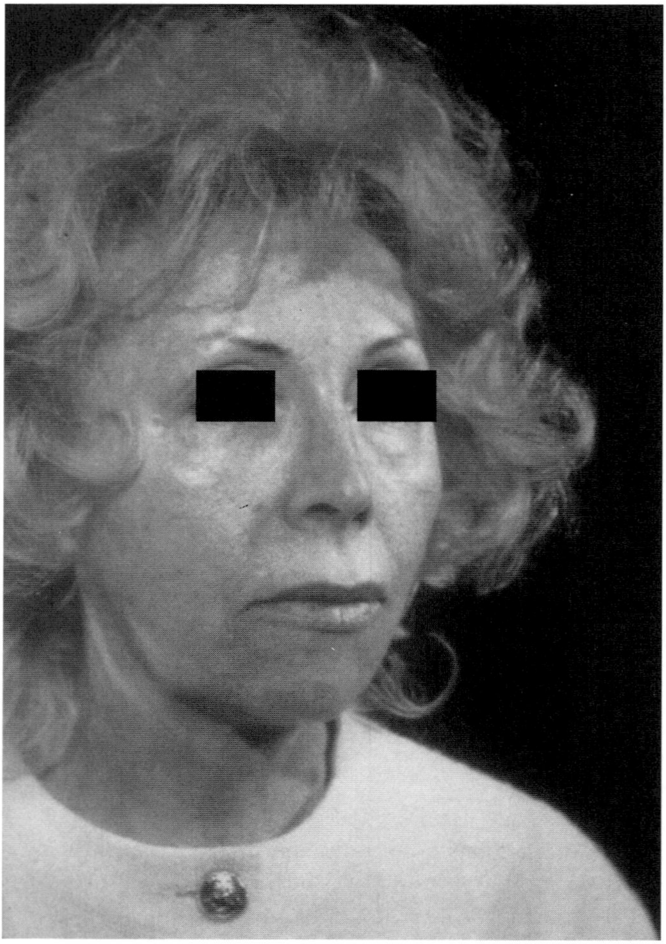

Figure 1a Preoperative.

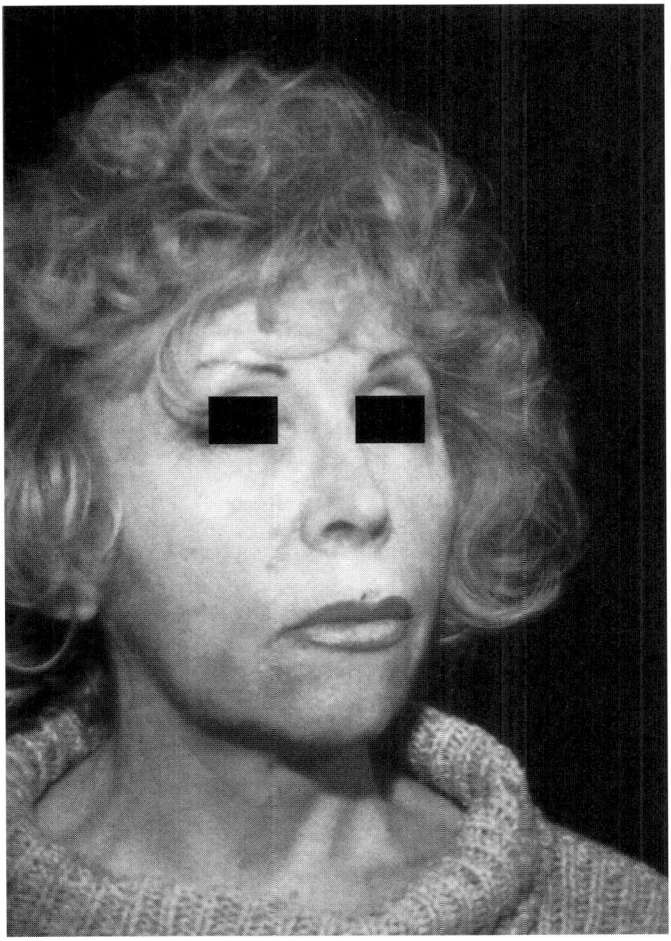

Figure 1b Postoperative eyebrow lift with fat transfer.

manipulation decreasing the chance of fat cell damage. Slow, constant, and gradual negative pressure is applied while the cannula or needle is directed in a spokewheel fashion to prevent depression in the skin overlying the donor site. The syringes with fat are placed in an upright position.

The aspirate is allowed to settle into two distinct layers over a period of 5–6 mins. The bottom layer, consisting of lidocaine and Ringer's solution, is decanted and discarded.

The recipient site is marked and prepped in sterile fashion. The entry sites, along the margin of the hairline, are injected with 2% lidocaine containing epinephrine 1:100,000. Nerve blocks are useful to prevent distortion of the soft tissues caused by infiltration of anesthetic into the recipient area. Small amounts of fat are deposited, in herring bone fashion, in tunnels which are oblique and lateral to the medial part of the forehead and oblique to the temporal region. The cannulae used are curved to follow the contour of the frontal bone. When contraction occurs during the healing process, an upward lift and shrinkage of the skin result. The grafts are placed up to 1 cm above the eyebrow, since a downward pull occurs if the fat is placed in close proximity to the eyebrow. Three separate sessions are usually necessary, injecting 2–3 mL of fat each time at 3-month intervals. Patients over 40 years of age require the larger amount of fat.

It is advisable to fill the dark shadows under the eyes with 0.5 mL of fat to obtain a better overall cosmetic result.

II. POSTOPERATIVE CARE

The implants in the forehead are not massaged, although the implants under the eyes are massaged gently for a few seconds each day to avoid contraction of the implant.

III. DISCUSSION

Fat transfer to the forehead is an excellent method to lift the eyebrows. With the use of small quantities of fat, approximately 20% reabsorption has been noted.

Ptosis of the eyebrow does not occur with the use of Botox if the injection is in the medial and upper portion of the forehead. Combining fat transfer with Botox prevents muscular contraction, which tends to pull down the skin of the eyebrow and forehead. Filling the lower eyelids to improve the dark shadows improves the total result.

14

Carbon Dioxide Laser Resurfacing and Autologous Fat Injection

Steven L. Harlan
Laser Resurfacing and Vein Center, Des Moines, Iowa

Combining laser resurfacing and autologous fat injection (AFI) presents special challenges and considerations. For rejuvenation of the aging face, experienced carbon dioxide (CO_2) resurfacing cosmetic surgeons are accumulating experience in augmenting with autologous fat injection where laser resurfacing results tend to be limited. Besides rolling back the clock a few more years than laser alone, fat injection makes patients more attractive people. Prior to the use of CO_2 resurfacing, if someone had an unattractive chin, cheeks, or lips, they still were unattractive, although younger looking, after laser treatment. This is an exciting adjunctive therapy milestone, and the majority of relevant experience at this time is with autologous fat injection either weeks before laser resurfacing procedure or in the weeks after the healing from the laser resurfacing procedure.

In most cases, autologous fat injection is performed two or more weeks after laser resurfacing. Fulton[1] has demonstrated that for cosmetic surgeons with extensive experience managing the healing of both laser resurfacing and autologous fat injection some fat injecting may be performed on the day of full-face laser resurfacing. Tightening the skin and smoothing wrinkles with ultraspulsed CO_2 lasers do not address all of the changes that cause the face to look older. Standard remedies such as chemical peels and rhytidectomy do not correct the problem of subcutaneous fat atrophy, loss of supporting cutaneous structure, and facial dimensional problems. Laser resurfacing alone will correct deep nasolabial furrows, subcommissural furrows of the lateral chin, the horizontal mentolabial crease of the central chin, narrow mandibular width, and malar grooves and fossae. Particularly common in the older laser resurfacing patient is loss of facial volume in the brow, infraorbital malar, and perioral areas.

Laser resurfacing cosmetic surgeons will be able to offer patients a quantity of the most natural, inexpensive, and versatile filler substance—the patient's own fat. With same-day harvesting, the cosmetic surgeon and patient together may plan to freeze all of the fat for later treatments or to inject some of the fat immediately before the resurfacing procedure. It is recommended that this initial fat injection be performed in the malar areas of the face if a same-day injection scenario is planned.

I. PRINCIPLES OF CO_2 LASER RESURFACING

In the 1970s, the concept of selective photothermolysis was introduced and provided physicians and engineers with a goal of designing lasers that were highly selective and safe. CO_2 lasers became commonly used as a surgical tool in the 1980s. However, the original continuous-wave CO_2 lasers produced significant thermal injury to the facial skin and vermillion portion of the lips. As a resurfacing tool, these lasers unavoidably produced carbonized char, significant thermal injury, and unacceptable risk of scarring.[2–5]

With advancement in laser technology in the 1990s, ultrapulsed CO_2 lasers are able to produce high-energy outputs and exceedingly short-pulse periods providing for ablation and then thermal relaxation of the tissues. As long as char is not present to conduct heat to the adjacent tissues, thermal injury and risk of scarring is controllable and acceptable. Ultrapulse CO_2 lasers are ideal for resurfacing facial skin because of their long wavelength of 10,600 mm giving them strong water absorption to ablate without regard to color or pigmentation. The physical properties of this particular wavelength also provide the superior hemostatic properties of CO_2 lasers. A single pulse of ultrapulsed CO_2 laser will only penetrate to a precise and limited depth into tissue. Bloodless and precise vaporization of tissue provide direct visualization of endpoints that are unmatched by anything else. Visualization of the desired endpoints like smoothing a scar, smoothing a wrinkle, and removing solar elastotic skin is excellent.

When the physician is properly trained, understands minimum power densities for the ablation threshold, and observes appropriate density and number of passes, the risk of scarring is very acceptable with the ultrapulsed CO_2 lasers. Excessive passes or excessively overlapped passes will increase the risk of scarring. Dermal structures apparently will tolerate only a certain number of passes before scarring becomes likely. This varies of course with the thickness of the skin, location, and dermal appendages present and from patient to patient.

The ultrapulsed CO_2 lasers do produce enough thermal effect on the dermal collagen to cause a pleasing tightening of the skin. With healing and remodeling of the dermal collagen, it is hoped that some of the effects of tightening are long lasting. It is believed that heat generated from the ultrapulsed CO_2 lasers has a specific effect on type I collagen fibers. These collagen fibers respond to temperatures as low as 50–60°C by shortening and contracting.[6,7] Remodeling of dermal collagen during the weeks and months after initial healing provides the opportunity for some of this tightening effect to remain. Avoidance of sun exposure and wind exposure in the first 12 months after laser resurfacing is particularly important for the effects to be lasting. Long-term skin care programs utilizing topical alphahydroxy acids, topical retinoids, and topical antioxidants may be important. As with new innovations, many more new questions arise than are answered, and therefore clinical experience will be essential.

Laser research has demonstrated the production of new collagen and elastin fibers in the dermis after CO_2 resurfacing. The practitioner knows that the rejuvenation of skin from a single CO_2 laser resurfacing procedure is difficult to match with anything else. Most patients can obtain impressive and very natural results without the polished or plastic looking change associated with deep dermabrasion and deep chemical peel. The bloodless and precise procedure of ultraspulsed CO_2 laser resurfacing is now often preferred over dermabrasion and deep chemical peeling.

During the procedure, the laser resurfacer must constantly be aware of the amount of tightening occurring to the inferior orbital skin. Mild ectropion of the lower eyelid which resolves in the first week postoperatively is acceptable and is frequently the intended endpoint in the hands of very experienced operators. However, excessive tightening of the malar and lower eyelid skin may produce ectropion that lasts for months but generally resolves if scarring is not present. Contracture and ectropion can be permanent, particularly if scarring of the orbital skin occurs. All early signs of scarring like persistent localized erythema and induration should be addressed immediately with topical steroids used intermittently and dilute intralesional 3 mg/mL triamcinolone.

CO_2 laser resurfacing is extremely stimulating to dermal pain receptors. Unless at least excellent regional nerve block anesthesia is used, patients must be very deep under intravenous sedation. An anesthesiologist attempting intravenous anesthesia for this procedure often will have the patient anesthetized either too deeply or too lightly during such a significant amount of the procedure that it is quicker for the surgeon and assistant to anesthetize the facial area locally. The collimated CO_2 lasers do not defocus and have the potential to ignite gases in tubing anywhere in the room. Appreciation of lidocaine limits with local and tumescent anesthesia must be foundational in the knowledge of cosmetic surgeons performing these procedures.

Lidocaine mixed in 2000 mL of saline as a 0.1% tumescent solution can be introduced into the trunk subcutaneous fat without toxic serum levels of lidocaine developing. It must be understood that when 0.05–0.1% lidocaine solutions are used in a panniculus, the safety of 35 mg/kg for lidocaine established by Klein[8] and Lillis[9] hold true. When 1–2% lidocaine solutions are used in the face, the actual safety limits are much closer to the original 7 mg/kg limit established for lidocaine toxicity.

In addition to establishing a protocol plan for prevention, the CO_2 laser resurfacing cosmetic surgeon must be prepared immediately to culture and diagnose all infectious and allergic complications.

Except for the fairest individuals with type I skin, posttreatment hyperpigmentation will occur in 50% of patients without preventive measures. This tendency develops by the third week postoperatively and requires 2–3 months to stabilize. It is important to get any hyperpigmentation resolved in the first 6 months, as it may become more dermal and recalcitrant after this. Posttreatment hyperpigmentation is reduced when patients use a class V topical steroid cream starting in postoperative week 2. In postoperative week 3, it is also very helpful if the patient can begin wearing a sunscreen containing make-up which blocks all wavelengths of light. In all but the fairest of type I skinned blue-eyed individuals, hydroquinone cream and retinoid cream is resumed by the third postoperative week.

Finally, the potential but not guaranteed benefits of skin rejuvenation and a more youthful appearance must be more important to the patient than the 10–15% risk of hypopigmentation. Both men and women who are contemplating laser resurfacing must consider and accept the risk of hypopigmentation as a tradeoff for the benefits of erasing wrinkles and improving loose droopy skin. This is a particularly important consideration with women who strongly dislike the idea of wearing makeup. Delayed hypopigmentation after laser resurfacing is a delayed hypertrophic scarring phenomenon. A process of delayed fibrosis and scarring causes melanocytes to disappear and in some cases, there is a delayed disappearance of pilosebaceous follicles by this subtle process (Fig. 1). The incidence of delayed hypopigmentation can be reduced by the intermittent use of topical and intralesional steroids.

Provided that these principles are observed, the most common cause of problems with laser resurfacing is a patient who is unwilling or unable to follow directions.

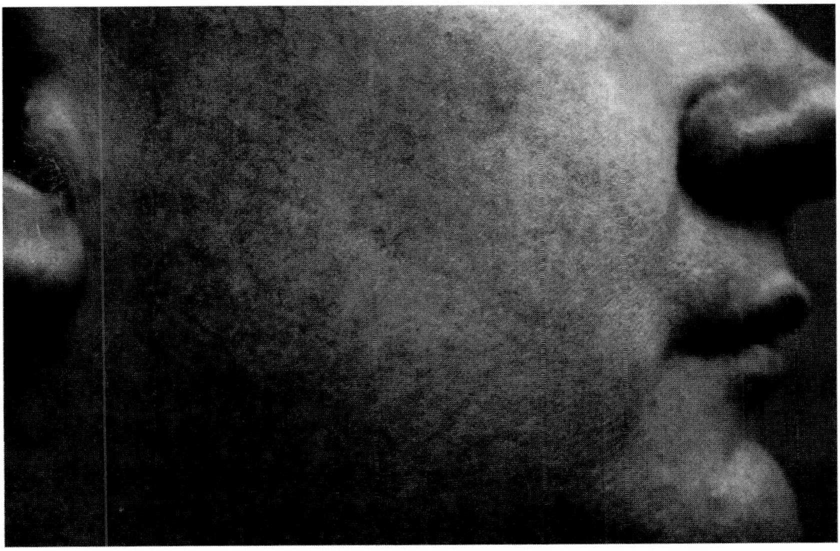

Figure 1 CO_2 laser induced delayed hypopigmentation showing early perifollicular disappearance of melanocytes and disappearance of follicles within larger areas of hypopigmentation.

II. PATIENT SELECTION AND INDICATIONS FOR COMBINED CO_2 LASER RESURFACING AND FAT INJECTION

Patients may be offered autologous fat injection at the time of a laser resurfacing, 2 weeks after, and several months after laser resurfacing.

Skin and wrinkles are not the only changes that cause the face to look older. Standard remedies like chemical peels, face lift, and laser resurfacing fail to address one of the primary problems with the aging face, subcutaneous fat atrophy. Loss and flattening of supporting brow fat contributes to brow ptosis and lid ptosis. Loss of malar fat accentuates jowls, nasolabial lines, and lower eyelid fat pads. Flattening of the lips and loss of perioral fat contributes to the aging lower face. Subcommissural depressions and diagonal furrows of the lateral chin are very resistant to laser resurfacing. Results with the horizontal mentolabial crease of the chin are also disappointing with laser alone. Subcision or aggressive pretunneling is important here, but after two fat injections, this mentolabial area responds nicely.

We believe that AFI to the nasolabial furrow area can be more effective than face lift or laser resurfacing alone. Patients are frequently very pleased even with subtle improvement of the nasolabial furrow with AFI. Often, even when the physician feels that the fat graft take was less than expected, the patients are pleased with the nasolabial furrow area. The fibrous connections between the nasolabial crease and the subcutaneous musculoaponeurotic system (SMAS) soften and become less harsh. With or without aggressive pretunneling or subcision, fat injection is especially helpful in this area. If the patient later has face lift surgery and the SMAS is pulled laterally, the previous nasolabial fat injection may be helpful in softening the SMAS-dermal connections in this area, providing a better result.

Acne scarred patients frequently have an asymmetrical loss of subcutaneous fat in the cheek and temple areas. Quite often, it is apparent that the destructive inflammation of acne contributed to irregular atrophy and fibrosis of the subcutaneous adipose tissue. Skin tightening from resurfacing CO_2 lasers can help this problem somewhat, but usually not adequately. The most logical natural replacement is autologous fat. Soft undulating acne scars respond to superficial fat injection. Deep injections are indicated for raising of the entire area and creating symmetry. Not infrequently, acne scarred patients report a more satisfactory improvement from the autologous fat injection than from their laser resurfacing (Fig. 2).

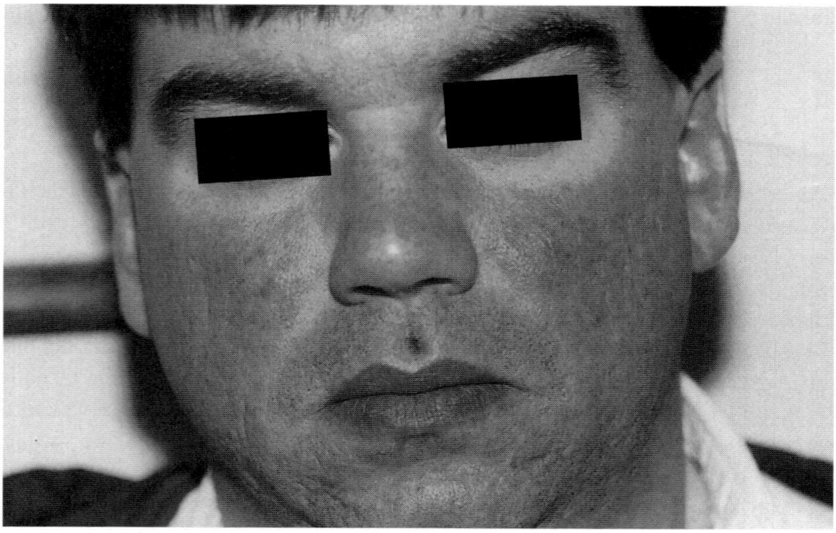

Figure 2a Acne-scarred patient before CO_2 laser resurfacing.

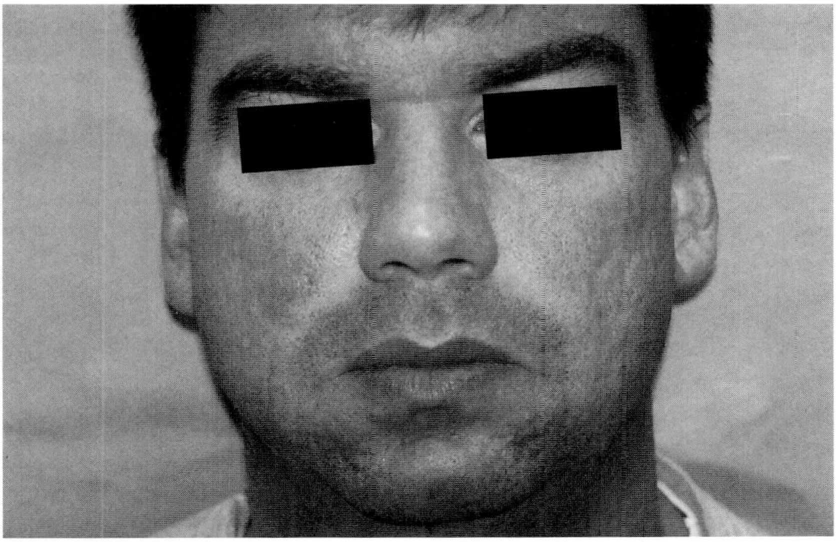

Figure 2b After laser resurfacing.

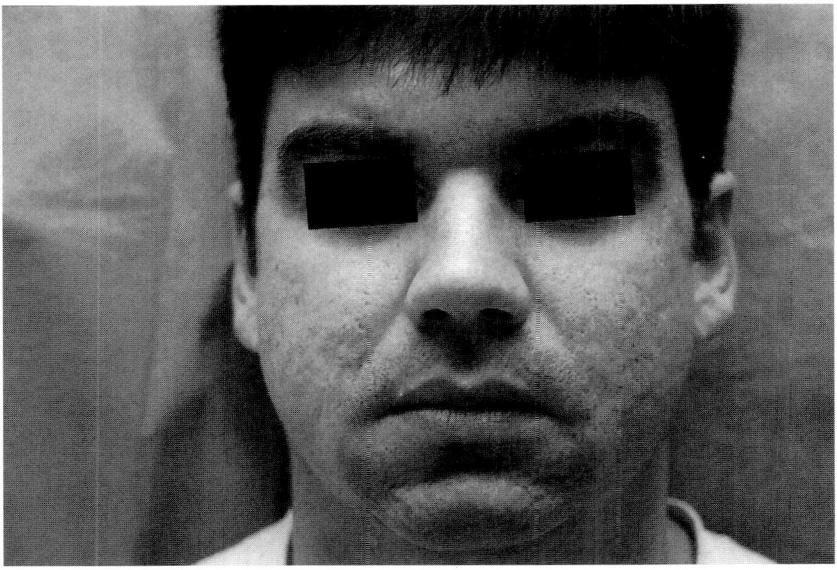

Figure 2c After autologous fat injection to cheeks and temples.

Angular, concave areas of the face must be softened with a deep-filling substance that addresses the problem. Superficial dermal fillers are not helpful for this. Patients seeking laser resurfacing to improve the face actually may have natural asymmetry or concavity of the facial fat that must be esthetically addressed. Patients with lipoatrophic or asymmetrical areas and deep dynamic furrows are likely to be disappointed after laser resurfacing alone. The cosmetic surgeon must point out these limitations early in the consultation. Fortunately, these patients generally quickly comprehend this and are receptive to autologous fat augmentation. Frequently, narrow mandibular width is part of the problem with asymmetry and esthetics.[10] Ideally, the horizontal width between the mandibular angles should be approximately 80–85% of the distance from the chin to the glabella for optimal esthetics. Augmentation with autologous fat addresses this problem nicely (Fig. 3).

These patients are likely to be disappointed after laser resurfacing alone unless the cosmetic surgeon points out to them during the preoperative consultation

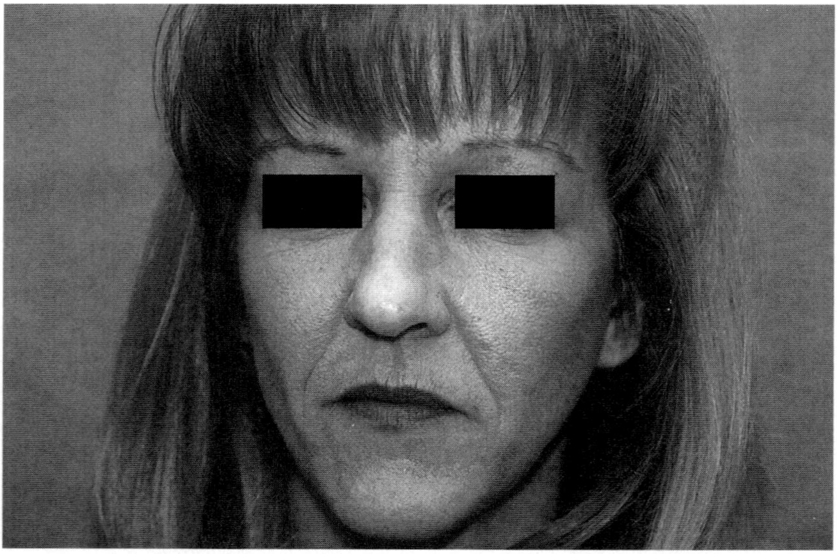

Figure 3a Patient after full-face laser resurfacing with only modest improvement because of narrow mandibular width and cheek fossae.

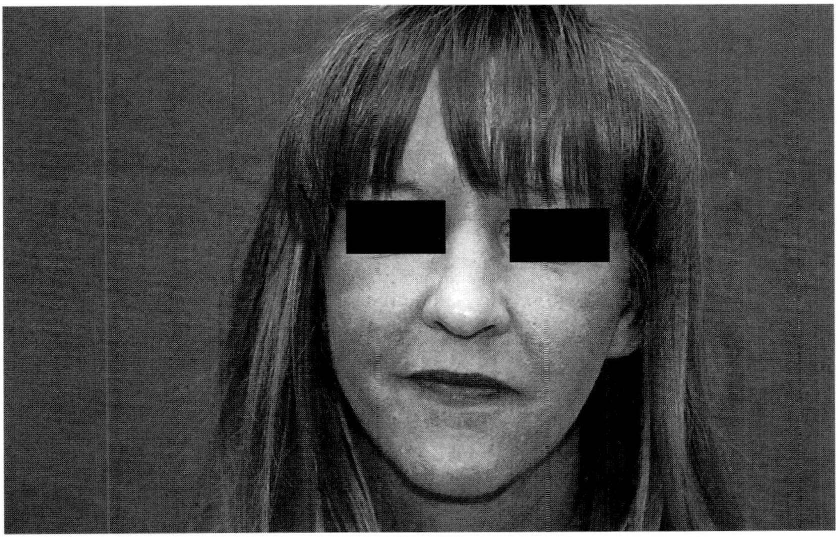

Figure 3b Appearance after autologous fat injection.

the appropriate expectations for laser results and the role of autologous fat augmentation. Addressing these unattractive features with fat injection will increase the likelihood of a happier and more attractive laser patient.

The psychology and the expectations of the autologous fat injection fit well with laser resurfacing. Patients seeking laser resurfacing often are looking for more natural rejuvenation alternatives to face lift. The patient who has seen examples of the "wind tunnel" look from facial surgery and does not like the idea of long incision scars at the periphery of the neck and face frequently is receptive to a combination of two more natural approaches. Multilevel rejuvenation of the skin with laser resurfacing and concomitant rejuvenation of the supporting adipose tissue fits well with this type of cosmetic patient.

Patients often are fat injection candidates and do not know it. They seek out the cosmetic laser surgeon because they believe they need something but are not sure exactly what it is.

Patients who want dramatic facial changes are better referred for face lift surgery. They are likely to be disappointed that autologous fat injection and laser resurfacing did not produce the changes that are part of their expectations.

Patients seeking laser resurfacing rather than face lift are often biased psychologically toward the most "natural" filler substance even if the results are less dramatic and less permanent than with synthetic or semisynthetic implant materials. This type of patient is generally more tolerant of the failure of a "good take" with AFI, because the physician and patient at least attempted the most natural replacement. These patients tend to be less unhappy with less than a good take. A partnership between the physician and the patient with the patient supplying the filler substance and the physician performing his or her best to make that filler substance work is a very positive start. Together the physician and the patient perform as a team to obtain the best results. Pleasing improvement is expected, but it is understood that the most common sequelae is failure to achieve a good take with fat grafting. Patients often feel a sense of accomplishment along with the physician by contributing the filler substance to the cosmetic procedure. This may account for some of the unexpectedly high acceptance by patients when fat take in an area is disappointing.

Many of these laser resurfacing–oriented patients will not tolerate as well a permanent synthetic implant that moves or shifts or has the slightest unnatural bulge. Whereas for these same patients, a small step-off or leading edge of fat graft that is not perfect seems to be much more likely to be accepted and overlooked by these patients. Making someone more attractive with their own fat is more exciting and pleasing to patients who are the best candidates for this procedure.

III. PSYCHOLOGICAL CONTRAINDICATIONS

It is especially important for the patient who is preparing to undergo combination laser resurfacing and autologous fat injection to be capable of following directions. We do not laser resurface anyone who cannot or will not successfully use a small amount of topical hydroquinone in the morning and a small amount of retinoid cream nightly for 4 weeks preoperatively. The preoperative regimen is as important as a device for psychological screening as it is a helpful preparation of the skin. Patients who fail the test of following these simple preoperative skin care instructions should definitely be reconsidered as laser candidates. However, a neurotic patient with reasonable expectations who follows instructions well is not necessarily a bad candidate for this procedure. Patients who have complaints or difficulties with these products like nonspecific irritation or allergies should be cautioned to reconsider laser resurfacing. They should be warned that dealing with postoperative hyperpigmentation may be a long and difficult road, and that they should expect that postoperative pain, discomfort, and itching could be a roller coaster experience for up to several months postoperatively. The goal of skin rejuvenation and looking younger must be a powerful enough motivating factor that the patient wishes to go ahead with the laser resurfacing.

Patients who have exceedingly high expectations for autologous fat injection results are probably better served by referral for face lift surgery. Also, if the patient has a fixation, an overly critical opinion of very slight jowl effect, or slight cheek ptosis, one should talk about the option of referral for face lift consultation.

Both men and women who are considering laser resurfacing must consider and accept the risk of permanent hypopigmentation as a tradeoff for the benefits of erasing wrinkles and improving loose droopy skin. This is particularly important with women who strongly dislike the idea of wearing make-up. We rarely resurface female patients who say that they cannot or will not wear make-up.

The physician should beware and proceed with caution when the problem the patient is seeking to improve can be covered up with make-up. This patient is unlikely to accept any kind of pigmentation imperfection postoperatively. Do not laser a woman who wants to be able to "go without make-up." This patient is seeking perfection, not likely to tolerate hypopigmentation, and may not follow instructions for covering up with make-up the first months post-operatively.

IV. PREOPERATIVE PREPARATION

Female patients should have a sunscreen containing make-up that they are able to tolerate wearing before the laser procedure is performed. Make-up is extremely important for resolution of postoperative erythema in women. Women who wear no make-up in the first 2 months following laser resurfacing have a more prolonged recovery from erythema. Make sure that men and women have found, tested, and tolerate an SPF 30 sunscreen before the day of laser resurfacing.

The patient is instructed in using a chlorhexidine wash to both the donor site and the face the night before and the morning of the procedure and a plan for prevention of infection is implemented (Form 1). The CO_2 laser resurfacing cosmetic surgeon must be prepared immediately to culture and diagnose all infectious complications. Prophylactic acyclovir and valacyclovir are used from 24 h preoperative to 10–14 days postoperative in all laser patients. Should suspicious areas of specifically localized pain or small ulcerations develop, intravenous acyclovir or at least 400 mg q 4 hr. po acyclovir should be considered. Reactivation of perioral herpes simplex virus (HSV) is the most common source of infection. However, transmission of HSV from contact with another individual during reepithelialization has occurred and makes prophylactic acy-

FORM 1
LASER SURGERY PREOPERATIVE INSTRUCTIONS

DO NOT TAKE ANY ASPIRIN CONTAINING PRODUCTS FOR TWO WEEKS PRIOR TO SURGERY. USE ONLY TYLENOL OR ACETAMINOPHEN. DO NOT TAKE VITAMIN E TWO WEEKS BEFORE THIS PROCEDURE.

1. For facial surgery, cleanse face with HIBICLENS skin cleanser thoroughly the night before surgery and the morning of surgery. For women, this includes removing all mascara and eye makeup. Do not wear makeup to the Surgery Center. It is suggested that you also shampoo your hair.

2. **PRESCRIPTION MEDICATIONS:** Antibiotics, pain medications, and/or tranquilizers/sleeping pills should be in your possession two days before surgery. If not, call the nurse.

3. START THE ANTIBIOTICS THE DAY BEFORE SURGERY.

4. Take a Halcion tablet for sleep the night before surgery and take one for relaxation the day of surgery 1 hour before arriving at the clinic. **YOU MUST HAVE A DRIVER!** Also arrange for someone to drive you to the Surgery Center for your initial postoperative check-up.

5. Purchase the following over-the-counter postoperative care items: Aquaphor ointment from Biersdorf – two packages of 4X4 inch or 3X3 inch rough wise mesh gauze – one box of Q-tips -one package of 4X4 inch TELFA pad dressing -Hibiclens antiseptic wash in the blue bottle - a 500 ml bottle of sterile saline for post-op cleaning - white vinegar from the supermarket.

6. For local anesthesia patients in the surgery center, eat light to normal breakfast and take your usual medication.

7. Please leave all valuables at home. Please do not wear earrings or necklaces. Do not wear make-up, nail polish, or HAIRSPRAY to the surgery center the day of surgery.

Form 1 Laser surgery preoperative instructions.

clovir or valacyclovir an important consideration for even those who have no history of HSV infections.[11–13]

Other infections such as bacterial impetigo are prevented with prophylactic antistaphylococcal antibiotics started the day before surgery and meticulous postoperative wound care. Itching and burning with nonspecific findings in the period from 4–14 days postoperatively should be considered either allergic contact dermatitis, *Candida* overgrowth, or *Pseudomonas* infection. *Pseudomonas* infection frequently produces a "cobblestone" appearance of swelling, crusting, and delayed healing usually over the lower face or perioral areas.

To prevent posttreatment hyperpigmentation, which will occur in 50% of patients without prevention measures, Lustra cream (Medicis) and Retin A 0.1% cream (Geneva Pharmaceuticals) is utilized. The hyperpigmentation tendency develops by the third week postoperatively and requires 2–3 months to stabilize. Except for the blue-eyed fairest of type I skinned individuals, the Lustra/RetinA combination is used daily or twice daily for 2–4 weeks preoperatively. Besides screening for compliance problems, this routine is intended to suppress postoperative melanocyte activity.

V. INFORMED CONSENT

An explanation of the benefits and risks of both laser resurfacing and tissue augmentation as well as alternative procedures is important. It must be determined that the patient not only understands the information and has appropriate expectations, but they are reliable and will follow instructions. The screening for these problems and for the uncooperative patient with a personality disorder occurs weeks earlier in consultation and preparation.

We require that patients desiring combined laser resurfacing and fat injection sign a consent form for each of these procedures (Form 2). Patients must understand that fat injection involves overcorrection in two or three sessions. Additional sessions are commonly needed for certain areas, such as the lips, to obtain lasting results. The surgeon should develop a feel for the patient's desire for more overcorrection and fewer treatment sessions or less overcorrection and more treatment sessions. This should be discussed and documented in the medical record. Patients should also be warned that the first autologous fat injection session may be the least dramatic result of all the sessions, and that they should not be discouraged by minimal fat take with the first AFI session. With each and every session of fat injection, they should be prepared for swelling and overcorrection, which improves greatly by the sixth postoperative day.

When AFI is performed the same day as laser resurfacing, the plan is to perform a conservative injection with less overcorrection. Some areas may not be injected with the initial session. However, a good harvesting of fat should be performed during this session with freezing of the donor fat.

Bruising and swelling may be increased when fat injection is added to laser resurfacing, especially when it is to be performed in the same session. Among special considerations in combining these procedures in one session, the fat injection should not be attempted immediately after laser resurfacing but rather immediately before the laser resurfacing. The markings and landmarks necessary for precise placement of fat are obscured immediately after laser resurfacing. There are also concerns about contamination of the fat injecting process in the oozy aftermath of resurfacing.

The patient must understand that there may be some slight irregularities from taking donor fat, particularly if the lateral thigh is used. Outer thigh fat is easy to anesthetize and to remove bloodless fat. This makes it an excellent donor site and often one thigh is larger than the other. Irregularities of the skin are more common and problematic on the lateral thigh versus the lower lateral abdomen, flank, or posterior superior hip areas.

FORM 2A

SPECIAL CONSENT OF LASER RESURFACING SURGERY

PATIENT:______________________________DATE:____________________

1. I hereby authorize Dr.______________________ to perform a laser surgical procedure known as laser resurfacing,

for__________________________with a plastic surgical procedure on

the________________________and surrounding structures, on ____________________________.
(Patient's Name)

2. The procedure in paragraph #1 has been explained to me by the doctor and I completely understand the nature and consequence of the procedure. The following points have been specifically made clear:

a. Some of the possible complications of the operations are infections and abnormal discoloration or pigmentation of the skin. Rarely, scars, inflammation, and allergic or other bad reactions to one or more of the substances used in the operation can occur.
b. Objectionable scarring is rare but scar formation is possible as a result of the surgery. Every effort will be made to prevent them or to make any incision as inconspicuous as possible.
c. That early postoperative skin changes are conspicuous, with redness changing to pinkness.
d. That there will be a pink discoloration of the skin for weeks and that in some cases this can persist for longer.
e. Due to the nature of the procedure, an exact end result cannot be predicted, and I have not been given any guarantee of specific results.
f. That hypopigmentation (prolonged or permanent white areas) results to some degree in 10% of patients.

Please initial_________

g. That wrinkles around the eyes and mouth caused by muscles of smiling, squinting, and frowning do not improve sometimes, and never improve as much as sun damage wrinkles.

Please initial________

h. That acne scars are expected to improve significantly but may not improve as much with one treatment as hoped. That some acne scars do not improve much with one or more treatments and other modalities must be considered. I understand that laser resurfacing does not "eliminate" acne scars.

Please initial________

3. I am aware that the practice of medicine and surgery is not an exact science and I know and acknowledge that no guarantees have been made to me as the results of the operation or procedure. I understand the goal of this surgery is improvement, not perfection;

Please initial________

4. I agree to keep the above doctor informed of any change of address so that he can notify me of any late findings, and I agree to cooperate with the above doctor in may care after surgery until completely discharged.

Please initial________

5. I have read the above consent and fully understand the nature of the procedure and the risks involved. I acknowledge and understand that no expressed or implied warranty has been given to me.

This operation has been explained to me and I fully understand the nature of the procedure and the risks involved. I acknowledge and understand that no expressed or implied warranty has been to me.

Patient's Signature__________________________Date_____________

Witness____Date_______

Form 2a Special consent of laser resurfacing surgery.

FORM 2B

CONSENT FOR AUTOLOGOUS FAT INJECTION

PATIENT____DATE___________

I hereby authorize Dr.___________and/or his or her associates to perform autologous fat injection on me. I fully understand that this procedure has limited application. No guarantee or assurance has been given to me by anyone as to the results that may be obtained. I am aware that the practice of medicine and surgery is not an exact science, and I acknowledge that no guarantee has been made to me as to the results of the operation or procedure.

Dr.____________has discussed in detail with me the information that is briefly summarized below:

Nature and Purpose of Autologous Fat Transfer:
Autologous fat transfer is a surgical procedure by which donor fat is taken by syringe liposuction technique from the tummy or hips and injected under the skin to attempt to replace or build up the layer of fat immediately under the skin.

1. **Risks:**
   I understand that among the known risks are bruising, lumpiness dimpling, scarring, numbness, minor depressions, and swelling. More procedures incorporating fat injection may be needed in the future.
   I am aware that in addition to the risks specifically described above, there are other risks such as loss of blood and infection that may accompany any surgical procedure, as well as injury to nerves that may lead to numbness and/or weakness of a facial expression.
   I recognize that during the course of the operation, unforeseen conditions may necessitate additional or different procedures than those set forth above. I therefore further authorize and request that the above named surgeon, assistants, or designees perform any such procedures that in their professional judgment are necessary and desirable.
2. **Anesthesia:**
   I understand that local anesthesia is normally required when AFI is performed and that this surgery will be performed under local anesthesia and light sedation. I consent to the administration of local anesthesia and sedation.
3. **Alternatives to Autologous Fat Transfer:**
   Alternative methods for removing excessive skin do exist and facelift, chemical peeling and/or dermabrasion may improve the wrinkles of the skin.
4. **Informed Consent:**
   I have had sufficient opportunity to discuss my condition and proposed surgery and all of my questions have been answered to my satisfaction. I believe that I have adequate knowledge on which to base an informed consent to the proposed treatment.
5. **Photographs:**
   I consent to be photographed before, during and after treatment; these photographs shall be the property of the above doctor and then may be published in scientific journals and/or shown for scientific reasons.
6. **Cooperation:**
   I agree to keep the physician and his or her staff informed of any change in my permanent address, and I agree to cooperate with them before, during, and after the surgery. I understand that I should not smoke or drink alcoholic beverages for at least 2 weeks before, both of which can jeopardize the surgical procedure and the healing afterwards.

(Please initial each paragraph and sign below)

_______________________________ _______________

Signature of Patient or Legal Guardian Date

____

Witness

Form 2b Consent for autologous fat injection.

VI. FINAL CHECKLIST FOR PATIENT SELECTION BEFORE LASER RESURFACING AND AUGMENTATION WITH AUTOLOGOUS FAT

1. Does the patient have skin pathology and subcutaneous support pathology suitable for laser resurfacing and augmentation with AFI?
2. Has the patient been evaluated sufficiently preoperatively for following instructions and cooperation with using make-up, sunscreens, topical retinoids, and topical bleaching agents?
3. Is the patient able and willing to wear make-up to prevent all exposure to light for 3 months after the procedure and possibly longer to correct hyperpigmentation?
4. Does the patient have a realistic expectation of both the healing time and the postoperative pain, discomfort, and swelling to be expected and an appreciation for the months of roller coaster events that can occur when patients do not follow instructions?
5. Does the patient have realistic expectations of the outcome and the amount of improvement to be expected?
6. Has the patient taken oral retinoids in the last 12 months?
7. Does the patient have pink or red permanent make-up in the lips which could turn black with laser?
8. Does the patient have any implanted silicone which can interfere with laser healing or cause fat injection inflammatory complications?
9. Is the patient a smoker or using oral vitamin E or other blood thinning agents?
10. Is there a history of perioral herpes simplex and, if so, should prophylactic acyclovir or valacyclovir be given earlier or longer than usual?
11. Do preoperative photographs with acceptable clarity and quality exist for the patient, as patients often forget their initial appearance?
12. Has the patient studied and indicated by questions that the consent form warning of any possible risks has been clearly stated and that the risk of hyperpigmentation and hypopigmentation have been significantly addressed and initialed?
13. Does the patient understand that demarcation lines can exist for a period of time after laser resurfacing?
14. Does the patient have a realistic appreciation that some loosening of skin and wrinkles may recur in the months after the procedure, particularly to the lower eyelid orbital skin and the areas of dynamic muscle rhytides?

VII. ANESTHESIA

Patients take a 0.25 mg triazolam (Halcion) and a hydrocodone (Vicodin) tablet orally 1 h before the procedure and make arrangements so that they are not operating an automobile. For a small touch-up fat injection with frozen fat, this is usually the total preoperative medication before local anesthetic is injected.

For laser resurfacing and fat harvesting, the patient is placed on pulse oximeter and either 5 mg of intramuscular (IM) prochlorperazine (Compazine) or 50 mg of IM hydroxyzine (Vistaril) is given for nausea prophylaxis. Then 2.5 mg of IM midazolam (Versed) is given and 25–50 mg of IM meperidine (Demerol) is given depending on the patient's size and age.

The periorbital nerve block sites and the perioral nerve block sites are the most painful to inject with 1% lidocaine, 0.25% bupivacaine (Marcaine), 1:100,000 epinephrine mixture. Therefore, we first preinject each of these sites with 0.3 mL each of buffered 0.5% lidocaine with 1:200,000 epinephrine containing 0.2 mL of sodium bicarbonate solution in each 3-mL syringe. For best control, the nerve blocks are injected with a 5-mL syringe with a 1 in. 30-gauge needle. For facial areas of fat injection, the local anesthetic is injected into the subcutaneous areas expected to be tunneled after the appropriate nerve block is placed. Vasoconstriction to prevent bruising, hematoma, and intravascular injection is the intended purpose. There is no information, anecdotal or otherwise, that fat graft take is compromised by lidocaine with epinephrine pretreatment of the recipient areas. Because of increased risk of intravascular injection, fat injection is not performed in the crows feet areas, globellar, and supra-orbital areas.

For CO_2 laser resurfacing, the usual scheme of nerve block anesthesia has been modified to provide the type of preparation that this pain-stimulating procedure demands (Fig. 4). A small 0.3- to 0.5-mL bolus is placed in each nerve block site. From the upper lip, lateral commissure, and mental sites, some of the nerve block mixture is extended or fanned into the upper and lower lips. The nerve block mixture is limited to a total of no more than 10 mL. Marcaine adds to the cardiotoxicity of lidocaine.

To prevent lidocaine toxicity, we wait until immediately before laser resurfacing each section of the face to place a small intradermal bolus of 0.5% lidocaine with 1:200,000 epinephrine every 1.5 cm. Fanning the anesthetic below the dermis does not give adequate anesthesia for CO_2 laser unless the tissues are tumesced and distended tightly. The unilateral chin, upper lip, cheek, and nose are first done as one unit. Stop at the level just below the lateral canthus so that the temple areas may be treated after the eyelids are resurfaced. This will help prevent accidental extra passes over the zygomatic area. The entire

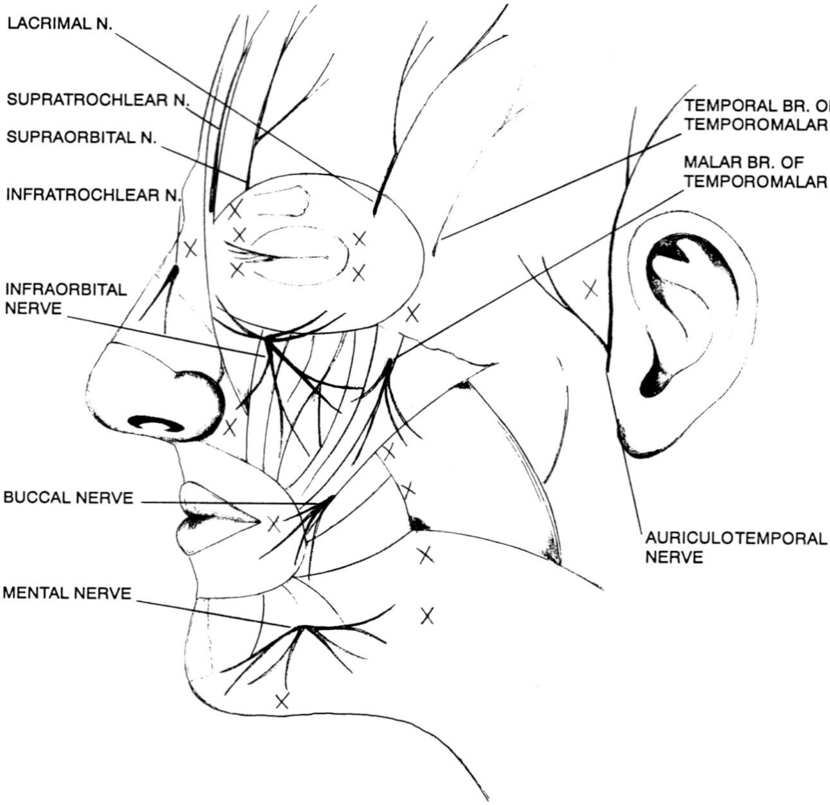

Figure 4 Modified nerve blocks for CO_2 resurfacing.

unilateral chin, lip, nose, and cheek are then lasered before proceeding to intradermal anesthesia of the opposite side.

The unilateral upper and lower eyelids are done as a unit. Finally, the entire forehead and temple areas are intradermally infiltrated with 0.5% lidocaine with 1:200,000 epinephrine and lasered as one unit. For best control and intradermal placement, 3-mL syringes with 0.5-in .30-gauge needles are used. The intradermal 0.5% lidocaine with 1:200,000 epinephrine is only maximal and adequate for CO_2 laser for approximately 10–15 mins. Observing these guidelines for conserving lidocaine allows full-face resurfacing with 40–50 mL of intradermal 0.5% lidocaine.

Additional use of 1 or 2% lidocaine could lead to central nervous system (CNS) toxicity[14] with a twitching or tingling sensation in the extremities or perioral areas. Toxic serum levels of lidocaine are reached much more rapidly with the 1 and 2% preparations. For these concentrations, the maximum safe lidocaine dose should be considered much closer to the original 7 mg/kg guidelines. A 50-mL vial of 1% lidocaine contains 500 mg of lidocaine, which is roughly 7 mg/kg for a 70-kg individual.

Marcaine is additive to the toxicity, particularly, to the cardiotoxic effects of lidocaine.

VIII. ANESTHETIC CONSIDERATIONS BY REGION

A. Forehead and Glabellar Areas

Prior to brow autologous fat augmentation, regional anesthesia is done for the forehead/brow area by introducing 1% lidocaine with 1:100,000 epinephrine at the supraorbital nerve, supratrochlear nerve, infratrochlear, and nasal nerve areas. Anywhere from 0.3 mL to 1.5 mL may be used in a particular region for a block. Prior to autologous fat injection in the brow, 1.5 mL of lidocaine is placed in each side of the glabellar area and each side of the supraorbital area to obtain desired vasoconstriction for safety reasons. For laser resurfacing anticipated to the forehead, 0.25% Marcaine is added to the nerve block mixture.

B. Cheek, Temple, and Perioral Areas

A careful deep regional anesthesia in the area of the infraorbital nerve serves more than one purpose. Besides contributing to the regional anesthesia for both autologous fat injection and laser resurfacing, this particular injection produces anesthesia and vasoconstriction in the facial muscles of expression in the infraorbital malar areas that lie medial to the lateral canthus. The zygomaticus and levator labii muscles are potential sites for fat-injection grafting and careful placement of tunnels of fat.[15] Vasoconstriction of the facial vein running under these muscles of facial expression in the malar area (Fig. 5) is also a desirable goal.

For the lower and lateral malar areas, several small boluses of the nerve block mixture at the anterior border of the masseter muscle produces helpful regional anesthesia prior to laser resurfacing or autologous fat injection. Before

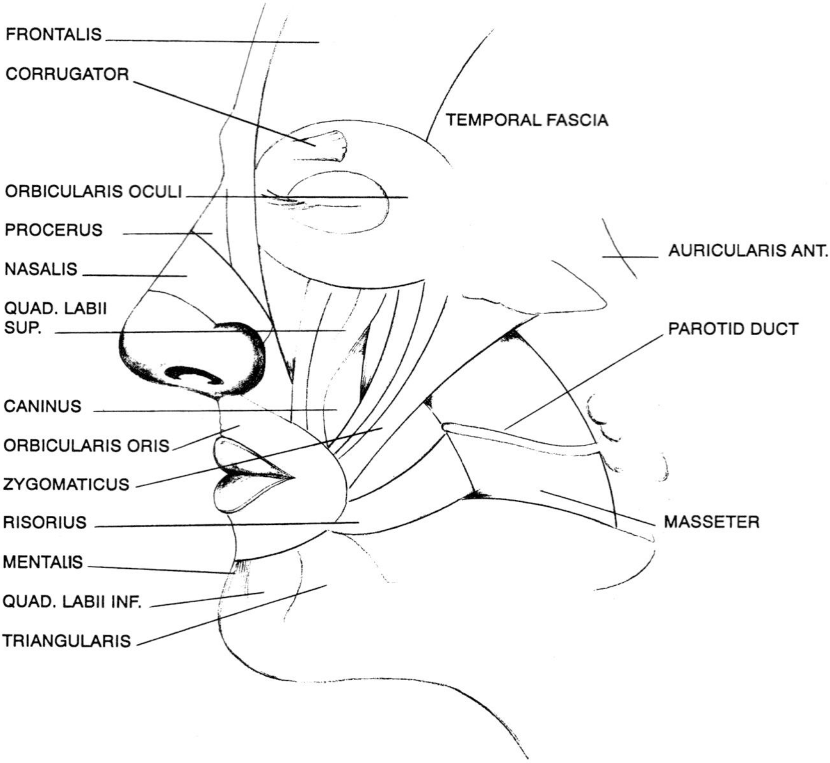

Figure 5 Muscles of facial expression of the medial face receptive to autologous fat implantation. Note parotid duct exiting over the masseter muscle.

laser resurfacing of the upper lateral face, the zyomaticofacial nerve areas, as well as the lateral orbital rim are approached with the modified nerve block anesthesia. The crease near the nares is blocked in a similar fashion as well as the region around the mouth approximately 6 mm from the lateral commissure with the modified nerve block anesthesia.

For laser resurfacing or autologous fat injection of the chin, the mental nerve and the inferior edge of the mandible immediately below the lateral commissures are blocked. From these sites, a small amount of the nerve block mixture is extended or fanned into the area near the lower lips.

C. Eyelid Anesthesia

Timing is important for conserving and minimizing the total quantity of lidocaine used. Eyelids metabolize 0.5% lidocaine with 1:200,000 epinephrine to levels suboptimal for CO_2 laser application in about 10 mins. Superficial use of the Marcaine-containing nerve block preparation at the medial and lateral canthal areas make subdermal anesthesia of the pretarsal skin less painful. Careful avoidance of skin vessels with loupe magnification will minimize bruising.

IX. AUTOLOGOUS FAT TECHNIQUE

A. Fat Harvesting

After chlorhexidine prep, the entry point as well as the extent of the donor areas are marked with indelible surgical marker and the entry points are injected with 1% lidocaine with 1:200,000 epinephrine. A small puncture is made with a #11 blade, angling toward the harvest area if possible. Using a 14- or 16-gauge Lamis or Zaki infusion needle (Byron Medical), 200–300 mL of 0.1% lidocaine with 1:100,000 epinephrine is infused for tumescence of the fat donor site. After vasoconstriction, the donor fat is harvested with a 12-gauge Coleman harvesting cannula using either 12- or 20-mL disposable plastic syringes. If blood in the donor fat is a problem, a 20-mL syringe half filled with fat is easier for washing two or three times with sterile saline in a closed system. The excess saline and infranatant is decanted, discarded, and the fat transferred to 3-mL plastic disposable syringes and decanted one final time. Some physicians prefer to use a sterilizable centrifuge system but it is not necessary for fat graft take. The fat is then ready for freezing with or without an initial injection session.

Outer thigh fat is easy to anesthetize and remove bloodless fat, which makes this an excellent donor site. However, irregularities of the skin are more common and problematic on the lateral thigh versus the lower lateral abdomen, flank, or posterior superior hip areas.

Although many physicians prefer the outer thigh for donor fat, harvesting in a thin person with less than 15% total body fat may be difficult without recruitment of an additional area such as the lower lateral abdomen or the posterior flank area. Both of these areas are resistant to fat wasting and produce an excellent filler substance. For leaving a minimal visible scar, the outer thigh can be approached from either the superior aspect or from the infragluteal crease. In a small thigh, postoperative irregularities are more likely with lateral tunneling from the infragluteal crease. More fat may be taken from a superior vertical tunneling approach in the small lateral thigh without leaving skin irregularities. For the posterior flank, a medial and a lateral puncture mark inferior to the area being harvested with both vertical and diagonal tunneling is preferred. The buttocks and iliac crest may be somewhat in the way using this approach to the flank from below. However, the diagonal tunneling provided by the dual medial and lateral puncture sites make this posterior flank harvesting much easier. In a small thin tummy, a puncture mark that is paramedian and halfway between the umbilical level and the suprapubic line is used. This allows more effective harvesting into the suprapubic area as well as the periumbilical fat.

B. Fat Injection

Only a few cosmetic surgeons have significant experience with full-face laser resurfacing and simultaneous autologous fat injection, and information remains anecdotal.[16] Survival or take of injected fat graft does not appear to be a problem. In fact, the most common complaint by cosmetic surgeons relates to survival of slightly misplaced grafted fat when injected on the day of laser resurfacing. This appears to be primarily related to tumescent anesthesia for laser resurfacing. Tumescent anesthesia for laser resurfacing utilizing a 0.25–0.4% lidocaine mixture with 1:400,000 epinephrine can make an accurate and even injection more challenging in the brow and infraorbital malar areas. When autologous fat injection is performed immediately after tumescent anesthesia, it appears to be more difficult to avoid uneven or slightly misplaced distribution of fat into the infraorbital malar area, the nasolabial fold, and subcomissural lateral chin. Alternatively, small amounts of 1% lidocaine with 1:100,000 epinephrine with a 10-min delay for vasoconstriction as the preinjection anesthesia may be more conducive to precise placement of fat in these areas. Marcaine is added if a delay time of over 30 min is expected before fat injection.

Besides reducing the volume of local anesthetic near the inferior orbital skin, the author also suggests switching to a larger caliber injecting cannula such as a 14-gauge for better control in the infraorbital malar areas. Deeper placement in the malar muscles of facial expression should be considered. An unwanted lump of fat in the inferior orbital skin is probably the most frequently troublesome of all complications from autologous fat injection. It can quickly become fibrotic and attach to the orbicularis muscle and be resistant to intralesional steroid intervention. Immediate removal with a 14- or 16-gauge cannula should be considered. Incision and dissection may be eventually necessary if the physician waits too long.

Tumescent anesthesia can also make even distribution in fat grafting of the forehead more challenging. Lumpiness and sharp step-offs at the edges of fat grafting in the forehead can be particularly noticeable.

In the glabellar and periorbital areas, the accidental intravascular injection of fat is the most feared complication. The preoperative infiltration of epinephrine containing local anesthetic probably reduces the risk of this accident. The use of needles, like the bullet needles for injecting, featuring a pointed but closed tip with side opening and the presence of good vasoconstriction makes accidental intravascular injection of fat theoretically almost impossible. However, because of lack of valves in the veins of the glabella and crows feet areas, fat may enter the veins. These areas are avoided by the author. Small amounts of local anesthetic and a meticulous precise visual and tactile observation of the injecting tunnel of fat during withdrawal of the bullet needle are also extremely important to reducing the risk of blindness and stroke that has been reported from intravascular injection of fat. The feel and accuracy of the fat placement is excellent with a 10-min delay after injection of epinephrine containing local anesthetic into the donor sites with vasoconstriction having taken place and the initial slight distention from the local anesthetic disappearing at 10 mins. For the first session of lip augmentation, two sterile tongue blades are placed under the upper lip. A small amount of fat, 0.5–1.0 mL, is placed into each side of the midline. Only the vermillion portion is injected with the upper lip rolled upward.

X. SAFETY CONCERNS OF ULTRAPULSED CO_2 LASER

Facial laser resurfacing with ultraspulsed CO_2 lasers creates special considerations and safety requirements. Certainly not the least of these is fire in the operating room: fire of a patient's hair containing hairspray, fire of surgical draping materials, fire in the patient's endotracheal tube or oxygen line. It is our general policy to have all oxygen-containing tubing out of the surgical field area by the time the laser resurfacing commences. Preoperative O_2 via nasal cannula during the initial stages of IM or IV sedation while regional anesthesia is being established is an important part of the procedure. One or 2 L per minute of O_2 per nasal cannula in the lightly sedated patient is an excellent idea during establishment of local anesthesia, and it allows the patient's CO_2 levels to increase slightly, augmenting the patient's natural respiratory reflexes. A healthy candidate for cosmetic surgery should be able to come off supplemental nasal cannula oxygen by the time laser CO_2 resurfacing commences and maintain a stable O_2 saturation in the 90s on the pulse oximeter.

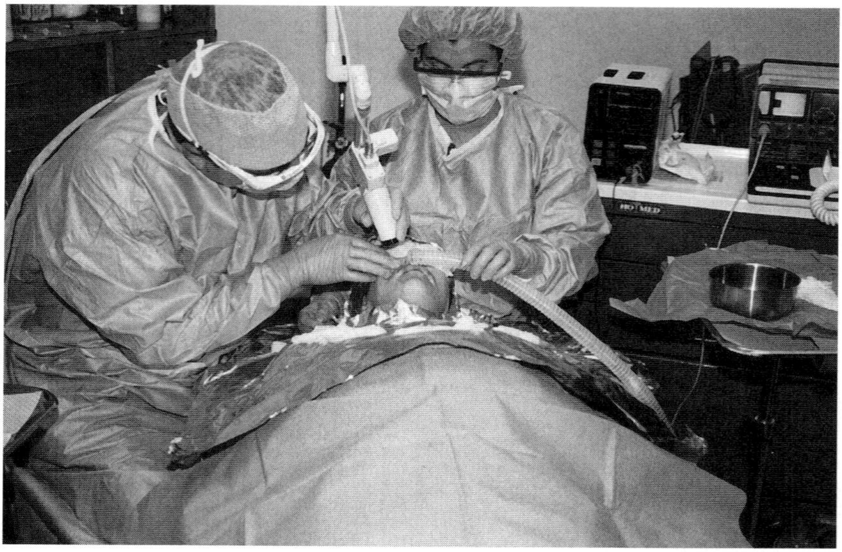

Figure 6 CO_2 laser resurfacing safety with autoclaved aluminum foil draping and water bowl close at hand.

A bowl of sterile water or saline next to the surgical field always must be present (Fig. 6). Draping materials should be either wet fabrics or aluminum foil. Hairspray is extremely flammable, and the patient should not be allowed to come into the operating theater with styling products in the hair.

Besides draping to prevent accidental burns of the patient outside the intended surgical areas, the eyes and teeth must be protected with special considerations created by laser resurfacing. Saline-soaked cotton gauze should be used at all times to protect the teeth, as inadvertant CO_2 laser strike will blacken tooth enamel. The surgical assistant is vital in maintaining wet gauze over the patient's eyes during this procedure where there is often frequent head movement and position change. When the eyelids are being treated with the laser, the globe must be protected with a laser-safe metal protection system. The burnished metal Jaeger plate system (Byron Medical) is our choice, and it is well tolerated by the patients. For the staff and the operating room environment in general, special safety considerations must include appropriate and adequate signs warning of laser use. An established system of laser in the standby mode when staff are moving in and out of the room and the appropriate safety eyewear for CO_2 laser resurfacing are necessary.

The laser plume is a potential health hazard for all operating room personnel. Masks should be the appropriate type for laser. A smoke evacuator must be in top condition and have regular filter changes so that suction is maximized. The assistant must keep the vacuum tubing within 1–2 cm at all times during periods of heavy plume. Working around the eyelids with the vacuum tubing directly connected to the laser handpiece is very adequate in this low-plume situation, and it creates an extra hand which is frequently important when working on the eyelids.

XI. LASER RESURFACING CONSIDERATIONS BY REGION

A. Brow and Temple Areas

Removal of solar elastosis, ablation leveling of the shoulders of rhytides and creases, and general tightening of the forehead area with the laser produces very positive results. The vertical glabellar dynamic rhytides are less responsive but still improve. The brow tightening, lifting, and improvement to the appearance of the eyes is quite remarkable in some patients. In other patients, particularly men, the type I dermal collagen responds only slightly to the effects of CO_2. In the temple areas, the tightening effect of the laser can be used with vertically oriented rectangular patterns to lift and tighten the lax skin of the lower lateral eyelid. This also reduces the immediate postoperative ectropion effect from resurfacing the lower eyelids and the malar areas.

During full-face laser resurfacing, the chin, cheek, and nasolabial furrow areas are generally lasered first. The laser pattern is rotated between the first and second or the second and third passes to obtain smoother postoperative results. Wet gauze is used gently to wipe away the white fuzzy debris after each and every pass of the CO_2 laser. Going over the same area twice without removing debris will create brown char which can conduct thermal injury to the skin.

Patients with a lax lower lid are checked frequently for excessive tightening of the inferior orbital skin. The pattern used in this area is smaller and oriented to minimize downward pull on the lateral lower eyelid. The physician should frequently perform a traction test of the lateral lower eyelid. A small amount of ectropion immediately postoperatively is generally acceptable as long as the safety range for number of passes has been carefully observed. The perioral rhytides as well as the cheek rhytides dramatically improve from tightening the entire cheek and lateral face with CO_2 laser resurfacing. Cheek ptosis and mild jowl problems improve.

The nasolabial furrow is resistant to laser resurfacing alone. However, after laser resurfacing, the line is smoother and less sharply demarcated.

B. Eyelids

Experienced CO_2 laser resurfacing cosmetic surgeons may laser the entire orbital skin to the lash margins. Wet wooden Q-tips are used to bend the lashes away from the laser to prevent accidental singing of the lashes. The patient should be warned preoperatively, however, that an eyelash can be temporarily singed during the procedure and that it will regrow. The pretarsal skin, which is the 5 mm closest to the lash margin, should receive less passes with the laser, with two probably being the maximum. In patients with thin skin, the millijoule power setting and overlap densities are reduced on the eyelid skin. The surgeon stops and repeats the lower lid traction test after every pass in the patient with lax lower eyelids.

XII. POSTOPERATIVE CARE

Immediately after laser resurfacing, the last bit of laser debris is wiped away and Flexzan (extra sticky) semipermable dressing is applied to each cheek and temple areas as one unit. With any of the semiocclusive dressings, the postoperative pain is dramatically reduced in the first 5 days, which makes patients happier and much more likely to recommend laser resurfacing to their friends. A strip of Flexzan is then applied to the entire forehead, to the upper lip, and to a 1-cm strip of the chin at the jaw line only. This creates three bridges between, and overlapping onto, the two large side pieces. The lower lip and superior two thirds of the chin are left open and free of Flexzan, because this is an extremely mobile area. A small piece of Flexzan can be placed on any exposed fat injection puncture marks at this time.

Liberal amounts of Aquaphor ointment from Beiersdorf are then applied to the eyelids, chin, and all other open areas. The nose can be covered with two or three horizontal strips of overlapping Flexzan. Stretchy tube netting is used to hold the Flexzan in place. Individual TELFA pads are used over the ointment-covered chin and to protect it from direct contact with the netting. Small pieces of TELFA pad are used to help cover any other exposed edges that the netting is resting against.

The patient should clean frequently the open areas with saline-soaked wide mesh gauze and Q-tips and reapply the Aquaphor ointment. At 24 h, the patient returns for new netting and Flexzan as needed and eyelid cleaning is reinforced. Loose Flexzan is replaced with TELFA over ointment on postoperative day 3. On days 4 and 5, the patient is allowed to remove loose Flexzan and replace with ointment and TELFA.

The use of Aquaphor ointment for the first week postoperatively and then switching to emollient creams in week 2, has reduced milia and erythema. Using ointments after the first week prolongs erythema.

Short-acting IM corticosteriods are helpful to minimize the postoperative swelling in the first 4 days, particularly when eyelids are resurfaced. Celestone, 1 or 1.5 mL, is given IM at the end of the surgical procedure.

Patients definitely do better with diazepam 5 mg orally three to four times daily for the first 3 or 4 days postoperatively. This helps the patient to tolerate all the semiocclusive dressings and netting, as well as the discomfort associated with healing and postoperative cleaning. It also helps to prevent symptoms from mild lidocaine neurotoxicity. An uncomplicated case will require pain medication like acetaminophen with hydrocodone four times daily for 5 days. Patients are encouraged to clean more vigorously four times daily 1 hour after taking the pain medication. Patients continue cephalexin 500 mg qid and valacyclovir 500 mg bid for the prescribed 10 days. Prophylactic valacyclovir 500 mg bid is used from 24 hs preoperatively to 10–14 days postoperatively.

All patients, except for the blue-eyed patients with type I skin, 2 weeks after laser resurfacing, are started on Lustra hydroquinone cream in the morning and RetinA 0.1% Micro cream at night. We instruct patients at first to apply these very sparingly to the lateral temple and cheek areas only to avoid irritation. They may increase Lustra to twice a day if necessary.

The other important factors in preventing hyperpigmentation include the use of a class V topical steroid cream for 1 week, from postoperative day 7 to postoperative day 14. Also, on day 7 postoperatively, patients begin wearing sunscreen-containing make-up, Total Block with Toner from Fallene. We believe that these measures reduce both erythema time and incidence of hyperpigmentation. The patient is given written postoperative instructions (Form 3).

XIII. COMPLICATIONS WITH COMBINED FAT INJECTION

In addition to establishing a protocol plan for prevention, the cosmetic surgeon must be prepared immediately to culture and diagnose all infectious and allergic complications. Infections such as bacterial impetigo are prevented with prophylactic antistaphylococcal antibiotics like cephalexin and meticulous postoperative wound care. Itching and burning with nonspecific findings in the

FORM 3
POSTOPERATIVE INSTRUCTIONS FOR LASER RESURFACING PATIENTS

Day number 0

1. Sleep elevated on three pillows or in a chair for the first four nights after the procedure.
2. Apply ice in a plastic bag with a clean dry cotton towel over the bag, or a bag of frozen peas in a clear cotton towel to the eyebrow and cheek areas for 15 minutes every hour or two the first 24 hours to minimize swelling. You may have some increased swelling three days after the procedure related to the Celestone medication wearing off. Use ice again for 24 hours if you notice increased swelling.
3. **For patients with eyelid surgery in addition to simple resurfacing:** Make the doctor aware of any one sided swelling, discomfort, bruising, or change in the vision. If necessary, the doctor will meet you at the Emergency Room to evaluate the possibility of blood vessel bleeding and causing swelling behind the eye. Patient should also make the doctor aware of any vomiting or retching from nausea from medication as this can also cause bleeding behind the eye. Absolutely, do not bend over during the first five days, bend at the knees to pick something up.
4. Wherever the skin is open and not covered by the Flexzan bandage, clean the skin with wet Q-tips followed by dry Q-tips every hour to prevent build up of yellow crusts. Apply Aquaphilic ointment immediately after each cleaning procedure to keep the areas moist.

DAY NUMBER 1 (First post-op day after the procedure)

1. Taper off of the ice compressing but continue the hourly cleaning while awake to all exposed skin not covered by the Flexzan dressing.
2. Call the office and tell the staff how you are doing or be seen for a check-up.
3. Sleep elevated on three pillows or in a chair.

DAY NUMBER 2,POST OP

1. Come to the clinic and have an examination by the surgical staff if you were not seen on day number 1. The staff will make sure that the unbandaged areas are getting cleaned properly and will help you with any difficult to clean areas.
2. It is important to take your pain pill four times per day with a little food. Approximately 45 minutes after the pain pill, do a thorough cleaning with the water soaked Q-tips followed by dry Q-tips to work all of the white or yellow crusting until pink clean skin is established. Continue to apply the Aquaphilic ointment to the clean pink skin immediately after cleansing.
3. Sleep elevated on three pillows or in a chair.

DAYS NUMBER 3 AND 4, POST-OP

1. Instead of cleaning with water soaked Q-tips, begin cleaning with diluted vinegar soaked Q-tips. Mix 2 tablespoons of white distilled vinegar into a glass of water and use this dilute solution as your cleaning and compressing solution. Soak gauze squares in the solution to compress larger areas prior to cleansing, and soak the Q-tips prior to cleaning in the vinegar dilute solution. Complete the cleaning with dry Q-tip and the re-application of the Aquaphilic ointment. The vinegar soaking is soothing and prevents the resistant bacteria from growing that sometimes occurs after three days post-op.
2. Make sure that either you or the staff has trimmed away any Flexzan dressing that does not stick to the skin and prevents you from cleaning the area. Loose Flexzan is OK as long as the area of skin can be cleaned with Q-tips. If the Flexzan is not sticking and it prevents the cleaning of the area, it must be trimmed away.
3. The Flexzan dressing continues to be held in place where it sticks well. The Flexzan speeds healing and prevents the pain of the air getting to the skin.

DAY NUMBER 5, POST-OP

1. The Flexzan dressing is removed and the dilute vinegar solution is continued on any areas that tend to crust. An Aquaphilic ointment is used more sparingly and primarily on areas that are still crusting a bit, four times a day only. Use the TELFA over ointment on any areas still healing.
2. Moisturel lotion is begun on healed (pink) areas to keep the skin soft.

Form 3 Postoperative instructions for laser resurfacing patients.

DAY NUMBER 6 THROUGH 7, POST-OP

1. Moisturel lotion is used to keep newly healed areas from feeling dry.
2. Dilute vinegar cleaning and Aquaphilic ointment is used for the last few days only for the areas that are still healing.

DAYS NUMBER 7 THROUGH 14, POST-OP

1. A cortisone lotion, like DesOwen lotion is provided by the staff. You are to use this twice daily, applied to all newly healed areas to help speed the resolution of the pinkness. **THIS IS IMPORTANT!** Use the cortisone lotion only for one week, then discontinue.
2. Let the staff know if you are experiencing any increase in itching. There is a small amount of itching that goes on as part of the healing process and this is normal, however, we want to know about a change or an increase in itching.

DAYS NUMBER 14 THROUGH 40, POST-OP

1. Begin every day wearing sunscreen containing **make-up** like Total Block with Toner from Fallene, Clinique Citybase, Iridelle, or similar, to all healed areas. **THIS IS IMPORTANT.** This prevents small amounts of ultraviolet light, (even from fluorescent lighting) which may keep the pinkness from fading as early as possible.
2. If you plan to be outside at all, apply SPF 30 sunblock, (Hawaiian Tropic, Pre-sun Ultra with parsol, Ombrelle) sunblock lotion prior to applying make-up containing sunscreen. Ideally, you should not be outside unless you have both sunscreen and the UV blocking make-up on. Avoid all wind exposure!
3. If you are a person with brown eyes, or a person who tans easily, at day 15, you must begin applying one of the bleaching creams that prevents hyper pigmentation and irregular tanning, such as Solaquin Forte, Eldoquin Forte, or Lustra cream. **THIS IS IMPORTANT.** If you are prone to hyper pigmentation, you should be using RetinA 0.1% micro cream with the Lustra cream twice daily by day 15, or you should be back on the Obaji Nu-Derm cream system by day 15. Begin theses products **sparingly** at the outer edges of the temple and cheek areas, to get used to them gradually. Call the office if you experience any problems.
4. **Six weeks after** laser resurfacing, approximately days 40-47, **all patients** are instructed to apply **Clobetasol cream**, supplied by the staff, sparingly to all lasered areas except for the eyelids. Do this twice a day for one week and then discontinue. **THIS IS IMPORTANT** to stabilize final skin color and to reduce the risk of area losing skin color (hypopigmentation). If you see areas of hypopigmentation developing, call the office so that we may give you additional instructions.
5. Ten weeks after laser resurfacing, to prevent hypopigmentation (areas losing color and appearing whiter). We may consider treating the areas with the Temovate cream twice daily for one additional week for stabilizing skin color.

Form 3 *continued*

period from 4 to 14 days postoperatively should be considered either allergic contact dermatitis, *Candida* overgrowth, or *Pseudomonas* infection.

Fibrosed small nodules of fat can occur anywhere with autologous fat injection. For unknown reasons, an area of grafted fat develops a fibrous capsule and rounds up into a soft to firm nodule. This can be treated with intralesional triamcinolone 5 mg and increasing to 10 mg. Ultimately, the small nodule may require local anesthetic, rupturing with an #11 blade, and gentle pressure.

In the glabellar and periorbital areas, the accidental intravascular injection of fat is the most feared complication. Preoperative infiltration of epinephrine containing local anesthetic probably reduces the risk of this accident. The use of the bullet needle featuring a pointed but closed tip with a side opening and injecting on withdrawal makes accidental intravascular injection of fat theoretically unlikely. Small amounts of vasoconstricting local anesthetic and meticulous precise visual and tactile observation while injecting the tunnel with fat is extremely important to reduce this rare complication.

Overinjection of fat from excessive take or a slightly misplaced take of grafted fat can occur when the fat injection is performed on the day of laser resurfacing.

XIV. COMPLICATIONS OF LASER RESURFACING

A. Scarring

Early scarring is usually indicated by an area of persistent redness and induration which is noted by the second or third postoperative week. This area should be addressed immediately with intermittent topical steroid used for brief periods and then discontinued. Clobetasol cream may be used on an area twice a day for 7–14 days and then discontinued. It is also helpful to recheck at this time and use intralesional triamcinolone 3 mg/mL on any areas which do not appear to be softening and losing erythema. Intralesional triamcinolone injections are helpful every 3 weeks until the area is stable. At 4 weeks postoperatively, patients may begin using a small patch of silicone gel sheeting such as Epiderm from Biodermis, with slight pressure over the area. Wearing silicone gel sheeting 6–12 h daily is helpful to improve scars. We also allow our acne scar patients to use this beginning 4 weeks postoperatively on the remaining acne scars to obtain additional smoothing.

Ultrapulsed CO_2 laser scarring is usually caused by inadvertent extra passes over the same area, inadvertent back-up and lasering over an area of debris, or inappropriately high-density overlap settings. The physician must also be absolutely aware of the minimum ablation threshold energy using each particular spot size for each laser being used. Turning down the power density per pulse or millijoules per pulse to below this level may cause scarring. Below the threshold of ablation, the tissue does not vaporize and becomes a heat sink transmitting thermal injury deep into the tissues. The larger the spot size, the higher the energy necessary. This relationship of energy to spot size is an exponential increase in energy required with increased spot size.

With each pass of the CO_2 laser, the ablated tissue debris must be removed gently with sterile saline-soaked gauze. The physician should pay particular attention not to make the frequency of pulses too rapid or the density of pulses too overlapped to cause char. Even with scanner-type attachments, the physician must be careful not inadvertently to move backwards over an area being scanned.

Repeating too many passes and lasering too deeply is a cause of scarring. For thin skin, the safety endpoint may be two passes with Ultrapulsed CO_2 laser, and for thick skin it might be four. Generally, more than three passes are not performed unless the physician is extremely cognizant of endpoints. Once the reticular dermis has been entered, particularly over the lower half of the face, the skin becomes yellowish and has a slightly roughened or wool-like appearance. On the upper part of the face, this yellow color is not always as reliable. In the eyelid and temple areas, the skin generally will not develop the characteristic yellowish hue, and this cannot be used as an endpoint to help avoid scarring.

Appendages in the reticular dermis must be preserved to contribute to reepithelialization and healing without scarring Patients who have taken isotretinoin therapy in the previous 12 months should not have laser resurfacing. Patients who have had extensive electrolysis for permanent hair removal in the perioral and jaw line areas also may have compromised pilosebaceous units. These patients should be lasered more conservatively, as they are quite likely to have an increased risk of scarring with three passes.

During the first postoperative week, failure to clean the skin properly and allowing excessive crusting or allowing dessication and delaying epithelialization may contribute to scarring. Wound care should include cleaning open areas every 2 h while awake the first 3 days and reapplication of Aquaphor ointment and TELFA pads. TELFA pads are used with Aquaphor ointment until the sixth postoperative day for all treated areas except the eyelids where ointment only is used.

B. Infection

Secondary bacterial infection with *Staphylococcus aureus* is prevented with prophylactic antistaphylococcal antibiotics and with wound care. A complication or delay in healing should be immediately cultured, or this infection could lead to scarring. *Pseudomonas* infection is not an uncommon problem, causing stinging, itching, and delayed healing in the perioral areas or lower half of the face, usually from days 4–10 postoperatively. When a late *Pseudomonas* infection is present, the perioral areas frequently develop a reticulated pattern of crusting with a "cobblestone" appearance. Dilute vinegar compression soaks (1 tbs of white vinegar in an 8-oz glass of water) applied to the areas is soothing and also suppresses the *Pseudomonas* infection. Patients with a postoperative

Pseudomonas infection may require an oral quinolone antibiotic which covers *Pseudomonas*. We have our patients begin dilute vinegar soaks as part of their cleaning program on the third postoperative day.

Another source of burning pain or itching lasting beyond 5 days postoperatively is *Candida* overgrowth. Clinically, the physician sees no apparent problem and potassium hydroxide mount is not helpful. Diflucan 200 mg in a single dose is used if this is suspected.

Herpes simplex virus infection is the most scarring of all the infections and all patients should be considered candidates for prophylactic acyclovir or valacyclovir. If the patient has a history of perioral herpes simplex and the laser procedure involves perioral areas, we begin valacyclovir 500 mg bid 2 days prior to the procedure. All other patients will begin 24 h before the procedure along with the cephalexan 500 mg qid. An asymmetrical painful eruption of small clustered ulcerations should immediately alert the physician to herpes simplex virus infection. A culture for herpes simplex virus should be performed and the patient should be placed on acyclovir 400 mg q 4 hr. If an intestinal absorption problem is suspected, intravenous acyclovir should be considered.

C. Contact Dermatitis and Hyperpigmentation

Contact dermatitis and hyperpigmentation are grouped because the CO_2 laser resurfacing appears to be unique regarding these two problems. Empirically, CO_2 laser resurfacing seems to enhance both T-cell–mediated contact dermatitis and melanocyte-mediated postoperative hyperpigmentation in a way that is not seen in any other peel or resurfacing situation. Topical erythromycin seems to be the only topical antibiotic which does not quickly induce allergic contact dermatitis in healing CO_2 laser patients. It has been reported that 65% or more of patients will become sensitized with allergic contact dermatitis to bacitracin- and neomycin-containing antibiotic ointments.[11,12] Unlike an infection which tends to be asymmetrical and oriented to one portion of the face, allergic contact dermatitis tends to be general and symmetrical and is mostly associated with itching. Airborne contact allergens are a problem for laser resurfacing patients for at least 3 months after the procedure. The laser patient should avoid wind exposure and avoid exposure to airborne animal danders, pollens, molds, and chemicals if possible.

From 3–12 weeks after laser resurfacing, hyperpigmentation will occur in up to 50% of individuals without preventive measures. When it does occur, it is aggressively treated with a combination of hydroquinone, retinoids, glycolic acid, azelaic acid, and even light trichloracetic acid (TCA) peeling. If hyperpigmentation is allowed to persist longer than 6 months, it has a tendency to become dermal and more recalcitrant.

D. Permanent Hypopigmentation

Probably the most disturbing of all the sequelae of carbon dioxide laser resurfacing is delayed permanent hypopigmentation. It tends to occur more in older patients, patients who have had previous face lift, and people who are lasered with higher density overlap. An important factor is the tendency to develop scarring, which appears like a delayed scar formation phenomenon, without the hypertrophy, similar to what dermatologists refer to as scarring alopecia. The fibrotic process begins with the follicle and extends outward causing melanocytes to disappear. On close inspection, in some cases, even some of the follicular orifices disappear.

To prevent hypopigmentation, a strong class I topical steroid cream such as Clobetsol is applied twice daily for 1 week from day 40 to day 47, and day 70 to day 77, postoperatively by all laser resurfaced patients. Patients are instructed to watch closely for development of delayed hypopigmentation and to see the physician immediately for intralesional 3 mg/mL triamcinolone injections. Clobetasol cream is used twice daily for 2 weeks on and 2 weeks off for 3 months to specific hypopigmented areas while observing for return of pigmentation. Patients psychologically tend to deal better with hypopigmentation when there is something they can and should be doing, and the physician is wise in having them take some responsibility as well in this situation.

Once the hypopigmentation is permanent and no longer responds to intermittent topical and intralesional steroids, the patient is placed on the Obaji Nu Derm system with RetinA Micro cream. The patchy pigmentation that remains is lighter and more even with this therapy. Microdermabrasion is also helpful.

E. Itching and Purpura

Eczema and intermittent pruritus with streaks of purpura from scratching are relatively common postoperatively and usually occur between postoperative weeks 2 and 6. For patients who are scratching and bleeding in their sleep, a few days of wearing thin Duoderm dressing can be helpful.

Occasionally, new fragile dermal capillaries break for no apparent reason, causing a patchy purpura over the sides of the face between weeks 2 and 6. Patients are reassured and instructed to sleep with extra pillows under the legs and head to reduce movement during the night.

F. Acne and Milia

Mild side effects of ultrapulse CO_2 facial resurfacing include milia formation and acne exacerbations. Expression of milia and careful gradual increasing amounts of topical retinoid work well for the milia formation, and topical erythromycin preparations are added for exacerbations of inflammatory acne.

G. Permanent Make-up Complications

Laser resurfacing can smudge permanent make-up eye liner and should be avoided. Some patients have skin-tone permanent make-up tattooed in facial areas that could turn black from oxidation by the laser. Pink and red permanent lip make-up is common now. The edge of this pigment could be turned black by facial resurfacing.

H. Implanted Silicone Complications

Implanted deep dermal and subcutaneous facial silicone has been utilized in the past. Lasering over this may lead to prolonged swelling and erythema of the area. Implanting fat adjacent to silicone may also cause prolonged inflammatory complications.[17]

I. Recurrence of Wrinkles and Scars

Edema in the first 2 weeks after laser resurfacing hides some remaining acne scars. Patients must be warned that some of their acne scars will become more visible as the postoperative edema dissipates. This will help prevent a surprise and tearful psychological letdown if and when this occurs.

It is important to warn patients that some wrinkles may similarly be more visible as the immediate edema dissipates. Most of the tightening of the skin will be seen in the first month after the procedure; however, some degree of tightening and remodeling in fine wrinkles and in acne scars will occur for up to a year in some patients. Lower eyelid rhytides are notoriously prone to redevelop, particularly with wind and sun exposure. Patients are told before their first laser procedure that they may want to consider a touch-up procedure with local anesthetic for lower eyelids or for the upper lip after 12 months. The acne scar patients are told to consider having a second laser resurfacing performed at 6 or 12 months to the areas of most severe scarring, and to consider zyplast collagen adjunctive therapy.

J. Texture Changes

Laser resurfacing entire cosmetic units is important in considering texture changes. Treated areas will often have smaller pores, smoother texture, and lack solar elastosis with a definite texture difference creating a potential line of demarcation.

K. Ectropion

Ectropion is more likely to occur in patients who have had previous lower lid blepharoplasty surgery and in people with a lax lower lid.[13,18] This is a disturbing complication detracting from the natural rejuvenation effects. For the infraorbital malar skin of these patients, laser patterns should be small and oriented in a way that minimized downward traction on the inferior orbital skin. A small amount of ectropion immediately postoperatively which resolves in the first week is the endpoint for some patients in the hands of an experienced laser resurfacing cosmetic surgeon. Most ectropion usually resolves in 6–8 weeks as long as scarring of the orbital skin has not occurred. Any area suspicious for scarring in the orbital skin must be addressed immediately with steroids and silicone gel sheeting to prevent delayed contracture and delayed ectropion.

XV. CONCLUSIONS

Ultrapulsed CO_2 laser resurfacing results augmented with autologous fat injection can be thrilling for both the patient and the cosmetic surgeon. There is a particularly natural aspect to this type of facial improvement and rejuvenation (Fig. 7). The very best results possible and a pleased patient are the goals. This requires a well-selected and meticulously prepared, educated, and informed patient.

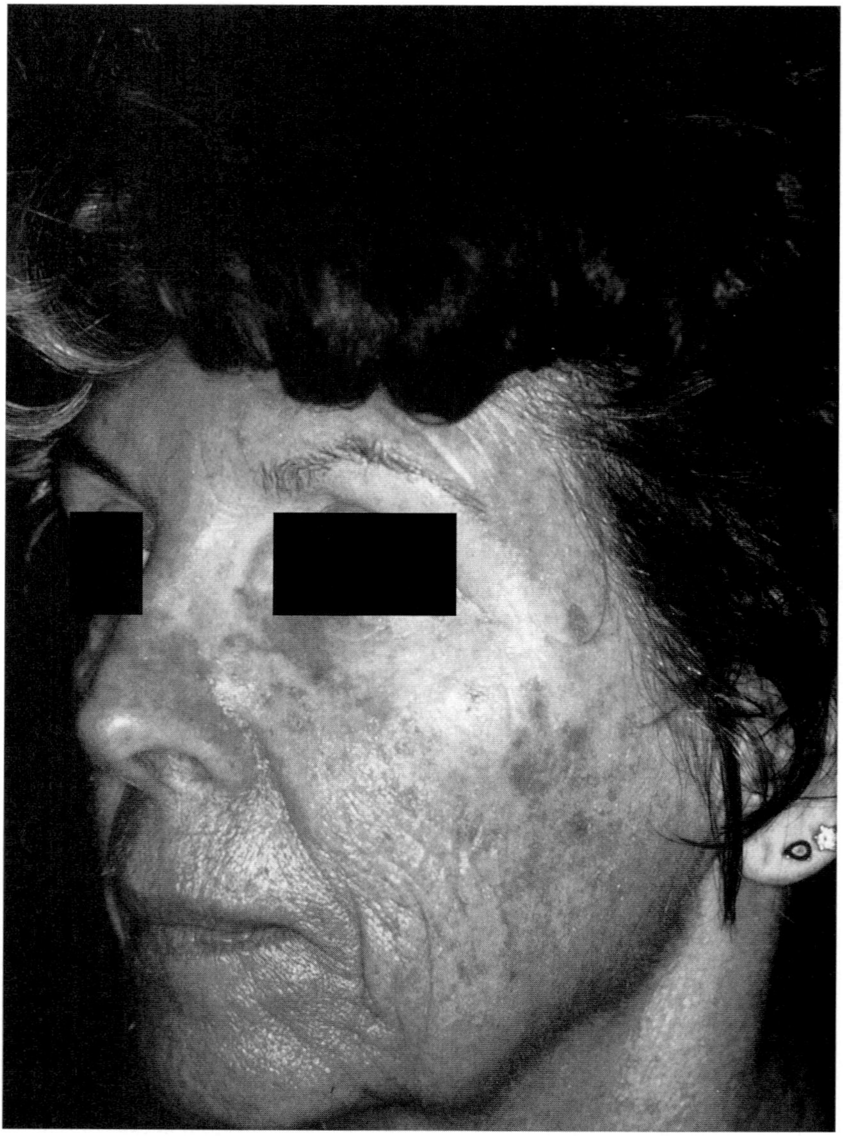

Figure 7a Preoperative. (Photo courtesy of James Fulton, Jr., MD, PhD.)

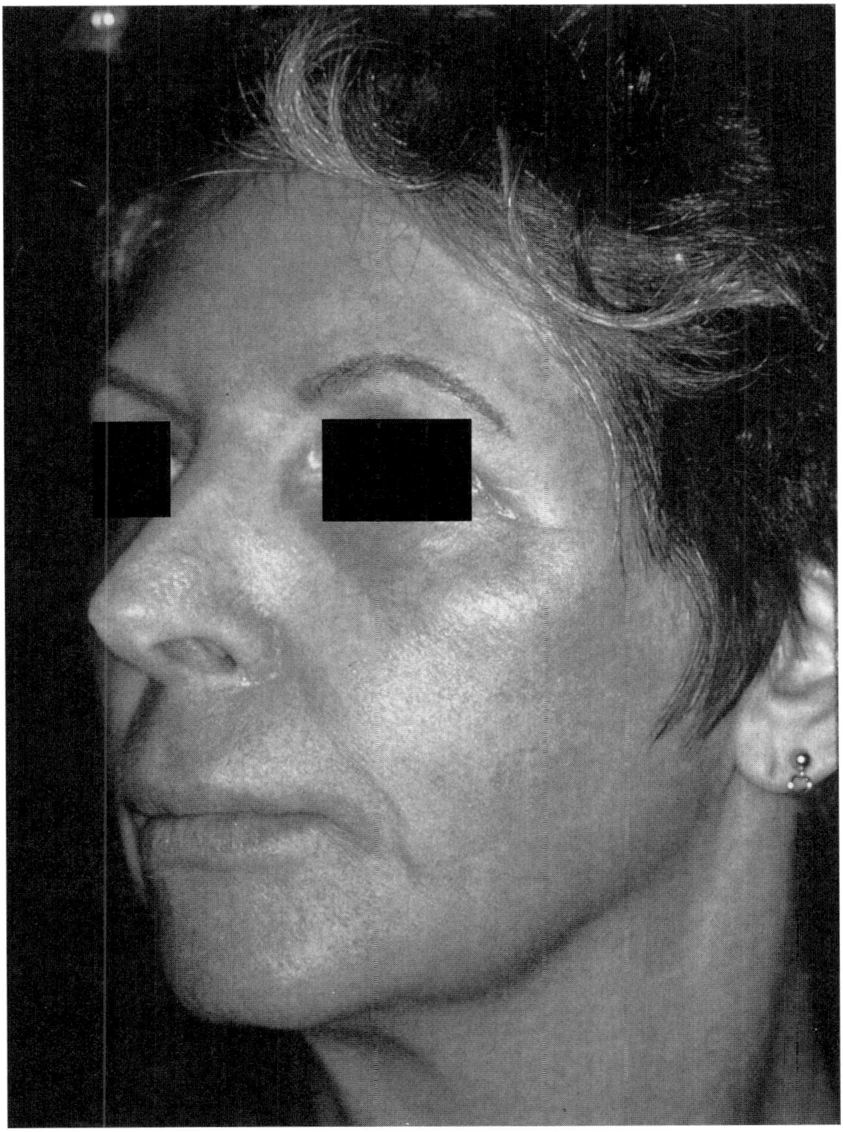

Figure 7b Postoperative simultaneous CO_2 laser resurfacing, autologous fat injection and rhytidectomy in a single session. (Photo courtesy of James Fulton, Jr., MD, PhD.)

REFERENCES

1. Fulton JE Jr, Suarez M, Silverton K, Barnes T. Small volume fat transfer. Dermatol Surg 1998;24:857–865
2. Polanyi TG, Bredemeier HC, Davis TJ Jr. CO_2 laser for Surgical Research and Medical Biological Engineering. New York: Pergamon Press, 1970
3. Slutzski S, Shafir R, Bornstein LA. Use of the carbon dioxide laser for large excisions with minimal blood loss. Plast Reconstr Surg 1977;60:250
4. Baker SS, Muenzler WS, Small RG, et al. Carbon dioxide laser blepharoplasty. Ophthalmology 1984;91:238–243
5. Weinstein C. Carbon dioxide laser resurfacing. In: WP Coleman, CW Hanke, TH Alt, S Asken, et al. eds. Cosmetic Surgery of the Skin: Principles and Techniques. 2nd ed., St. Louis: Mosby, 1997
6. Henriques FC. Studies of thermal injury. Arch Pathol 1947;43:489
7. Anderson RR. Laser tissue interactions. In: MP Goldman, RE Fitzpatrick, eds. Cutaneous Laser Surgery. St. Louis: Mosby, 1994
8. Klein JA. Tumescent technique for regional anesthesia. J Dermatol Surg Oncol 1990;16:248
9. Lillis PJ. The tumescent technique for liposuction surgery. Dermatol Clin 1990; 8:439
10. Louis PJ, Waite PD. Lateral augmentation of the angle of the mandible with porous hydroxyapatite block. Am J Cosm Surg 1998;15:125–35
11. Waldorf HA, Kauvar AN, Geronemus RG. Skin resurfacing to fine to deep rhytides using the char-free carbon dioxide laser in 47 patients. Dermatol Surg 1995;21: 940–946
12. Fitzpatrick RE, Goldman MP, Satur NM, Tope WD. pulsed carbon dioxide laser resurfacing of photaged facial skin. Arch Dermatol 1996;132:395–402
13. Nanni CA, Alster TS. Complications of CO_2 laser resurfacing: an evaluation of 500 patients. Laser Surg Med 1997;9(Suppl):242
14. Lillis PJ. Liposuction surgery under local anesthesia: limited blood loss and minimal lidocaine absorption. J Dermatol Surg Oncol 1988;14:1145
15. Fulton JE Jr. Personal communications, 1998
16. Yoho A. Personal communications, 1998
17. Coleman SR. Skin & Allergy News, June 1998, p 33
18. Weinstein C. Ultrapulse carbon dioxide laser removal of periocular wrinkles in association with laser blepharoplasty. J Clin Lasers Med Surg 1994;12:205–209

15

Rejuvenation of the Hand: The Hand Fill

Pierre Fournier
*Private Practice, Aesthetic Plastic Sugery,
Paris, France*

It is important to think of rejuvenating and embellishing the hands as it is to think of rejuvenating or embellishing the face (Fig. 1). However, very few techniques have been developed for this, and few patients at this time request esthetic improvement. Probably people do not know that such an improvement is possible, because as soon as they have been informed that certain procedures are available, they become very enthusiastic and ask to have them done.

Following facial or face and neck rejuvenation, it should be normal practice to prescribe similar treatment for the hands; after all, why have hands 10 years older than one's face?

The hands, like the face, often look much older and worn than the rest of the body, which is protected from the sun and wind for most of the year. A woman's hands may be damaged by housework, because they are not always protected by gloves, and in addition to the more or less corrosive external agents, the aging process takes its toll. Face make-up can camouflage aging, but the hands have no such artifice. Hands are like faces—some stay young a long time and their shape and color do not change. There are full hands, just as there are full faces, and there are thin hands, just as there are thin faces which have aged inside and out. Most often, as in any aging process, the hands shrink, the veins appear more bulging and more colored than in youth, and the complexion darkens. Hyperkeratoses appear, multiply, and spread. The back of the hand does not contain a lot of fat tissue and thus cannot reabsorb it. One should note that the back of the hand and the fingers age above all and not the palm or the palm side of fingers.

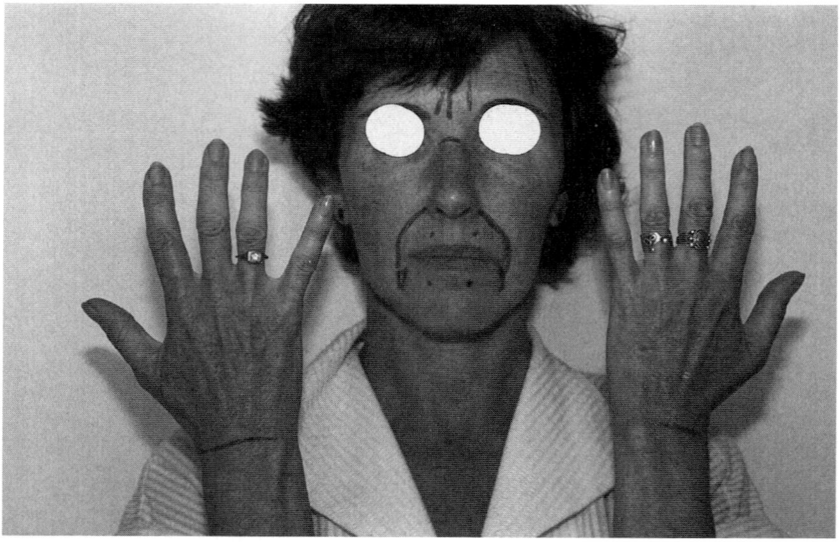

Figure 1 The hands as well as the face can benefit from rejuvenation.

The previous degeneration, apart from hyperkeratoses and wilted complexion, is due to the atrophy and the involution of the other components of the hand, mainly of the bones and intrinsic muscles associated with the alteration of the normal skin texture. The skin crumples, wears, and becomes darker just like the inward aging of the face. Tissue involution is thus taking place.

I. HAND REJUVENATION

A. Hand Lift

A hand lift has been proposed on the ulnar side of the hand with a greater or lesser undermining of the back of the hand and intradermal suture. Another technique proposes skin undermining at the wrist level on the back side of the hand. The scar is more visible, but of course the skin excess will be more even. This skin "lift" results in some improvement, and in order to better the results, there has been no hesitation in adding a resection of the skin excess on the fingers at the joint between the first and second phalanges. These techniques, at best, only treat the skin excess and not the complexion of the skin. Therefore, other means have been proposed to correct the aging complexion.

B. Dermabrasion

The back of the hand reacts remarkably well to mechanical dermabrasion. This technique poses no technical problems. However, the same inconveniences and complications as those of facial dermabrasion are possible: hyperpigmentation, hypopigmentation, and a sharp border with the untreated skin. We have performed a few of these dermabrasions with varying results; although the results were generally good. Long-term results are not known.

C. Skin Peel

Phenol peel has been done by a few surgeons. One hand and then another must be done at a 6- to 8-day interval, so that the patients will not be to handicapped. The follow-up is not as simple as that of mechanical dermabrasion. Hypertrophic scars can occur, which may be improved with cortisone treatment.

Some operators use the classic phenol solution which has aged for several years. After this aging it is weaker and can be used regularly, even on Orientals, with reportedly good results. The border between the treated and nontreated areas is not very visible, and thus can be on the wrist or on the forearm. Although lifts and chemical or mechanical abrasion treat the problems of excess skin or pigments, they do nothing for the veins which show up even more than usual. This characteristic disturbs some women, who may consciously or unconsciously keep their hands held high during a conversation, which makes prominent veins decrease or disappear. Artists and advertisers never show a hands congested or unesthetic hands.

D. Vein Stripping

Compared to the previous procedures, vein stripping at one time was relatively popular. The operation is very simple because of the short length of the veins and the practice of stripping the varicose veins on the lower limbs helps make the operation easier but this experience is not indispensable. A necessary and sufficient "internal stripper" can be a simple plumbline (e.g., a 1.5-mm fuse wire), a hair pin, a thick thread, or even a fine hemostatic forceps used like an external stripper which will allow resection where the veins are concerned. Not all the veins of the back of the hand are to be automatically stripped. It is remarkable to see the patient herself choose: "Take that one away, it makes me look old, but not that one which looks normal."

After local anesthesia, using lidocaine (Xylocaine) at 1% with epinephrine (adrenaline), of the skin where the incisions are to be made and where the vein crosses, the vein is approached at the level of the metacarpophalangeal joint in a crease. The incision is 4–5 mm long. The vein is exposed using hemostatic forceps and tied with fine catgut or 5 zero resorbable suture. The ligature acts in turn as a tracking wire, and a longitudinal incision is made in the subjacent area of the ligature and the stripper is inserted.

The author particularly likes to use a hairpin with a blunt end, once it has been bent into shape. At the wrist level, a small horizontal incision is made on the projection of the exposed vein and its upper part is tied up. The blunt end of the stripper is then exposed. A plastic bead can be put on the end of the hair pin as if one were making a necklace and then the pin can be bent at a right angle to obtain a solid construction. Soft traction of the upper part of the pin brings the vein back without problems. The vein is telescoped on the plastic bead as in a varicose vein stripping of the lower limbs. A pressure dressing is applied for 5–10 min, since the tributaries are of course torn during the procedure. This bandage is applied by an assistant and will not prevent the operator from stripping a second vein as far away as possible from the first.

After sewing the incision with a fine suture, a moderate pressure dressing is applied and the patient is advised to keep the hand in a raised position for the rest of the day. This can be accomplished by a scarf or sling tied around the neck. This bandage is removed after 24 h.

Bruising and edema are variable, but this is usually part of the operation and will slowly be resorbed. The sutures are removed in 4 days, and the hand regains its normal appearance in 6–8 days.

Once the patient can use the operated hand normally, the other hand can be treated or other veins on the operated hand can be stripped. The author has never seen persistent edema of the hand or exaggerated venous circulation in the remaining veins. Satisfactory results are always obtained. This could also be due to the fact that patients are rigorously selected or highly motivated.

E. Autologous Fat

Autologous fat grafting is an established method for tissue augmentation.[1–6] The author now injects adipose tissue in order to counter the exaggerated appearance of the veins on the back of the hand. Fat is removed with a syringe and a 4-cm long needle having a 2-mm external diameter (14 gauge) and 10 mL is injected into the entire back of the hand, although sometimes 12–15 mL is used. The hand fill is as justified as the face fill, since it also corresponds to loss of tissue. The hand lift or stripping only indirectly and imperfectly correct the problem with a heavy scar price to pay if a lift is done. It is obvious that a fill of the back of the hand cannot cause pigmentary blotches to disappear, but the skin darkens because of wrinkling and bunching of the skin over the responsible deep elements. It is obvious that the skin changes which are due to aging will not be altered, but those caused by the loss of volume of the internal structures will be remedied.

Even if complete rejuvenation is not accomplished, at least the faded complexion problem will disappear or decrease greatly and the color of the dorsal veins will be lightened and their overall appearance will be less noticeable. Once the result has been obtained, it is up to the esthetic surgeon to decide if a light peel, mechanical dermabrasion, or electrocoagulation of the pigmentary spots is to be advised. Hydrogen peroxide applied by the patient repeatedly over a period of many weeks has good results. Electrocoagulation of the pigmentary spots on the back of the hand is an excellent treatment if it is carefully done.

F. Silicone

Hand fill has been practiced in the Philippines and in Japan using injectable silicone. The use of injectable silicone is illegal in France and many other countries even though we know that, in some cases, if silicone is injected in very small amounts, at long intervals, it can obtain good results (e.g., the small wrinkles of the upper lip or forehead can thus be treated this way). The silicone must also be very pure. It has the drawback of migrating in some cases and can cause unesthetic mounds on the face. These mounds are impossible to treat because of the possibility of facial paralysis during the treatment. Surface defects caused by their ablation, which is necessary for localized hardening, inflammation, or tissular necrosis are also possible. The dangers caused by the incorrect use of silicone are much greater than the good results that can be obtained. After a certain amount of time, the silicone can also become yellowed. This is not a problem for Orientals, but it is difficult for whites. In addition, for some unknown reason, Orientals tolerate silicone injections much better than whites. This explains why there has been a wave of injections of small doses of silicone in the Orient, which eliminates the need for stripping or lifting the hands.

II. HARVESTING ZONE MARKINGS

The extraction zone surface will vary depending on the quantity of adipose tissue to be extracted. Obviously, this extraction should not leave any surface irregularities.

The chosen zone should be marked while the patient is standing if this zone is located in the trochanteric region or the hip. It should be done lying down if it is in an exterior buttock or in the abdominal wall, since morphology is modified by the position.

The markings should be either a circle, an oval, a square, or a rectangle. The interior of the marked zone should be subdivided in small rectangles approximately 4 cm long and 3 cm wide. Their number should be approximately equal to the number of syringes necessary to do the hand filling. With one syringe, 5 mL of adipose tissue can be extracted. A regular extraction can be performed easily and without leaving a residual deformation in the totality of the zone if only 5 mL of adipose tissue is removed from each one of the rectangles, which thus serve as measuring units.

III. OPERATIVE ZONE DISINFECTION

Rigorous asepsis is necessary. The integument is usually disinfected with either 90% alcohol or any other product, according to the surgeon's preference. The surgeon will be gloved and masked. Draping of the operative zone should be done as for any other operation before the anesthesia is given.

When using nonsterilized external refrigeration as anesthesia or as a complement to local anesthesia, disinfection should be done and the drape should be placed when this is finished.

IV. OPERATIVE ZONE ANESTHESIA

Anesthesia in the operative zone should be different than that used for reduction microlipoextraction. Recall that:

1. In the case of microlipoextraction for reduction lipoplasty when reinjection of tissues is not foreseen, "The most important thing is not what one extracts, but what one leaves."
2. In the case of microlipoextraction to reuse the tissues obtained during the extraction, "The most important thing is not what is left, but what one extracts."

In both cases, the tissues must be respected by the surgeon, but the tissues to be respected differ in each case.

The fraction that must live is what matters, not the fraction that will be eliminated. A noninfiltrated "dry" extraction is preferable to an infiltrated extraction because of the hydraulic traumatism of the injection added to the chemical traumatism caused by the injection of epinephrine and Xylocaine. Therefore, in the case of microlipoextraction and reutilization of the extracted tissue, utmost care should be taken by the surgeon for the tissue to be reinjected.

A. Anesthesia of the Harvesting Zone

Anesthesia of the donor zone should be performed first. The following should be done to extract the tissue to be reinjected.

1. Local Anesthesia

Even if local anesthesia is not theoretically desirable for the above-mentioned reasons, it is often used. As in the case of a nerve block or regional anesthesia, 1% lidocaine with 1:200,000 epinephrine is used. For anesthesia through infiltration, the greater the surface to anesthetize, the weaker the dilution. In general, a dilution to the fourth is used. The dilution is done using 2°C saline solution. A weaker dilution is perfectly possible at a fifth, an eighth, or a tenth).

All marked regions should be infiltrated at the same time; it will simply take longer for the anesthesia's effect to be felt. The infiltration should be made in all the thickness of the layer of adipose tissue, but especially in the depth. External refrigeration with ice bags should always be used for 20 min. The purpose is to obtain an extraction with as little bleeding as possible and to reinforce the anesthesia.

When it is possible to create a thick fold with the left hand (e.g., on the abdomen), one can use a long needle (spinal needle type but finer), anesthetize exclusively the depth of the adipose tissue in this fold and obtain anesthesia of the overlying tissues where an extraction can be easily obtained in a noninfiltrated zone.

2. Cryoanesthesia

The best method is external cryoanesthesia.

3. General Anesthesia

Sometimes patients want general anesthesia rather than local. This is inconvenient, since the patient is kept in the clinic longer, but it is the ideal technique for large extractions. Twilight anesthesia can also be used. Two or three zones can be refrigerated simultaneously, and the surgeon can perform the extractions satisfactorily. This types of anesthesia is for ambulatory patients.

4. Observations

External refrigeration of the areas to be operated on should be done before any of anesthesia—general, twilight, block, or local as well as before any kind of extraction or injection of adipose tissue. It should be maintained for about 20 mins.

V. PURE HAND LIPOFILLING TECHNIQUE

A. Instruments

The amount of equipment necessary for a hand fill is very small:

1. The "classic" filling needle (2 mm outside diameter and 4 cm is a useful length); disposable (14-gauge) ones now exist (14 gauge)
2. 30- and 21-gauge needles for local anesthesia
3. 10-mL disposable plastic syringes
4. 20-mL saline solution ampule at room temperature or at 2°C, which will serve to dilute the local anesthetic, to avoid all dead space in the syringe, to diminish the impact, facilitate the transportation, and to wash the adipose tissue
5. 2-mm external diameter transfer needle
6. Wooden or metal test tube holder

We no longer use 2-mL insulin syringes or 5-mL syringes, which do not permit washing of the extracted adipose tissue. If the face fill is done at the same time as body liposculpturing, the adipose tissue extracted with the help of single-orifice cannulae of 3- or 4-mm outside diameter and 50 or 60-mL syringes can be reused if it is transferred to 5 or 10-mL syringes on which the usual 2-mm outside diameter (14-gauge) filling needle has been mounted.

B. Extraction Technique

The syringe allows the extraction, preparation, and reinjection of the adipose tissue in a closed space, whereas respecting the integrity and sterility of the extraction. It also serves as a reservoir to conserve the extracted tissue in the freezer if necessary.

C. Extraction of the Adipose Tissue

Using a 2-mm outside diameter filling needle (14-gauge) mounted on a 10-mL syringe, 2 mL of saline solution at room temperature or at 2°C is placed in the syringe. The left hand makes a fold on one of the small rectangles marked on the extraction zone, and the needle is pushed through the skin in the middle of the adipose tissue. The left hand is then placed flat on the extraction zone, and will stay that way throughout the operation. The right hand pulls out the syringe plunger completely, and the surgeon begins the extraction. During the whole extraction, the plunger will be kept in this same position.[7]

Back and forth movements are made in a fan-shaped way almost the length of the needle. Four to five of these movements are made in one direction before going to the neighboring region, without taking out the needle, which must remain under the skin. Between 30 and 50 s are needed to obtain 5 mL of adipose tissue.

Once 5 mL has been obtained, the needle is removed, and 3 mL of saline solution is drawn into the syringe. The latter is placed vertically on a test tube holder with the plunger up.

The needle that was used for this extraction is mounted on another syringe for use in another cylinder extraction. This takes place while the adipose tissue in the first syringe is going through a period of decanting, which is the first phase of washing.

Once 5 mL of adipose tissue has been extracted with the second syringe in the same way, 3 mL of saline solution should be added, and the syringe should then be placed vertically on the test tube holder. The predetermined amount of adipose tissue will be removed in each one of the rectangles marked on the extraction zone.

D. Washing

When the extraction is finished, the surgeon washes the adipose tissue with the saline solution. Once the 3 mL of saline solution has been drawn into the syringe and it has been placed vertically on the test tube holder, one can see the mixture separate into two parts: the adipose tissue on top and the bloody saline solution on the bottom. This separation takes place after a few seconds. The washing also can be done with a 5% glucose solution.

This hydrohematic fraction is emptied into a cup, and 5 mL of saline solution is again drawn into the syringe. The syringe should once again be placed on the test tube holder in order to obtain the separation of the two fractions while performing the same maneuver on the other syringes.

There should be as many washings as necessary to obtain bloodless adipose tissue. In general, two to four washes are necessary, sometimes more if the sample is very bloody or if not enough time is allowed between each wash. A transfer needle will allow adipose tissue to pass from one syringe to another in order to get equal amounts for symmetrical regions. The reinjection can be done when all the syringes have been thus prepared.

Locally there should be an occlusive, slightly compressing dressing made of a few compresses kept in place by several Elastoplast bands or a layer of Elastoplast. This dressing should be removed after 24 or 48 hs. It may have to be changed after a few hours if bleeding persists.

E. Reinjection

Once the dressing is done, the injections should be performed. The needle used for the reinjection should be of the same caliber as that of the extraction needle; namely, 2-mm outside diameter.

One must inject with caution and make sure of not injecting into a vessel. This maneuver can be performed if desired with the needle mounted on an empty syringe. Once this is done, the syringe full of adipose tissue is used and the injection begins.

Unless the patient is under general anesthesia or twilight anesthesia, a wrist block using 5 mL of 0.5% Xylocaine with epinephrine is used. After waiting 15 mins (the block is accomplished after the fat tissue removal so that no time is wasted), the upper part of the back of the hand is injected with the 10 mL of removed adipose tissue. The veins are fewer and less apparent on the upper part of the hand. The lump thus created is easily spread out by being softly pushed. The fat is gently spread throughout the back of the hand as the initial opening is kept covered with one finger so that no fat comes back out. Before doing this, the patient is asked to make a fist, which makes the operation easier (Figs. 2 and 3).

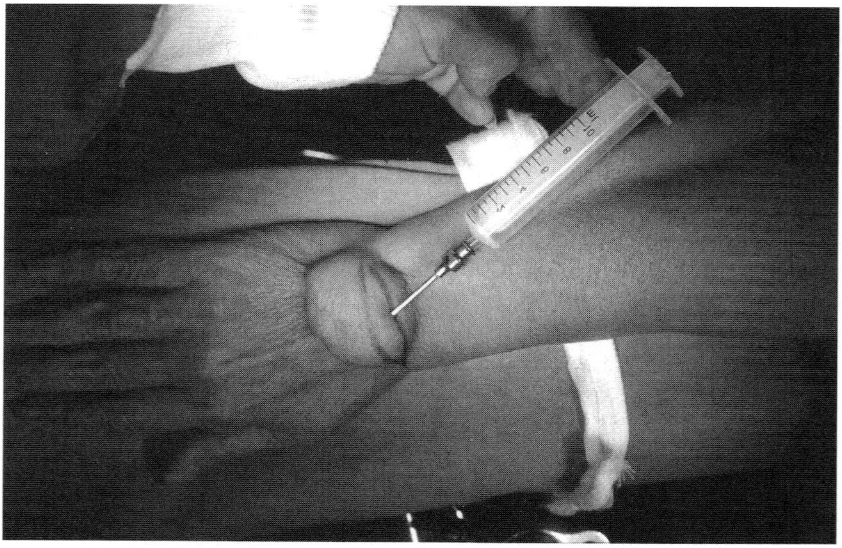

Figure 2 Fat injection into the hand.

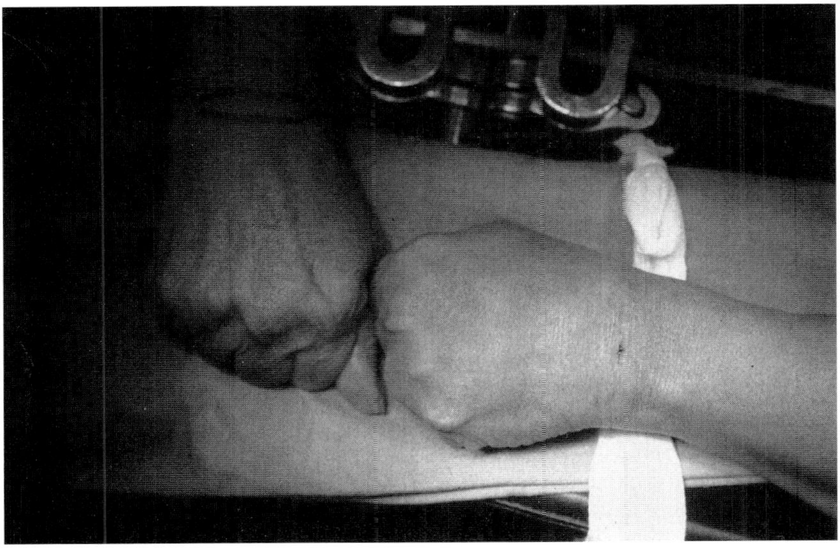

Figure 3 After the injection, the patient is asked to make a fist to spread out the fat tissue evenly.

One must not forget to push a little fat tissue to the base of the fingers. Our first patient told us, a while after the operation: "I am very happy with my hands, but my fingers have not changed and look even older in contrast. The next time I come to get an injection, please do not forget to inject my fingers."

Bandaging is not used and the patient can go back to his or her normal life the day after the operation. The only advice is to place the hand in an inclined position using a scarf knotted around the neck. Edema and some bruising are part of the operation. The bruises can be covered with foundation or Cover Mark until they disappear.

When the hand filling is done on an ambulatory basis without general or twilight anesthesia, the whole operation, including preparation, operation, and patient rest, takes about 1.5–2.0 hs. Vibramycin for 5 days is prescribed, more for medicolegal reasons than for risk of infection, which is minimal. Anti-inflammatories (Reparil) are, on the other hand, prescribed for 15 days. Analgesics are not necessary.

VI. "PREPARED" FAT GRAFTING: DELAYED HARVESTING

A few years ago, we started to prepare the donor area to obtain better results and decrease the number of injections necessary for a long-lasting result. The aim is to create an inflammatory condition of the fat tissue and to graft a "healing fat tissue" instead of a normal fat tissue. This possibility is explained to the patient, who will choose between "fresh fat grafting" or "prepared fat grafting."

In this latter case, we give a conventional local anesthesia and/or a cryoanesthesia (external), and with a spinal needle or an ordinary 19-gauge needle, repeatedly stroke the whole thickness of the fat layer of the harvesting zone for 1 or 2 mins. No fat is extracted. A dressing is applied, and the patient returns home. The future receiving area may be prepared the same way. After 10 or 12 days, the patient returns, and a conventional fat harvesting will be performed and the first injection accomplished.

The remaining fat is kept in the freezer for future injections. Histological studies have shown that the specimen of the fat obtained with this delayed harvesting has many new vessels, new collagen fibers, and a great number of fibroblasts. Out clinical experience confirms that this "prepared fat" decreases the number of injections necessary for a long-lasting result.

VII. IMPLANT CONSERVATION

A. Conservation of Implants in the Freezer

Conservation of all types of implants coming from adipose tissue is easily possible through cold. The results do not differ from those with the fresh implants normally used. The purpose of this conservation is to save the excess implant extracted or to avoid a new extraction when repeated injections are foreseen.

Tissue conservation by refrigeration is a classic technique, as in the case of other cells (sperm), and the innocuousness and efficiency of the technique also have been proven. All types of implants coming from adipose tissue have been kept in the freezer and reused by the author; some after a whole year. Types of frozen implants include:

1. Adipose tissue cylinders extracted with a 2-mm diameter needle (washed with saline solution)
2. Microimplants extracted with 12/10–mm (19-gauge) or an 8/10–mm (21-gauge) needle (washed with saline solution)
3. Implants extracted with 3- to 4-mm diameter cannulae and 60-mL syringes that can be used with 5- or 10-mL syringes and the classic 2-mm diameter fill needle (washed with saline solution)

4. Autologous collagen extracted with a 8/10– or 12/10–cm diameter needle (washed with distilled water and kept in the freezer); this collagen can be used as a dermic implant (using an 8/10–cm needle or dermojet) or as a subcutaneous implant.

B. Technique

After washing with either a saline solution or distilled water to rid the implant of all traces of blood or hemoglobin, in every case the sterile cap of an injection needle is placed on the hub of the syringe. All extraction syringes are placed on a test tube rack and are then placed in the freezer or refrigerator.

To reuse the implants thus kept, one need only take them out of the freezer compartment 1 h before reutilization. This defrosting of the implant is done at room temperature (22°C) in a test tube holder. It is well known that the freezing should be rapid, whereas the defrosting should be slow.

A quicker freezing than that obtained using the freezer can be obtained with liquid nitrogen. The syringe is placed for a short time in a liquid nitrogen–filled container before being placed in the freezer. A quicker defrosting can be obtained in warm water (28 or 30°C). The syringe or syringes with the cap are placed in a container with faucet water warmed to this temperature.

The results obtained with the help of these conserved implants are no different than those obtained with fresh implants (on the face or hands). The repeated operations necessary are more easily accepted by the patients.

VIII. COMPLICATIONS

Edema is not a complication but rather a part of the operation. It varies from one patient to another. Sometimes edema is a small amount but sometimes it deforms the area.

In most cases, after 3 days, the deformation of the treated area is no longer visible except perhaps to the patient. Less often, the treated regions can have abnormal edema for about 8 days.

Ecchymoses are rare, because the adipose tissue is washed. There can, however, be small ones related to the anesthesia or the adipose tissue injection. Sometimes they are delayed, and appear on the second or third day. Ecchymoses can be hidden with make-up.

Pain and discomfort are rare complications. The author has never seen liquefaction of the graft or infection. Often a localized, sensitive induration may be noticed in the treated areas, which is more palpable than visible, disappearing spontaneously in a few weeks, but sometimes lasting longer.

In order to avoid significant reactions, it is important not to overcorrect in volume the region treated. One wants the patient to return to a normal appearance within a few days. Necessary overcorrection can be done with repeated injections spread out over time. These are more easily accepted than an overcorrection in volume that leaves the patient marked for a much longer time as a consequence of the operation. Moreover, small quantities of fat take better than larger one.

The only unfavorable problem is the disappearance of the results obtained owing to total reabsorption of the graft. This is noted in a certain number of cases. In about 10% of the cases, the patient has cysts in one or several places on the dorsum of the hand. The surgeon can either crush them or remove them under local anesthesia (Fig. 4).

IX. DURABILITY OF THE RESULTS

The patient should be warned that multiple injections may be necessary. In the hand fill, 25–50% of the injected specimen is viable. Never try to overcorrect, since there is some uncertainty as to the degree of resorption and there can be an

Figure 4 Removal of cysts.

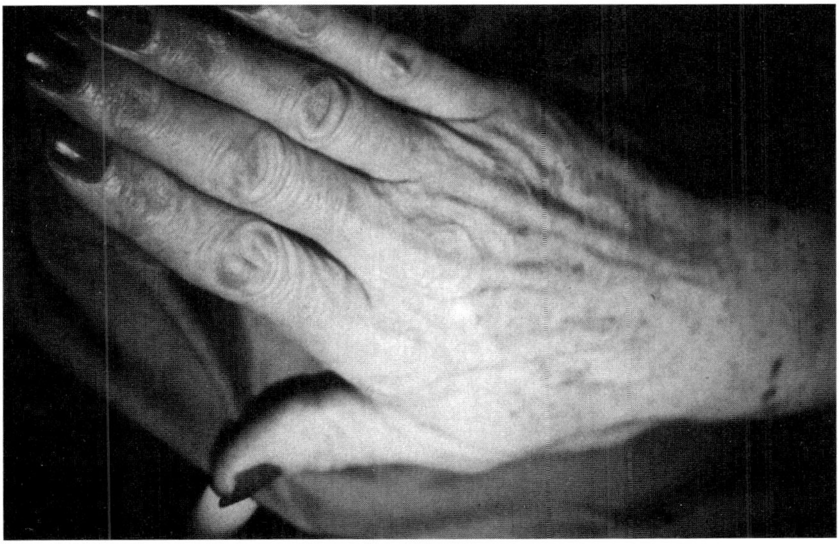

Figure 5a Preoperative condition.

unesthetic deformation of the area for several weeks. We have had to reinject the hands in a few cases where the resorption occurred after several weeks. The patient sees a notable change quite soon after the operation but a permanent result follows (Figs. 5 and 6). About 10% of patients have to be reinjected after 3 or 6 months in order to obtain a lasting result. The durability of the results is due to the particular structure of the injected area. The rigidness of this area, due to the carpal and metacarpal bones, creates a veritable inner splint. This limits the transfer's absorption. This sort of splint is not found in the face or even provided by the muscles, because these speed up the absorption of the injected specimen.

Even if the result on the back of the hand fill were only temporary, the simplicity of this operation makes it a must in the esthetic surgeon's armamentarium.

A person's age cannot only be judged by their face but also by their hands.

X. CONCLUSIONS

After studying different classic operations to rejuvenate or embellish the hand, we propose the fill of the back of the hand and fingers whose durability is greater than that of the face fill. This has been one of the most satisfactory operations

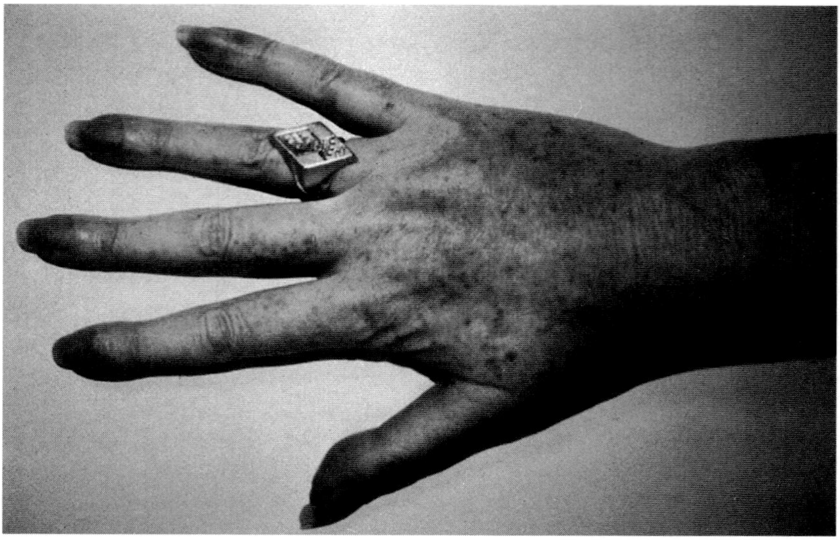

Figure 5b Postoperative condition.

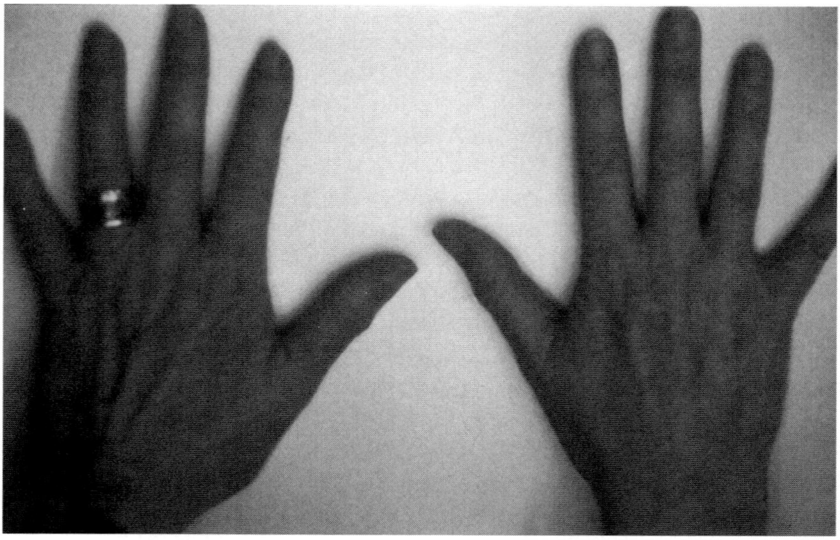

Figure 6a Preoperative.

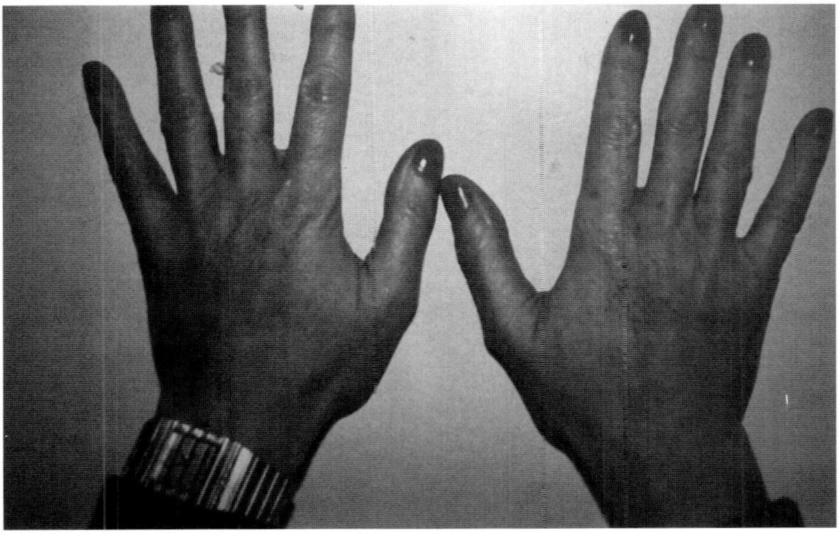

Figure 6b Postoperative: results after 30 months.

of fat grafting for the patient and the surgeon. It can be combined with a light peel and electrocoagulation of senile keratosis.

REFERENCES

1. Gurney CE. Experimental study of the behavior of free fat transplants. Surgery 1938;3:680
2. Munir C, et al. Greffe du tissu adipeux autologue. Etude experimentale. Rev Chir Est 1990;58:23
3. Peer LA. Transplantation of tissue Baltimore: Williams & Wilkins, 1955
4. Stevenson TW. Fat grafts to the face. Plast Reconstr Surg 1949;4:458
5. Skouge JW, Canning, DA, Jefs RD. Long term survival of perivesical fat harvested and injected by microlipo injection techniques in a rabbit model. Presented at 16th Annual American Society for Dermatologic Surgery Meeting, Ford Lauderdale, Florida, March 1989
6. Wetmore SJ. Injection of fat for soft tissue augmentation. Laryngoscope 1980;99:50
7. Fournier PF. Liposculpturing: the syringe technique Paris: Arnette Blackwell, 1991

16

Fat Transplant to the Buttocks and Legs for Enhancement and Deformities–Large Volumes of Fat Transplant

Lina Valero de Pedroza
Nubell Clinic for Aesthetic Surgery and Unidad Medica Clinica Del Country, Bogota, Colombia

I. HISTORY

Neuber, in 1893,[1] used small pearls of fat taken from the arm to fill depressed facial scars for bone loss following facial trauma. Since then free fat graft has been used successfully for various problems such as facial scars, facial hematrophy, saddle nose, first branchial cleft disorder, jaw defects, zygomatic arch loss, postparotidectomy defects, periodontal cysts, and chin augmentation. Grafts have been effective in preventing cerebral adhesions, covering nerves to prevent neuromas, and treating dural defects and adhesions. Orthopedic use includes treatment of stiff humeral-ulnar joints and finger joints, filling ostomyelitis cavities, replacing the lunate carpal bone, and preventing tendon adhesions. In general surgery, fat is used for closing ventral hernias and stopping persistent bleeding in the kidney and liver.

With the introduction of liposuction in the 1970s, there followed autologous fat transplantation utilizing aspirated fat.

Augmentation of tissue to improve the esthetic appearance or to refill atrophic regions is now common practice. Very satisfying results can be obtained from fat transplants to the buttocks and legs.

II. PREOPERATIVE PREPARATION

The patient is carefully evaluated and the results are explained using computer imaging. All details of surgery and possible complications are discussed with the patients. Patients who are smokers are asked to quit smoking at least 20 days before surgery, and they are also forbidden to take aspirin in any form.

Measurements are made over the chest, waist, hips, ankles, or any other area that is going to have a fat graft. Body weight is measured preoperatively and postoperatively.

Patients are given a prophylactic antibiotic, cephalexin (Keflex) 500 mg, orally every 12 hs starting 2 days preoperatively. The skin of the patient is sterilized for 3 days before surgery with surgical soap while taking showers twice a day. Before entering the operating room on the day of surgery, one more shower is taken.

Instruments to be used during surgery such as cannulae and Pyrex and glass flasks are vapor sterilized.

III. TECHNIQUE

The areas for fat retrieval are tumesced with Klein's solution.[2] The fat graft is obtained by suctioning areas all over the body such as the abdomen, waist, hips, back, arms, trochanteric region, inner superior third of legs, and knees. These patients usually need general body contouring as well as filling of buttocks and legs.

The fat graft is placed in sterile glass flasks which have been vapor sterilized (Fig. 1). The flasks are located between the suction cannula and the suction machine. The fat is retained in at least two or three flasks while normal liposuction contouring is continued.

The fat graft does not need any special treatment other than keeping it in sterile glass receptacles and immediately placing it in a sterile 10- or 60-mL syringe (Fig. 2).

In thin patients, any possible body area which has enough fat is suctioned for the grafts. Some patients may need 3–4 kg of fat weight in order to obtain enough graft. The patient is his or her own laboratory for growing a fat graft.

Since the Brazilian silicone prostheses for the gluteus is able to project approximately 3.0–3.5 cm posteriorly, this is the goal for the fat transplant.

In thin ankles with a circumference of 17 cm, the transplanted graft circumference should augment the circumference from 3.0 to 3.8 cm.

The fat graft is injected subcutaneously over any area needed to be filled. Minimum tunneling is performed and the fat is molded with finger and hand pressure while injecting to give a smooth appearance.

Figure 1 Fat for graft in sterile Pyrex flask following liposuction.

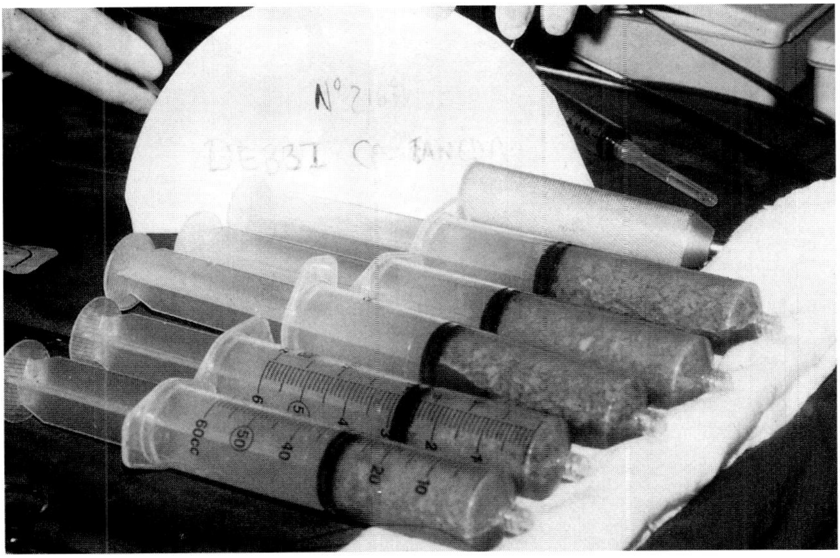

Figure 2 Storage of autologous fat in syringes.

A 12-gauge needle is used to transplant the fat into the buttocks and a 2-mm blunt cannula is used to fill ankles and legs. This will avoid artery or vein puncture with bleeding or injury to nerves and has not been associated with fat embolus.

In the buttock, measurements are taken from the intergluteal line to show the projection from 3.0 to 3.5 cm with the injected fat volume. Adding more volume results in the fat running over into lateral areas but not giving more projection. A similar occurrence is observed in the ankles. When the graft volume is enough, the procedure is stopped. Overfilling may cause skin necrosis due to excessive pressure.

From 200 to 350 mL is used to perform buttock augmentation and 80 to 110 mL is used for ankles.

IV. POSTOPERATIVE CARE

All patients remain 1 night in the hospital. Compression garments are applied immediately after surgery, and the patient should wear them for a period of 30 days postoperatively.

The areas of the body suctioned are massaged starting 15 days after surgery. Three massages of 1-h duration are performed every week for 2 months.

The grafted areas do not receive any treatment other than the normal care of a surgical scar where the skin puncture is present. Peroxide is applied to the wounds starting 3 days after surgery.

V. COMPLICATIONS

The authors have treated 1350 patients with ages ranging from 18 to 65 years. The patients were followed for up to 6 years (Table 1).

Table 1 Follow-up Time Period and Percentage of Patients Followed

Time	Patients followed (%)
2 months	92
3 months	87
6 months	68
1 year	55
2 years	36
3 years	28
6 years	13

The treated areas' circumferences were measured. Buttock augmentation had 0.5–1.0 cm loss at 2 months, which persisted at that measurement for years. This minimal decrease occurred if the patient had 3–5 kg of weight loss from the preoperative weight or of 3–5 L of fat was liposuctioned from the body. Patients losing 8–10 kg have no loss of buttock circumference. Patients who had gained 4–6 kg at the 6-year follow-up showed an increased circumference of 2 cm for each kilogram of weight gained. For those who underwent ankle or leg augmentation, there was a 0.5-cm loss in diameter at 1.5 months, which remained the same thereafter. The graft is, therefore, well preserved after 1.5–2.0 months.

One case of erysipelas appeared on the tenth postoperative day. The patient reported she was riding a bicycle and had a scratch over the left ankle producing a skin infection with edema, redness, and pain. The patient was treated with penicillin, 1,200,000 units, every third day for 12 days. The problem resolved quickly without loss of the fat graft.

Twelve cases of skin vesicles occurred because of the contact with Micropore over the skin of the ankles. These patients were treated with local antibiotics and the skin healed in about 5 days. The vesicles dried and resolved without damage to the graft.

Cellulitis with inflammation of the skin, which included redness and pain, occurred in one patient over the right gluteal area. The patient was treated with an oral antibiotic, Veracef (Cefradine) (2 g daily for 10 days). After healing there was no loss of the graft and no irregularities were observed.

All of the patients developed edema of the feet from the third postoperative day until the end of the first month. This was managed with the daily use of compression stockings (pantyhose) for 30 days, and at night the legs were elevated while resting. Five patients had foot edema for 3–5 months postoperatively. Two of these patients had peripheral vascular consultation and no pathology was found. Edema of the foot appears because when a thin structure like the ankle which is full of vascular structures is suddenly filled with a fat graft, pressure over the area occurs. Reduced venous flow results and produces the edema.

VI. DISCUSSION

Buttock prostheses cause prolonged incapacity ranging from 15 to 30 days. Mostly this is caused by extreme pain and the impossibility of lying on the gluteal area. In addition, patients are forbidden to have intramuscular injections into the buttocks. These gluteal prostheses cause small irregularities easily viewed through the skin, and many such patients were not able to wear a swimsuit. The buttock fat graft looks natural (Figs. 3 and 4), is well located, and the patient can lie in the supine position without restriction and can ambulate early on the day of surgery. Similar results are observed with leg fat augmentation (Fig. 5). Fat augmentation can be performed for reconstructive purposes such as polio sequelae (Fig. 6), agenesis of the gemellus muscles of the leg (Fig. 7), and facial hemiatrophy.

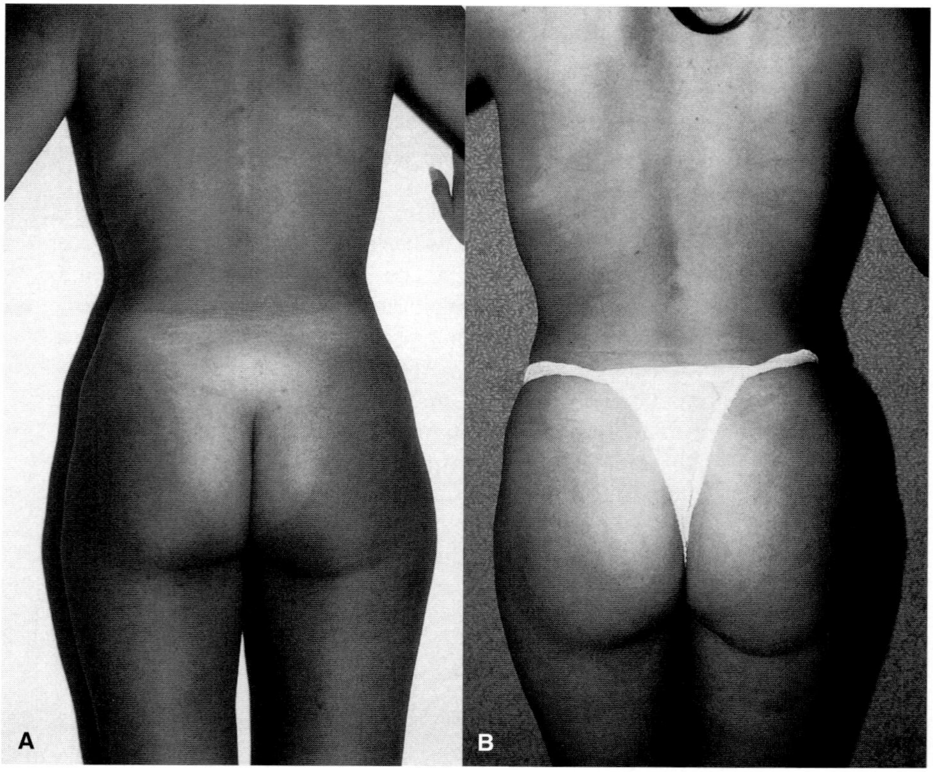

Figure 3 (**A**) Preoperative 19-year-old patient with small buttocks. (**B**) Postoperative 1.5 years. No weight gain.

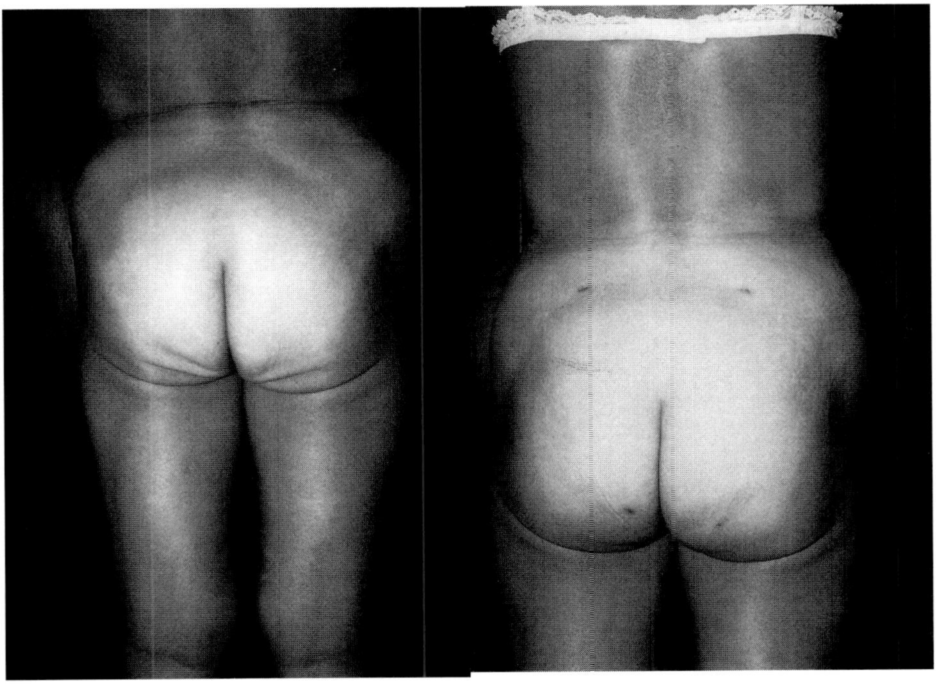

A B

Figure 4 (**A**) Preoperative 58-year-old female with small buttocks and loose skin. (**B**) Postoperative at 8 months after 600-mL fat graft to buttocks.

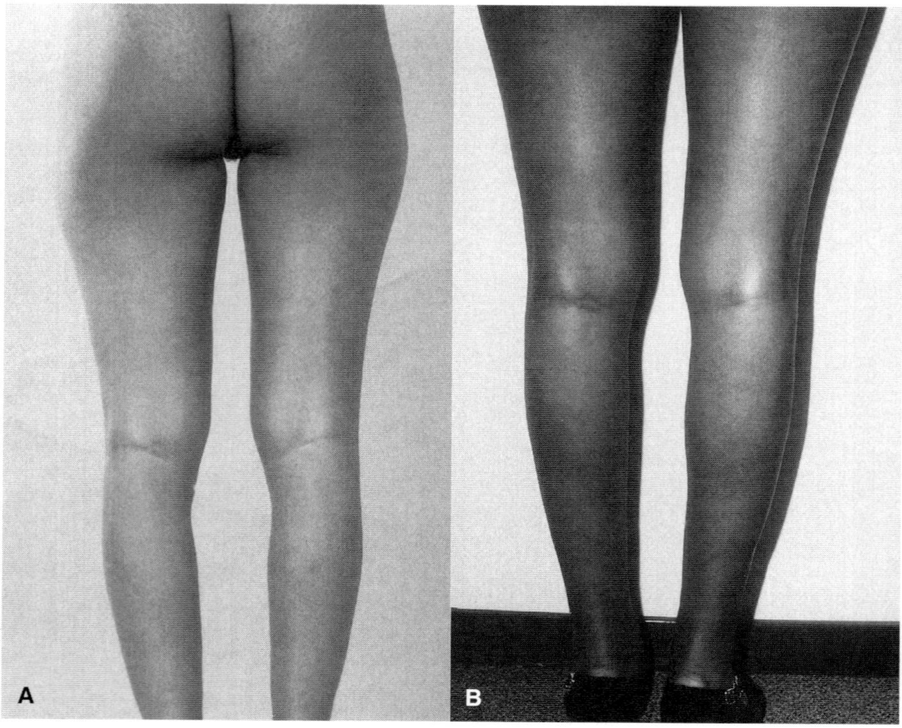

Figure 5 (**A**) Preoperative 21-year-old female with thin legs. (**B**) Two years post-operative 2 years. No weight gain.

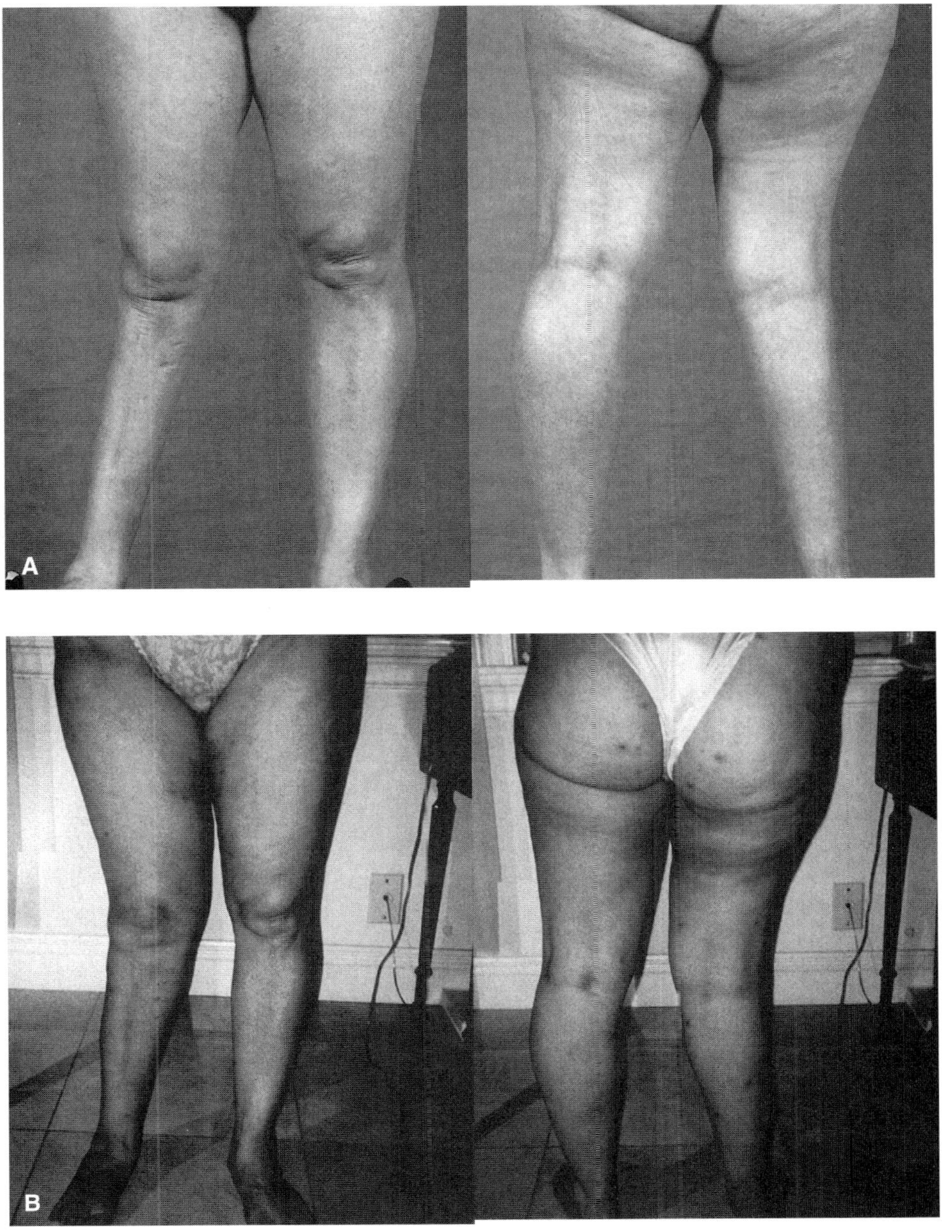

Figure 6 (**A**) Preoperative 44-year-old female with polio sequelae of right leg. (**B**) Postoperative 9 months after 800-mL fat transplant.

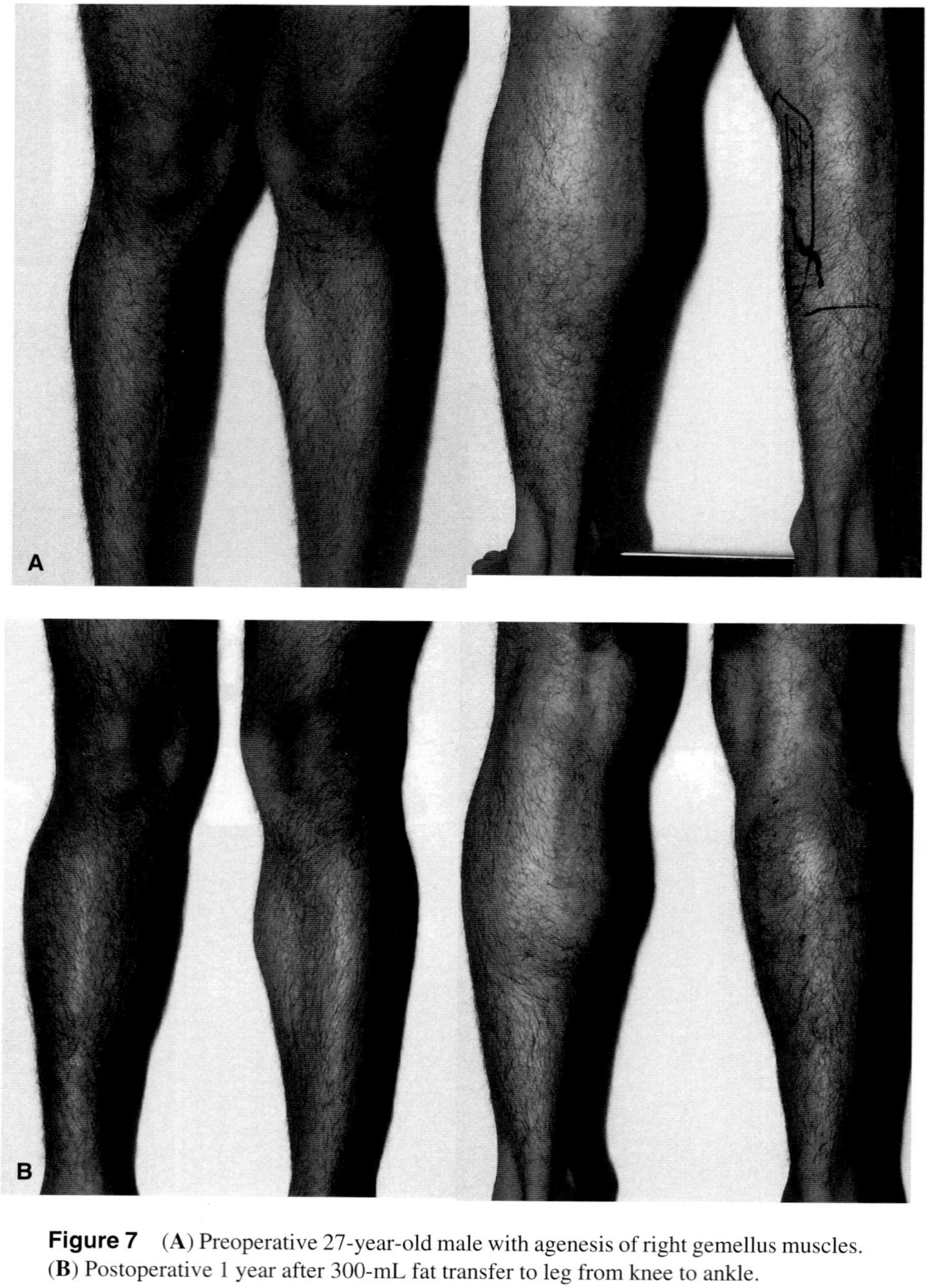

Figure 7 (**A**) Preoperative 27-year-old male with agenesis of right gemellus muscles. (**B**) Postoperative 1 year after 300-mL fat transfer to leg from knee to ankle.

Fat graft to the buttocks and legs has been shown to last at least 6 years in the follow-ups and is most likely permanent. The transplant is physiological and safe.

REFERENCES

1. Neuber F. Fettransplantation. Chir Kongr Verhandl Denstsche Gesellsch Chir 1893;22:66
2. Klein JA. The tumescent technique for liposuction surgery. Am J Cosm Surg 1987; 4(4):263–267

17
History of Breast Augmentation with Fat

Melvin A. Shiffman
Private Practice, Cosmetic and Reconstructive Surgery, Tustin, California

There has been some controversy about the use of autologous aspirated fat for mammary augmentation.

Czerny (1895)[1] reported the first use of autogenous fat in breast reconstruction. Lexer (1931)[2] described removal of the glandular breast in a patient with chronic cystic mastitis and reconstruction with fat rotated from the axilla. May (1941)[3] performed bilateral breast reconstruction using a free fat autograft on one side and a free fascia-fat autograft on the opposite side.

Bames (1953)[4] utilized fat from the buttocks denuded of epidermis in the breast and found 40% loss of volume because of fat liquefication. By placing the graft with its dermal side in contact with the breast tissue and the fascial side in contact with the pectoral fascia, there was about 90% graft survival. Reversal of the placement resulted in 60% survival.

Peer (1956)[5] stated that dermis-fat grafts provide a readily available transplantation material for establishing normal contour in small breasts instead of foreign implants.

Schorcher (1957)[6] reported autogenous free fat transplantation to treat hypomastia. He noted that the connective tissue elements remained intact with fat shrinkage to 25% of the original size by 6–9 months. He believed that if the graft was in several pieces, it would receive better nourishment for the recipient site.

Bircoll (1984)[7] was the first to inject autologous fat from liposuction around the breast. This helped avoid the complications of breast prostheses.[8] He used small droplets of fat (1 mL for each deposit), with a maximum of approximately 130 mL.[9]

Bircoll and Novack (1987)[10] described the use of insulin to pretreat the fat in order to increase survival percentage. Their report concerned one patient who had autologous fat injection into one breast to allow better symmetry with the opposite reconstructed breast (removed for cancer and reconstructed with a transverse abdominal rectus flap).

Asken (1987)[11] reported the use of autologous fat obtained by liposuction to augment the hypoplastic breast. He advised the injection of fat to be performed on withdrawal to avoid a large amount of fat in one location. Johnson (1987)[12] reported more than 50 augmentation mammoplasties by macroinjection from liposuctioned fat. Krulig (1987)[13] reinjected fat into the breast area and noted more than 50% resorption.

In a panel discussion at a meeting of the American Society for Aesthetic & Plastic Surgery, Bircoll[14] reported calcifications in three patients following breast augmentation with fat. All were biopsied and showed normal fat. Ousterhout[15] felt that the audience did not agree with the procedure in light of the calcifications and stated, ". . . to create calcifications in the breast by fat injections could cause considerable dilemma in mammogram diagnosis and possibly a missed malignancy diagnosis if each area of calcification was not biopsied."

There followed a flurry of letters in the journal *Plastic and Reconstruction Surgery*. Linder[16] stated that fat necrosis may require removal of some of the breast. Calcifications may occur in areas of fat necrosis, inflammatory reaction may produce a tumor-like mass and ". . . extreme reservation should be exercised in performing this procedure in the breast." Ettelson's[17] impression was that the ". . . preoperative and postoperative results both in the anterior and lateral views are virtually interchangeable as far as I am concerned. I am frankly shocked that the Journal would publish an article with such an unimpressive result." Hartrampf[18] had the impression that the fate of nonvascularized fat in the breast is fat necrosis, and this causes thickening and hardening which are clinically indistinguishable from carcinoma. Also, calcifications can occur in areas of fat necrosis. He stated, ". . . injection of any material into the breast, including autogenous fat, should be condemned." Baibak[19] complained about Bircoll's report[14] that there was no documentation of any change in the overall appearance of the patient.

Bircoll[20] responded by stating that no patient developed hard lumpy breasts. Calcifications in two patients occurred within 3 months of fat transplantation, and the calcifications were in the fat tissues surrounding the breast, which is distinguished by comparison with preoperative mammograms. Calcifications that remain stable can be observed rather than removed, and the calcifications in areas of micronecrosis are distinguishable from carcinoma.

Microcalcifications occur following breast augmentation with implants, open and closed capsulotomy, and breast reduction. As a new technique, the problem of microcalcifications in fat augmentation of the breast should be fully evaluated and understood before discarding the procedure. Bircoll[21–22] reported an estimate of 1.4% microcalcifications following fat augmentation of the breast which are periparenchymal and were not multidensity, rod-like, punctate, or branching spicules. Miller et al. (1987)[23] noted calcifications on mammogram in 3 (12.5%) of 24 patients who had undergone reduction mammoplasty. These postsurgical calcifications all had a benign appearance. Brown et al. (1987)[24] reported asymmetrical densities of mammograms in approximately 50% of patients who underwent reduction mammoplasty. Of 42 patients, a few developed calcifications on mammogram during the first year and over 50% had calcifications after 2 years. The calcifications were coarse and usually occurred in the periareolar and inferior portions of the breast. Only one patient with calcifications required biopsy. They concluded, "knowledge of the expected mammographic alterations and their time course may help to distinguish between postoperative change and carcinoma and thus eliminate the need for biopsy in some cases." Mitmick et al. (1990)[25] studied mammograms of 152 patients who had reduction mammoplasty. The incidence of calcifications was 24.3%. The calcifications were mainly rounded and well formed with an occasional lucent center. Finely rounded and slightly irregular microcalcifications adjacent to the larger calcifications were identified in 2.6% of patients.

Hin (1988)[26] performed autologous fat mammary augmentation with subcutaneous and retromammary injection. Illouz (1990)[27] was of the opinion that fat injection for breast augmentation has ethical and technical problems. Calcifications induced by fat resorption may resemble calcifications seen in cancer. Too much fat injected into the same location can result in necrosis of the fat and infection may occur. He injected into the breast only 5 mL of fat at a time and in different sites in a radial manner. Skouge (1991)[28] stated that early results of autologous fat injections for breast augmentation are encouraging but long-term results are not well documented. The degree of breast enlargement is limited, and fat necrosis has been shown to produce benign focal calcifications within the breast. "However, calcium deposits are common after reduction mammoplasty and other standard procedures involving the breast, and many radiologists believe that the distinction between benign and malignant deposits is usually clear." Yoho and Brandy-Yoho (1997)[29] stated that a good candidate for transplantation of fat into the breasts might be a postnursing woman who has lost some volume and whose skin is slightly sagging but not enough for more serious breast surgery. Fat transplant can fill the breast and sometimes restore or improve its original shape. In many cases, this may be preferable to artificial implants.

Although surgeons[26–32] have reported good results with autologous fat injection for mammary augmentation, opinions persist that are contrary to the facts. Gradinger[33] stated, "In my opinion, this procedure should be denounced as being potentially hazardous to one's health." Grazer[34] declared, "The use of the autologous fat injection for breast augmentation as advocated by Bircoll (1987)[14] is deplored because some patients can develop calcifications and esthetic results are marginal." Grazer (1992)[35] continued to denounce the procedure when he stated,

> Autologous fat grafting should never be used as a breast augmentation procedure. . . . Calcifications has been found to be a problem in some patients, masking malignant and premalignant changes. Biopsies must be performed in patients with nodules, pain, or calcification . . . the aesthetic results are marginal if large-volume augmentation is required.

Pitman (1993)[36] opined that the ". . . presence of calcifications in the breasts of injected patients makes the technique unacceptable."

The Society of Plastic and Reconstructive Surgeons (ASPRS) produced a Position Statement on October 1, 1992, which stated that the society (ASPRS) "strongly condemns the use of fat injections for breast enlargement," warning that the procedure may hamper the detection of early breast cancer or result in a false-positive cancer screening.

> . . . Furthermore, women who seek the procedure for breast enlargement are sometimes not informed that much of the injected fat will die, causing scar tissue and calcifications. . . . In a worst case scenario, a patient may face painful exploratory surgery or even mastectomy because of an uncertain mammography result.

It is this author's opinion that medically and legally it would be a breach of the standard of care to do a "mastectomy" because of an uncertain mammography result. Stereotactic needle biopsy is simple, accurate, and almost painless. "Exploratory surgery" would be virtually unnecessary in the hands of a capable knowledgable physician.

An article appeared in *Cosmetic Surgery Times* (1998) quoting Robert W. Alexander, MD, when he described satisfactory results with 60–70% fat retention from autologous fat injection for breast augmentation. There was a 3–4% incidence of small fat cysts or spherical calcifications which were easily differentiated mammographically from malignancy. What followed were letters objecting to the article and procedures. Lynch (President of the American Society of Plastic and Reconstructive Surgery) and Penn (President of the American Society for Aesthetic Plastic Surgery)[38] reiterate the Position Statement of the ASPRS (1992). Gorney[39] (Executive Vice President, Marketing Services, The Doctors' Company) stated,

> the installation of fat grafts in the breast is an extremely poor idea which will lead to highly undesirable consequences. . . . As any sensible surgeon knows, fat grafts to the breast are notoriously unreliable because the vast majority of the material reabsorbs within a matter of weeks to months, and what remains behind is merely the stroma of the cell, which very quickly calcifies and closely initiates the appearance of early cancer.

He states that there were four claims (presumably regarding breast fat injection) that were defended by the Doctors' Company. One was lost, one was tried and won, and two were settled. The Doctors' Company specifically excludes fat injection to the breast from coverage. Stone[40] stated that he would not incorporate the procedure into his practice, because ". . . injecting autologous fat into the breast will cause at some future date mammogram abnormalities which will lead to unneeded biopsies and my malpractice carrier specifically will not cover the procedure."

The letters regarding the article in *Cosmetic Surgery Times* were responded to by Schaefer (Editor-in-Chief).[41] She stated that the political scientist John Stuart Mills wrote,

> If all mankind minus one, were of one opinion, and only one person were of the contrary opinion, mankind would be no more justified in silencing that one person, than he, if he had the power, would be justified in silencing mankind. . . . If the opinion is right, they are deprived of the opportunity of exchanging error for truth: if wrong, they lose what is almost as great a benefit, the clearer perception and livelier impression of truth, produced by its collusion with error.

Those same surgeons who denounce fat augmentation of the breast because of calcifications performed breast augmentation with silicone gel implants which have up to 70% capsule contracture rate requiring surgical correction. The silicone-covered saline implants presently used have 10% incidence of contracture (smooth implants) and 9% incidence of rippling (textured implants), most of which will require surgical intervention. The fibrous capsule around the implant does develop calcifications which are readily distinguishable from malignant calcifications both by position and character. Breast reduction surgery is still performed with a small but significant incidence of stippled calcifications from fat necrosis within the breast parenchyma. Despite their appearance some months after surgery, these may need to have stereotactic biopsy, because they are within the breast. However, this does not contraindicate breast reduction by any surgeon's standard.

Patients who desire fullness of the breasts, not large breasts, are satisfied with the procedure. To some physician observers, this may appear to be unimpressive. Many patients fear the problems of implants such as scars, asymmetry, postoperative pain, deflation, capsule contracture, rippling, decreased mammography sensitivity, and possible autoimmune disease.

Fat injection into the breast results in very little discomfort. Lumps which appear are the result of oil cysts from too much fat injected into one area. Calcifications, which may occur months after the injections, can be sampled, if the physician and patient desire, by the simple expedient of stereotactic core needle biopsy. Open biopsy is unnecessary if the core shows benign tissue. These calcifications can then be followed by yearly mammograms just as is done with other surgeries causing microcalcifications.

REFERENCES

1. Czerny A. Plastischer ersatz der brustdruse durch ein lipoma. Chir Kongr Verhandl 1895;2:216
2. Lexer E. DeGesamte Wiederherstellungs-Chirurgie. Liepzig: J Barth, 1931
3. May H. Transplantation and regeneration of tissue. Penn Med J 1941;45:130
4. Bames HO. Augmentation mammoplasty by lipo-treatment. Plast Reconstr Surg 1953;11:404–414
5. Peer LA. The neglected free fat graft. Plast Reconstr Surg 1956;18:233–250
6. Schorcher F. Fettgewebsver pflanzuang bei zu kneiner. Brust Munchen Med Wochenschr 1957;99:(14):489
7. Bircoll MJ. New frontiers in suction lipoectomy. Second Asian Congress of Plastic Surgery, Pattyea, Thailand, 1984;80(4):647
8. Bircoll M. Autologous fat tissue augmentation. Am J Cosm Surg 1987;4(2): 141–149
9. Bircoll M. Cosmetic breast augmentation utilizing autologous fat and liposuction techniques. Plast Reconstr Surg 1987;79(2):267–271
10. Bircoll M, Novack BH. Autologous fat transplantation employing liposuction techniques. Ann Plast Surg 1987;18;327–329
11. Asken S. Autologous fat transplantation: Micro and macro techniques. Am J Cosm Surg 1987;4(2):111–120
12. Johnson GW. Body contouring by macroinjection of autogenous fat. Am J Cosm Surg 1987;4(2):103–109
13. Krulig E. Lipo-injection. Am J Cosm Surg 1987;4(2):123–129
14. Bircoll M. Panel on fat injection. American Society for Aesthetic Plastic Surgery, Los Angeles, March 26, 1987

15. Ousterhout DK. Breast augmentation by autologous fat injection (Letter). Plast Reconstr Surg 1987;80(6):868
16. Linder RM. Fat autografting (Letter). Plast Reconstr Surg 1987;80(4):646–647
17. Ettelson CD. Fat autografting (Letter). Plast Reconstr Surg 1987;80(4)646
18. Hartrampf CR Jr, Bennett GK. Autologous fat from liposuction for breast augmentation (Letter). Plast Reconstr Surg 1987;80(4):646
19. Baibak GJ. Liposuction. Plast Reconstr Surg 1987;80:647
20. Bircoll M. Reply (Letter). Plast Reconstr Surg 1987;80(4):647
21. Bircoll M. Autologous fat transplantation to the breast (Letter). Plast Reconstr Surg 1988;82:361–362
22. Bircoll M. Autologous fat transplantation: An evaluation of microcalcification and fat cell survivability following (AFT) cosmetic breast augmentation. Am J Cosm Surg 1988;5(4):283–288
23. Miller CL, Ferg SA, Fox JW IV: Mammographic changes after reduction mammoplasty. Am J Roentgenol 1987;149:35–38
24. Brown FE, Sargent SK, Cohen SR, Morain WD. Mammographic changes following reduction mammaplasty. Plast Reconstr Surg 1987;80:691–698
25. Mitnick JS, Roses DF, Harris MN, Colen SR. Calcification of the breast after reduction mammoplasty. Surg Gynecol Obstet 1990;171:409–412
26. Hin LC. Syringe liposuction with immediate lipotransplantation. Am J Cosm Surg 1988;5(4):243–248
27. Illouz Y-G. Fat injection: a four year clinical trial. In: GP Hetter, ed. Lipoplasty: The Theory and Practice of Blunt Suction Lipectomy. 2nd ed. Boston: Little Brown, 1990: pp 239–246
28. Skouge J. Autologous fat transplantation in facial surgery. In: WP Coleman, CW Hanke, TH Alt, S Asken, eds. Cosmetic Surgery of the Skin: Principles and Techniques. Philadelphia: Decker, 1991, pp 239–249
29. Yoho RA, Brandy-Yoho J. Fat transplantation. In: RA Yoho J Brandy-Yoho, eds. A New Body in One Day: A Guide to Local Anesthetic Cosmetic Surgery Procedures. Studio City, CA: First House Press, 1999, pp 79–87
30. Fischer G. Autologous fat implantation for breast augmentation. Workshop Liposuction and Autologous Fat Implant, Isla d'Elba, September 1986
31. Fulton JE. Breast contouring by autologous fat transfer. Am J Cosm Surg 1992;9(3):273–279
32. Silveira, J.N.W. Autologous lipograft for augmentation mammoplasty. A new treatment of breast hypoplasia. Rev da Soc Beras. Plast Est Reconstr 1993; 8(1,2,3):12–19
33. Gradinger GP. Breast augmentation by autologous fat injection (Letter). Plast Reconstr Surg 1987;80(6):868–869
34. Grazer FM. Body contouring. In: JG McCarthy, ed. Plastic Surgery, Philadelphia: Saunders, 1990, p 4021
35. Grazer FM. Special reconstruction and therapeutic procedures. In: FM Grazer, ed. Atlas of Suction Assisted Lipoectomy in Body Contouring. New York: Churchill Livingstone, 1992, pp 319–329

36. Pitman GH. Odds and ends: Lipomas, flap debulking, lymphedema, fat reinjection, and additional complications. In: GH Pitman, ed. Liposuction & Aesthetic Surgery. St. Louis: Quality Medical Publishing, 1993:447-469
37. Moyer P. Breast augmentation—results are satisfying using autologous fat as filler. Cosm Surg Times 1998;2(4):4
38. Lynch DJ, Penn JG. Fat injection to breast compromises safety. ASPRS and ASAPS do not recommend procedure. Cosm Surg Times 1998;2(6):3
39. Gorney M. Fat injection to breast compromises safety. Commenting on calcifications. Cosm Surg Times 1998;2(6):3
40. Stone A. Fat injection to breast compromises safety. Questioning the procedure
41. Schaeffer, I. Fat injection to breast compromises safety. Editorial response Cosm Surg Times 1998;2(6):4

18

Stabilization of Fat Transfer to the Breast with Autologous Platelet Gel

James E. Fulton
Newport Point Medical Center,
Newport Beach, California

Autologous platelet gels have been used since 1982 to stabilize wounds[1,2] and accelerate wound healing.[3,4] The conversion of fibrinogen to fibrin produces a clot which facilitates hemostasis and closes draining sinuses. The activation of platelet growth factors stimulates macrophages and fibrocytes.[5,6]

Bilateral augmentation of the breast by injection with fat requires large volumes of adipose tissue. Mixing the fat with autologous platelet gel takes advantage of the stabilization and growth factors.

I. PATIENT SELECTION

An examination is completed to determine if the patient is a candidate for the procedure. Candidates in good health, have no personal or family history of breast disease, and have sufficient disharmonious obesity are chosen. Mammogram and volumetric breast measurements are completed on each patient before the procedure.

The best results are obtained in (1) mature women over the ages 30–35 who have had several children and have breasts that are atrophic but not ptotic; (2) those patients with disharmonious obesity who will be content with a one cup–size increase in volume; and (3) those patients who desire moderate breast augmentation after mastopexy or the removal of other implants.

Avoid patients with ptotic breasts. Adipose augmentations have been disappointing in these cases. Beyond 2–3 cm of ptosis, the circulation and lymphatic drainage in the breast may be compromised. This dependent breast does not appear to support neovascularization of the implanted tissue as well as the nonptotic breast. These cases may need mastopexy at least 6 months before the augmentation.

II. INFORMED CONSENT

Patients must be told not to diet immediately after the procedure and that they may either lose or gain breast volume after the procedure. They are told to inform their radiologist reading their future mammograms that they had previous breast surgery. This will help explain any benign lipid droplets or benign eggshell calcifications to their surgeon and reduce the need for exploratory surgery.

The patient signs an Informed Consent document during the initial preoperative examination and again on the day of surgery. The possibilities of loss of breast volume, breast calcification, or masking of future malignancy are discussed with each patient.

III. TECHNIQUE FOR PREPARATION OF PLATELET GEL[7]

One unit of blood is drawn from the patient into a standard labeled blood collection bag containing citrate anticoagulant. Blood is initially centrifuged at 5600 rpm for 15 min to separate platelet-enriched plasma (PRP) from the red blood cell pack (Fig. 1). The PRP is best left at room temperature until the fat is ready for supplementation.

IV. FAT COLLECTION

The night before and the morning of surgery, the patient completes a total body scrub with povidone-iodine (Betadine). Before surgery, the areas of the body are marked for liposuction. The current breast volumetric measurements and projected implantation volumes are marked on each breast. A combination of midazolan, propofol, and ketamine is used to achieve dissociative anesthesia[8] while tumescent fluid is infiltrated into the areas of liposuction.[9]

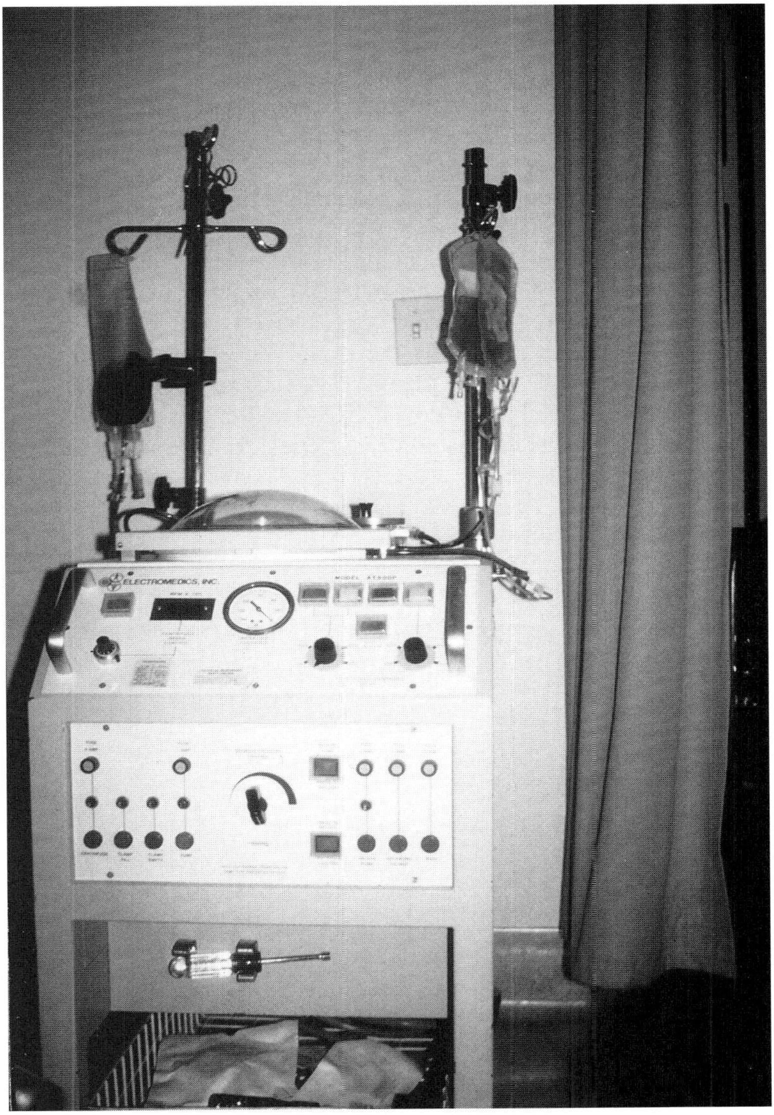

Figure 1 Preparation of the platelet gel: the unit of autologous blood is spun differentially to separate out the platelet gel.

Usually the thighs and flanks are infiltrated using a 30-cm long blunt 2-mm diameter stainless steel, multiple-port injection needle. After a 20-min pause to allow for vasoconstriction, the Tulip-syringe technique is used for fat harvesting.[10] Partial vacuum is gradually produced with a 60-mL Toomey syringe. The first 500–600 mL of adipose tissue is collected with a 30-cm blunt 3.7-mm liposuction cannula using a Mercedes tip (KMI, Anaheim, CA). The same procedure is repeated on the opposite thigh and flank until approximately 1200 mL of supernatant fat is obtained.

The aspirate is collected in a sterile glass container and washed several (usually two or three) times with lactated Ringer's solution to remove any broken fat cells or red blood cells. At this point, PRP (50–100 mL) is added. The mixture is agitated and allowed to "incubate" for 10 min. Then thrombin-calcium (thrombin 10,000 units in 10 mL of 10% calcium chloride) is added at the ratio of one part of thrombin to 10 parts of PRP (Fig. 2). The fibrin clot condenses and dehydrates the fat so only 50% or less of the supernatant fat remains for injection (Fig. 3). Free water and oil are decanted off through a sieve. The remaining fat (meshed in a fibrin clot) is transferred to injection guns.

Figure 2 Addition of thrombin-calcium: The thrombin-calcium converts the fibrinogen to fibrin within the platelet-gel-fat mixture. This results in clot formation.

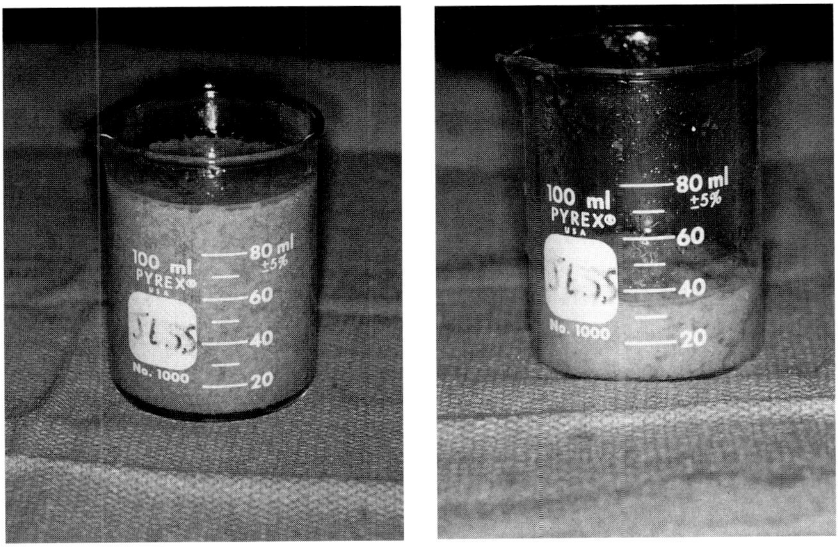

Figure 3 (**Left**) Fat prior to addition of thrombin-calcium. (**Right**) Dehydration of the adipose tissue: the fibrin mesh dehydrates the fat to 50% or less of its original volume.

V. FAT INJECTION

Technically it is important to collect fat at a reduced vacuum using a "vented" syringe technique and to inject small aliquots of stabilized fat throughout the breast area with special emphasis on augmentation of the upper quadrants. Better results are obtained from injecting 1-mL aliquots using a blunt needle than with uncontrolled fat injection using a syringe. The needle used has three ports, so 3 mm of fat and PRP is spread from each side of the needle as it is withdrawn. It is critical that each aliquot has a separate injection site among the collagen bundles of the breast area. Adipose tissue survival is maximized if the aliquots are spread under, into, and above the muscle. Augmentation is limited to 250–300 mL in each breast. After 90 days the implant volume is stable unless body weight is either added or lost. Then the implants enlarge or shrink correspondingly.

A total of 250–300 mL of this mixture is injected into each breast through a small stab incision below the inframammary fold. A ratcheted syringe delivers individual 1-mL aliquots through a triple-port blunt tip needle (KMI). Small 3-mm filaments are injected into the tissue through each port as the needle is with-

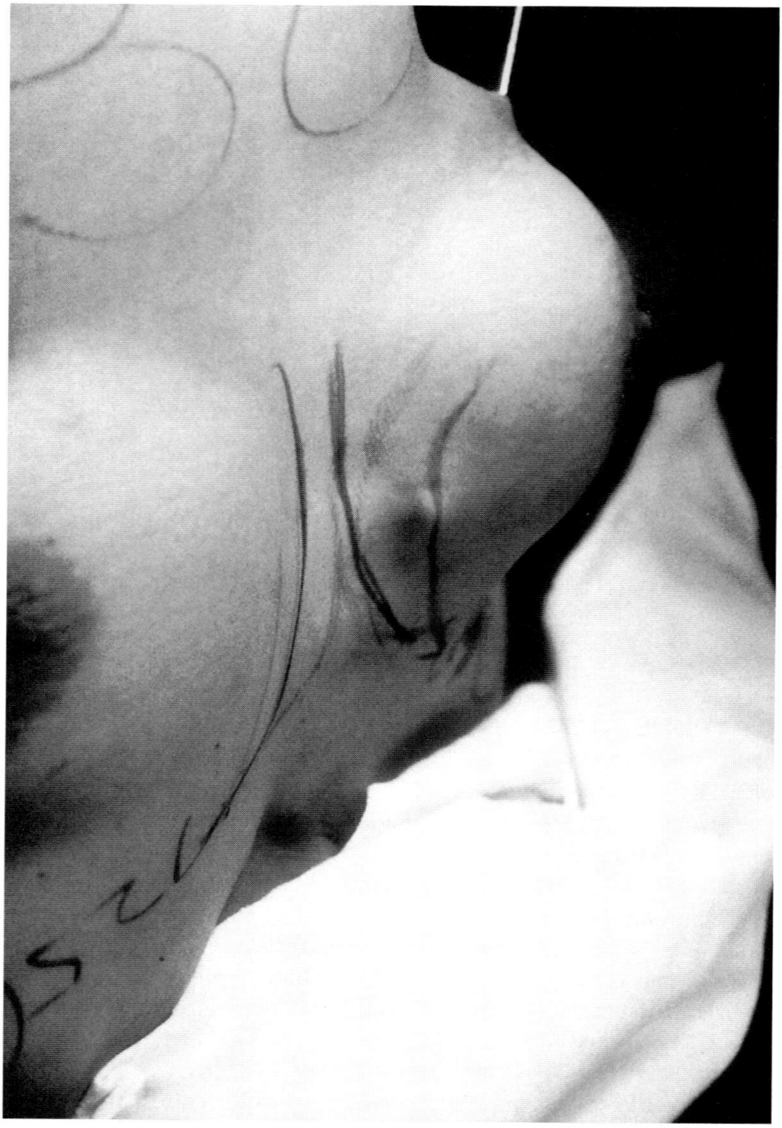

Figure 4A Infiltration of the adipose tissue: small aliquots of this "stabilized" fat is infiltrated throughout the breast area. Approximately 100 mL is placed under the muscle, 100 mL is placed into the area of the muscle, and 100 mL is placed above the muscle.

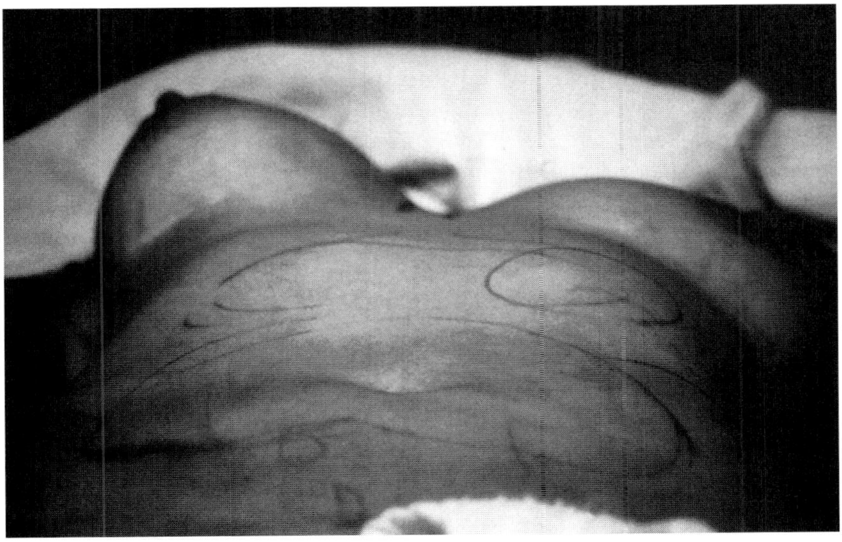

Figure 4B Following completion of injection right side.

drawn. Approximately 100 mL of adipose tissue is distributed into the subpectoral plane and another 100 mL is injected in a similar fashion into the plane of the pectoralis major muscle (Fig. 4). The final 100–150 mL is distributed in the plane superficial to the pectoralis muscle and deep to the breast. Between syringe injections, the breast tissue is gently massaged to disperse any fat globules. If one breast volumetric measurement is larger than the other, more adipose tissue is infiltrated into the smaller breast to adjust the size. The patients are given betamethasone (Celestone) to reduce inflammation and routine postoperative antibiotics, analgesics, and sleeping pills.

Following the procedure, the donor sites are covered with a compression garment.[11] The breasts are placed into a firm support bra, which is worn for approximately 1 month. The patients are seen at 1, 3, 6 months and annually thereafter for volumetric breast measurements. The patients also have mammograms annually after surgery.

VI. COMPLICATIONS

Complications are minor and infrequent. Fat collected without excessive trauma, washed free of debris with lactated Ringer's solution, stabilized with PRP, and transferred into the breast in small aliquots results in minimal problems. Patients rarely complain of breast discomfort. The use of the blunt cannula has been proven to be atraumatic. No hematomas or fat emboli have occurred. One patient developed an infection in one breast. The breast became indurated after 5 days and spontaneously drained approximately 100 mL of fluid through the injection site. The patient responded to ciprofloxacin and reinjection of adipose tissue 6 months later.

VII. DISCUSSION

The platelet-enriched fibrin clot dehydrates the adipose tissue, condenses the fat cells, and coats them in fibrin and platelet-derived growth factors. In the 23 cases studied, this produced a pleasing augmentation equivalent to a 200- to 300-mL silicone implant (Figs. 5–7).

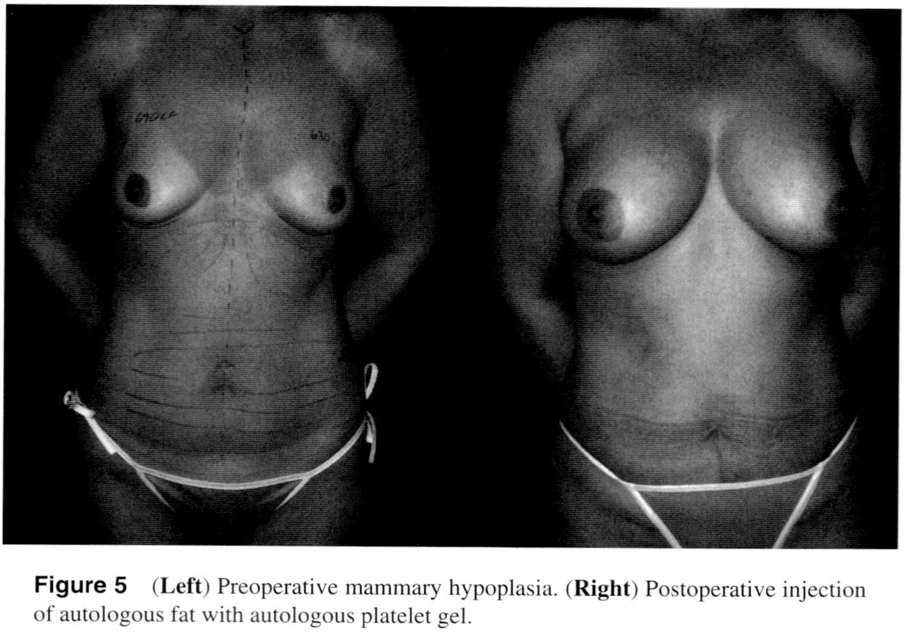

Figure 5 (**Left**) Preoperative mammary hypoplasia. (**Right**) Postoperative injection of autologous fat with autologous platelet gel.

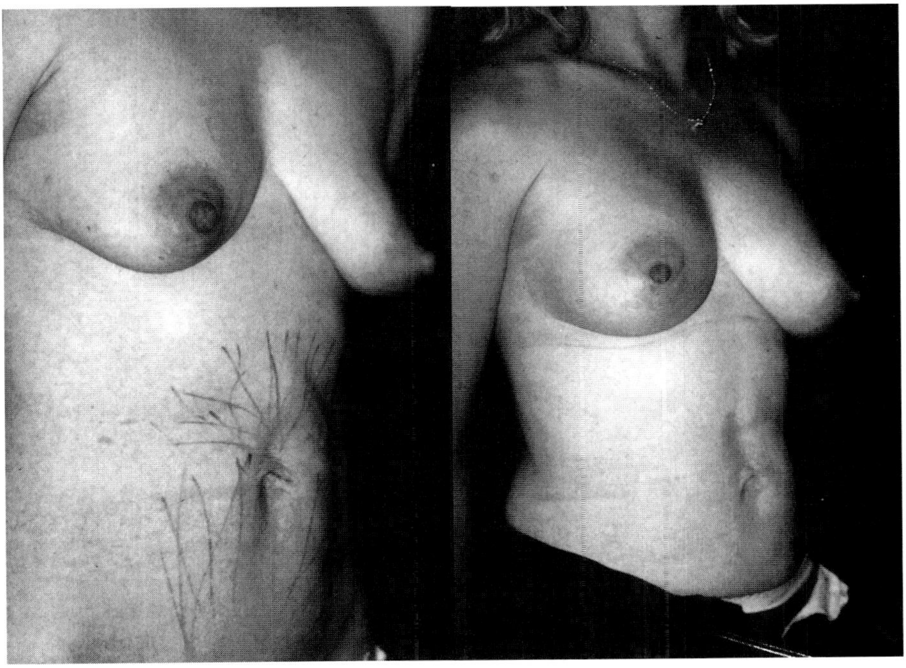

Figure 6 (**Left**) Preoperative small slightly ptotic breasts. (**Right**) Postoperative injection of fat with PRP.

No significant changes were seen on repeat mammograms except for the finding of fatty infiltration of the pectoralis muscle and prepectoral space. There was no masking or compression of the breast tissue. Calcifications did not develop in these cases.

The addition of platelet gel to autologous fat tissue transfer appears to stabilize the fat and increases retention rates. In a prior study, without the use of PRP, 65% retention of fat was achieved.[12] These results are improved to 77% with the addition of fibrin and platelet-derived growth factors. Since the platelet gel and transfer fluid will dissipate, higher retention rates are not expected.

Breast contouring with "gelled" autologous fat is a realistic addition to the breast augmentation armamentarium. Autologous tissue avoids the risk of capsule formation or systemic reactions from leaking silicone implants. The original concerns involving fat augmentation[13] have not developed. No distortion of

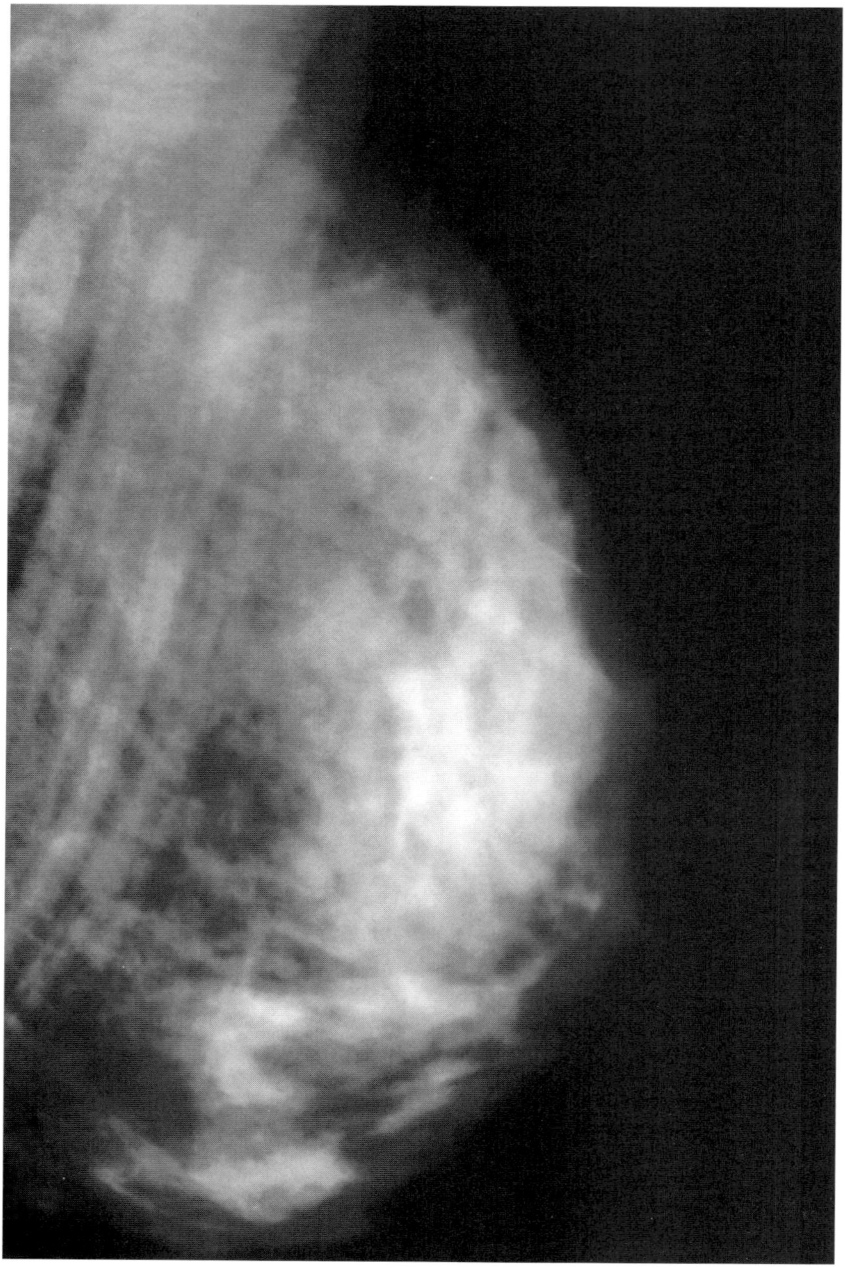

Figure 7 Mammographic findings: fatty infiltration of the pectoralis muscle. No calcifications present.

breast tissue or masking of breast cancer has been seen. Factors that minimize these distortions are blood-free harvesting methods using tumescent liposuction with a syringe, washing away any debris from the adipose tissue, using PRP to stabilize the implant, and spreading the implant into small aliquots in three planes throughout the breast area. This results in a "low-profile" augmentation that is natural in appearance.

VIII. GUIDELINES

1. Collect the fat predominantly from the outer thigh and flank where there is low vascularity and condensed fat. This fat is stable and not usually reduced with minor dieting.
2. Collect the fat atraumatically at a low pressure using a "vented" syringe technique.
3. Wash the fat as necessary to obtain a clear infranate. Add platelet-enriched plasma and clot the tissue after 10 min with a mixture of thrombin-calcium.
4. Inject controlled small aliquots of this stabilized fat with an injection gun throughout the breast area. Do not inject through a syringe, as a bolus of fat may be implanted inadvertently. Place approximately 100 mL under the muscle, 100 mL into the area of the muscle, and 100 mL above the muscle. Spread it out evenly as small 3-mm filaments through a triple-port injection cannula. Inject more into the upper quadrants of the breast, as the augmentation will settle down.
5. Wait 6 months before attempting repeat fat augmentation. This allows for neovasculization. Often the second implantation is more successful than the first.
6. Avoid the ptotic breast. These patients are not good candidates for fat augmentation. The survival of the implant and the lymphatic drainage system appear to be compromised.
7. Suggest that the patient maintain her weight after the procedure and not start a strenuous diet program immediately after surgery.

REFERENCES

1. Saltz R, Dimick A, Harris C, Grotting JC, Psillakis J, Vasconez LO. Application of autologous fibrin glue in burn wounds. J Burn Care Rehabil 1989;10:504–509

2. Oz MC, Jeevanandam V, Smith CR, Williams MR, Kaynar AM, Frank RA, Mosca R, et al. Autologous fibrin glue from intraoperatively collected platelet plasma. Ann Thorac Surg 1992;53:530–531
3. Knighton DR, Hunt TK, Thrakal KK, Goodson WH. Role of platelet and fibrin in the healing sequence. Ann Surg 1982;196:379–388
4. Nicholas JM, Dulchavsky SA. Successful use of autologous fibrin gel in traumatic bronchopleural fistula: case report. J Trauma 1992;32:87–88
5. Knighton DR, Fiegel VD, Austin LL, Cireski KF, Butler EL. Classification and treatment of chronic nonhealing wounds: Successful treatment with autologous platelet-derived wounds healing factors (PDWHF). Ann Surg 1986;204:322–330
6. Folkman J. Is there a field of wound pharmacology? Ann Surg 1992;244:1–2
7. Whitman DH, Berry RL, Green DM. Platelet gel: an autologous alternative to fibrin glue. J Oral Maxillofac Surg 1997;55:1294–1299
8. Friedberg BL. Propofol-ketamine technique. Aesth Plast Surg 1993;17:297–300
9. Klein J. Tumescent technique for regional anesthesia permits lidocaine doses of 35 mg/kg for liposuction. J Dermatol Surg Oncol 1990;16:248–262
10. Alexander R. Liposculpture with superficial plane: closed syringe system for improvements in fat removal and free fat transfer. Am J Cosm Surg 1994;11:127–131
11. Fulton JE Jr. Modified tumescent liposuction. Dermatol Surg 1999;25:755-766.
12. Fulton JE Jr. Breast contouring by autologous fat transfer. Am J Cosm Surg 1992; 9:273–279
13. Report of autologous fat transplantation. ASPRS Ad Hoc Committee on New Procedures. Am Soc Plast Reconstr Surg Sept 1987

19

Fat Transplants in Male and Female Genitals

Enrique Hernandez-Perez and Filadelfo Venadero-Albarran
Center for Dermatology and Cosmetic Surgery, San Salvador, El Salvador

I. HISTORY

Fat transplants have been used in reconstructive and cosmetic surgery for more than a century.[1–3] For many years, however, this procedure was practically forgotten. With the advent of liposuction as one of the most common procedures in cosmetic surgery, fat transplantation began receiving renewed interest approximately 15 years ago,[2] and it has been used with good results for the correction of facial wrinkles or for improving facial or body defects, either congenial (hemifacial atrophy) or acquired (usually traumatic).

There have been important developments over the years with fat transplants to various areas of the body. Satisfactory results have been obtained by injecting fat into female breasts, buttocks, and calves.[4,5] The use of autologous fat for penile enhancement was first reported in 1993.[4] Fat transplant to the female genitals was described in 1996.[6]

II. PREOPERATIVE PREPARATION

The operation is performed only in healthy adults who are highly motivated and to whom very realistic expectations have been carefully explained, assuming that each patient has genitals of normal size. Patients are informed that this is not an operation to treat impotence, frigidity, or anatomical deformities of any type. In the case of patients with redundant foreskin or true phimosis, a circumcision must be considered first.

The patient is prepared prior to surgery by obtaining a careful psychiatric history and giving him or her a thorough explanation of the procedure and possible problems. It is very important to avoid patients with significant psychiatric disorders.

Most patients are concerned with the diameter of the penis and degree of sexual satisfaction they can give to their partner. Although on examination the penis appears to be of adequate size to the physician, there is evidence of various degrees of individual insecurity. Proportion is very important in esthetic surgery. A patient may feel that his penis is smaller when a large deposit of fat exists in the pubis. It is necessary to inform him that aspiration of fat from this area must also be performed.

The evaluation of the size of the male genitals is certainly more difficult and possibly more subjective in females, but it is evident that fat atrophy (noticeable in the labia majora and pubis) occurs with aging and with loss of weight and in some instances may be genetic in origin. In these cases, the form of the vulva seems to have a more important role in the self-esteem of the female than the size of the organ.[6]

The ideal patient of this surgical procedure is an individual who, knowing that the size of his or her genitals is considered to be within the normal range, believes that the relationship with one's partner could improve if the size of the organ could be increased.

The patient is counseled about the fact that sexual satisfaction depends upon love and mutual understanding, as well as the use of caressing, and not specifically on the size of the penis or vulva. However, if the increase in diameter of the organ can improve the patient's security, the operation may be undertaken.[4,6]

Regardless of age, all patients are sent to a cardiologist for a complete cardiovascular check-up before surgery. For medical and medicolegal reasons, it is advisable to have written permission from the cardiologist to perform this type of surgery when using epinephrine. Other presurgical testing includes fasting blood sugar, human immunodeficiency virus (HIV), and coagulation. We prefer patients falling in categories I or II according to the criteria of The American Society of Anesthesiology.

Patients are instructed to wash the operative area with povidone-iodine or clorhexidine for 3 days prior to surgery. One day before surgery, they are given an oral wide-spectrum antibiotic. Patients are instructed to avoid aspirin and some antidepressants such as sertraline and beta blockers before the operation[7] and to stop smoking.[8]

III. SURGICAL TECHNIQUE

The patient is advised to take nothing by mouth starting at midnight prior to surgery. In the operating room, the areas to be treated are marked.

The operation is performed as an ambulatory procedure using local anesthesia and intravenous (IV) sedation (midazolam plus fentanyl). The authors prefer using sedation together with local anesthesia. The procedures are performed in a fully equipped operating room under the care of a certified anesthesiologist. Most cases of fat transfer are performed as a part of liposuction procedure or in combination with other cosmetic procedures.

Thirty minutes before starting the operation, a wide-spectrum prophylactic antibiotic (ceftriaxone, 1 g) is given intramuscularly (IM). The dosage is repeated if the procedure takes longer than 3 h. IV dexamethasone (8 mg) is given in an attempt to prevent the cascade of arachidonic acid, and metoclopramide (10 mg) administered IV, to avoid postoperative nausea.

Preoperative and postoperative photographs are taken of all patients.

A modification of Klein's formula[9] is used in the donor area. The solution consists only of lidocaine (0.05%) plus epinephrine (1:1 million) in saline solution. Sodium bicarbonate is not used, because the patients are under IV sedation and pain from the injection is not of concern. Sodium bicarbonate seems to increase the inflammation and lessen the duration of the anesthesia.[10] The local anesthetic used is at refrigerator temperature (cryoanesthesia).[10]

In certain body areas recognized for their excessive sensitivity, either larger volumes of the anesthetic solution are used or the amounts of the IV analgesics and sedatives are increased. The modified Klein's solution is infiltrated into the donor area up to tumescence.

For harvesting purposes, nonfibrous sites such as the lower abdomen, trochanteric areas, pubis, submental area, medial thighs, and inner knees are preferred.

With the patient monitored, the anesthesiologist administers IV sedation (midazolam plus fentanyl). The total dosage of midazolam varies from 15 to 20 mg (0.3 mg/kg). The total dosage of fentanyl varies from 300 to 500 μg (5 μg/kg–10 μg/kg). In the case of fat transfer alone, 50 μg of fentanyl, and 5 mg of midazolam are used as a starting dose. This is usually the only sedation needed. Both drugs are short lasting, and if administered over a prolonged period of time, there is no risk to the patient, especially with monitoring.

After antisepsis with povidone-iodine, tumescent anesthesia is used in the donor area, which is generally the lower abdomen. After waiting 15 min for the vasoconstrictive effect of the epinephrine, fat is harvested using 60-mL syringes with a snapper and small cannulae (Toomey or Tulip system) (Fig. 1). The fat

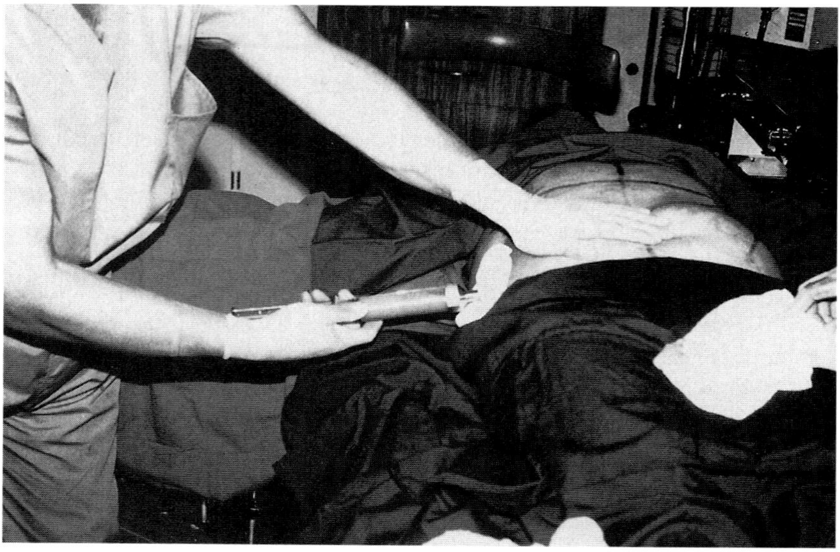

Figure 1 Fat harvested using a 60-mL syringe and small cannulae.

is manipulated very gently by decanting in the syringes after separation into the three classic levels. Triglycerides from the broken adipose tissue are in the upper part, pure fat is in the middle, and a small amount of blood mixed with modified Klein's solution is in the lower level[11,12] (Figs. 2 and 3).

In patients with an excessive deposit of fat in the pubis, it is aspirated at this time.

The authors prefer to manipulate the fat in a closed circuit to minimize the risk of contamination or to damage of the adipocytes. Insulin added to the fat has no benefit. When the fat has been harvested, the syringe is set with the needle downward, depressing the plunger to remove modified Klein's solution. The fat is not washed with saline, as is done by other surgeons.[13] This allows only the pure fat remaining to be used for the transplant. Standard operating room sterile technique must be used when harvesting and injecting fat to avoid infection (Fig. 4).

Frozen fat may be used for later injection; however, this is not used because of the possibility of infection and risk of overmanipulation with adipocyte damage.[8,14]

Anesthesia in the genital area is 1% lidocaine without epinephrine, using a disposable plastic 1-mL syringe with a 30-gauge needle. A minimal amount of this solution (from 0.25 to 0.50 mL) is injected into each of specific points.

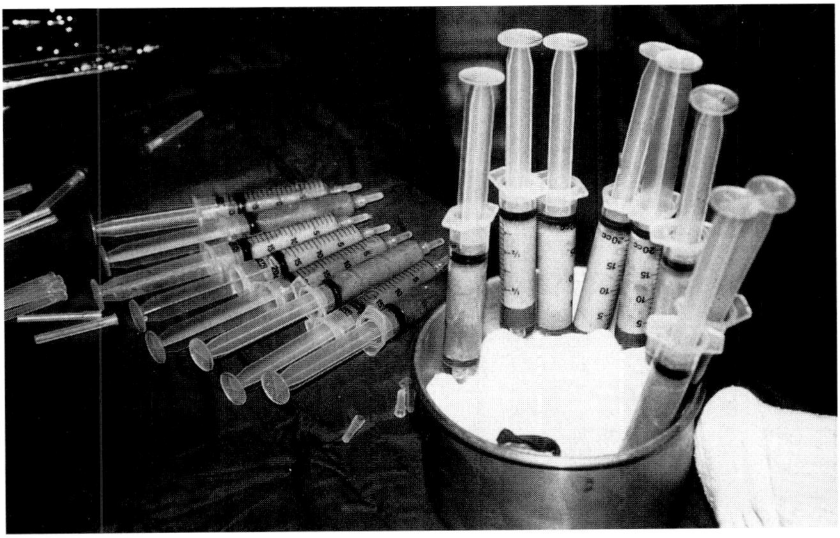

Figure 2 The preparation is made only by decanting.

Male: at 6 and 12 o'clock in the inner face of the foreskin. Female: at the upper pole of both labia majora and at the right side of the mons pubis. No more anesthetic is necessary.

IV. PENILE ENHANCEMENT

Two small incisions are performed with a #11 scalpel in the previously mentioned points and the fat is injected using thin cannulae (no larger than 14 gauge) of the same type used for breast or calf transplants. At this moment, the assistant must stretch the penis, taking the glans with the left hand while making a type of tourniquet at the base of the same with the right hand to prevent dispersion of the fat (Fig. 5).

The injection is made in a radial and retrograde form using a total of 60 mL of fat, which is adequate. The two small incisions are closed with 6-0 fast-absorbing plain gut and the penis is kept elevated to reduce inflammation and excessive swelling. A bandage is applied similar to that used after a circumcision. This position is maintained for 24 h after the operation. The patient returns on the fourth and eighth days. There is a surprising absence of pain and only slight swelling. The bruises are either slight or nonexistent (Fig. 6).

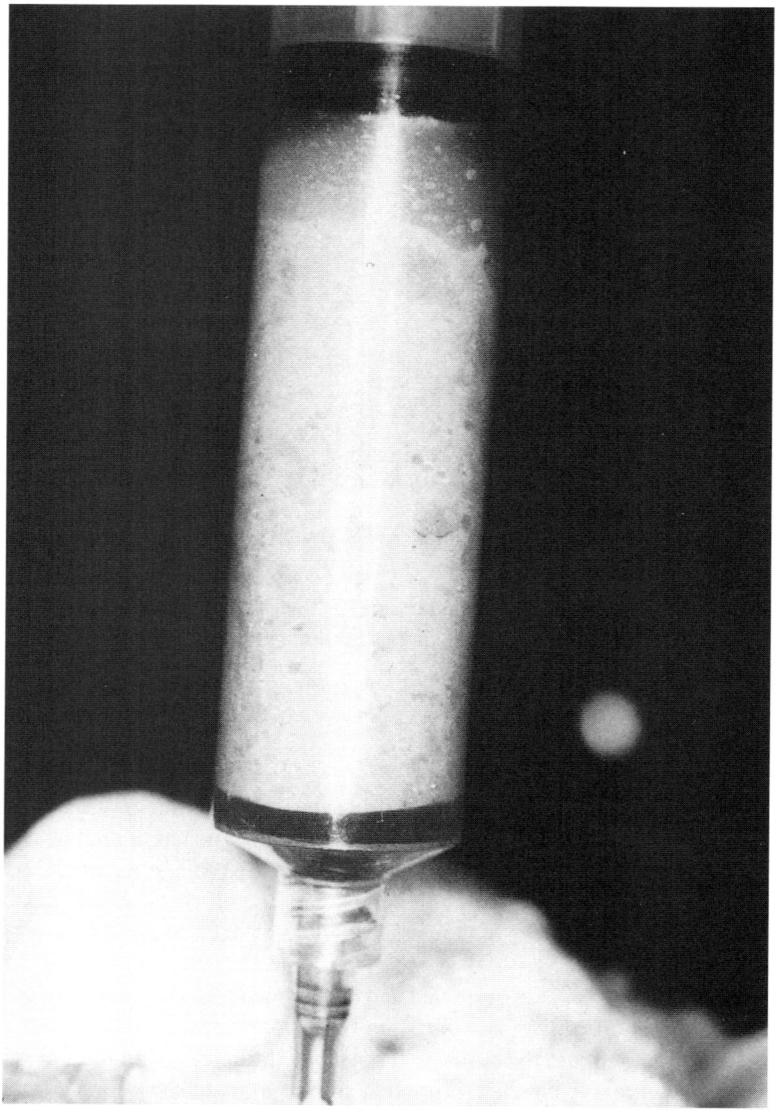

Figure 3 There are three classic levels. In the middle there is almost pure fat.

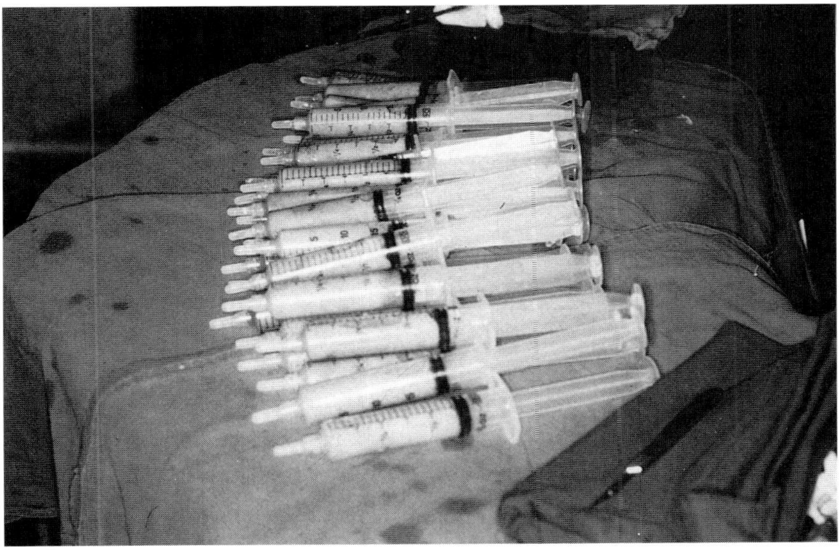

Figure 4 Fat ready for injection.

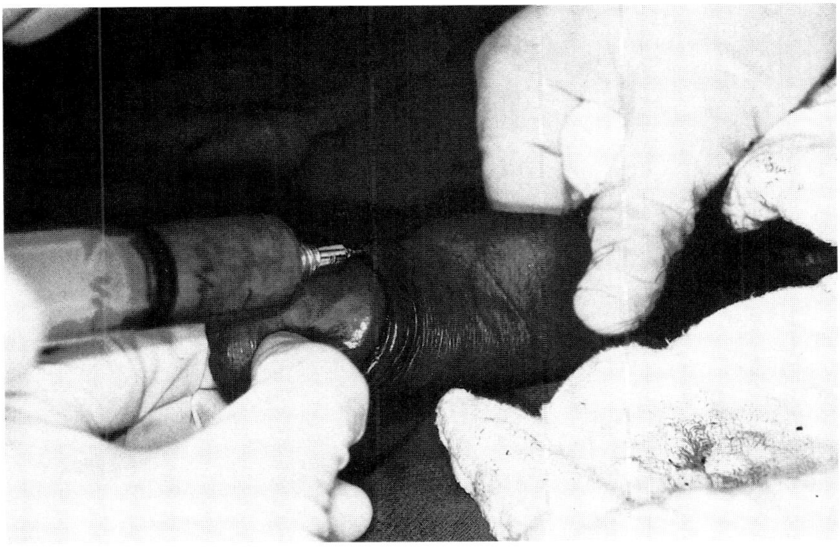

Figure 5 When the fat is injected into the penis, the assistant must stretch the organ between the hands.

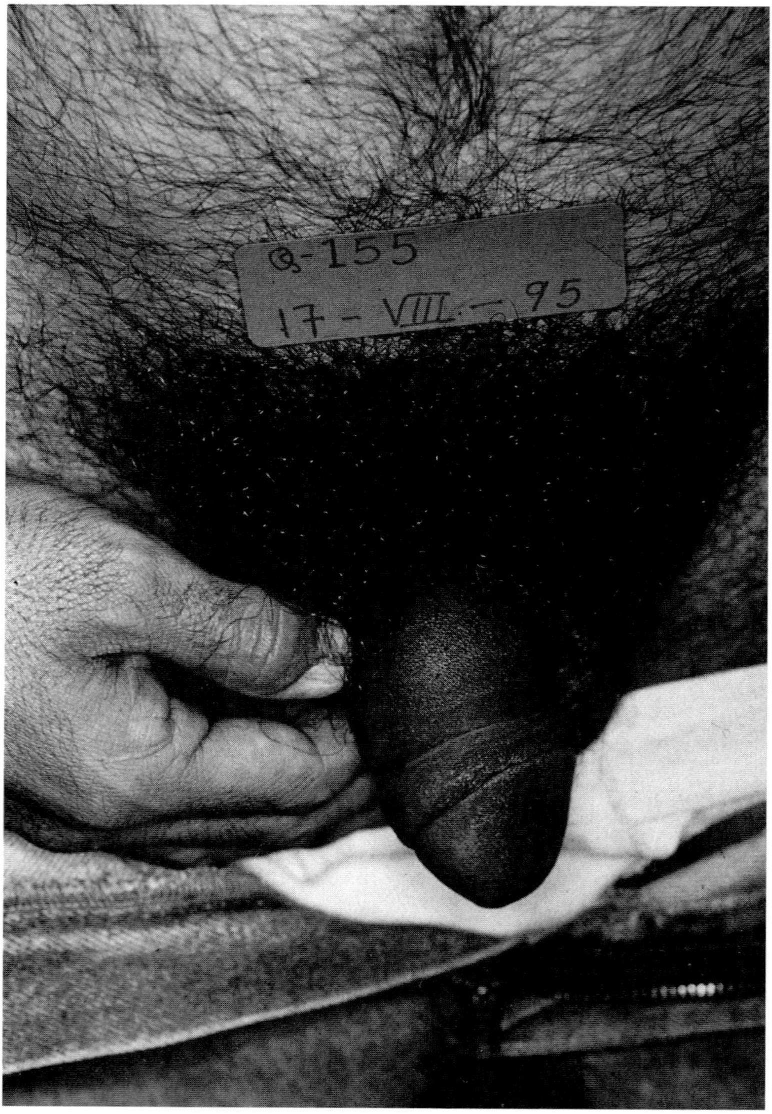

Figure 6 After injection of conservative amounts of fat, the penis appears enhanced.

V. VULVAR ENHANCEMENT

Using a #11 scalpel, three small incisions are performed in the previously anesthetized sites. Thin cannulae are used to inject the fat. First, the cannula is inserted in a caudal direction up to the lower third of the labium majora and the transplant is performed as the cannula is withdrawn (Fig. 7). Since many adipocytes die during surgery, some degree of overcorrection is necessary to compensate in the case of women.[8]

Once both labia have been injected, fat is transplanted into the pubis (Fig. 8). An average of 100–120 mL is used. Incisions are closed using 6-0 fast-absorbing plain gut (Fig. 9). Pressure bandages are used only on the abdominal donor area, and they are removed after 4 days.

One of the most important points in the injection technique is to place the fat in small segments and in different layers into the subcutaneous tissue to increase adipocyte survival.[1,15,16] The adipocyte survives through the blood vessels furnished by the adjacent tissues. If placed in large segments, such blood vessel nutrition may not occur.

Fat is not absorbed when injected in small amounts in the penis. It is possible that its anatomical conformation (rich vascularization with an almost total

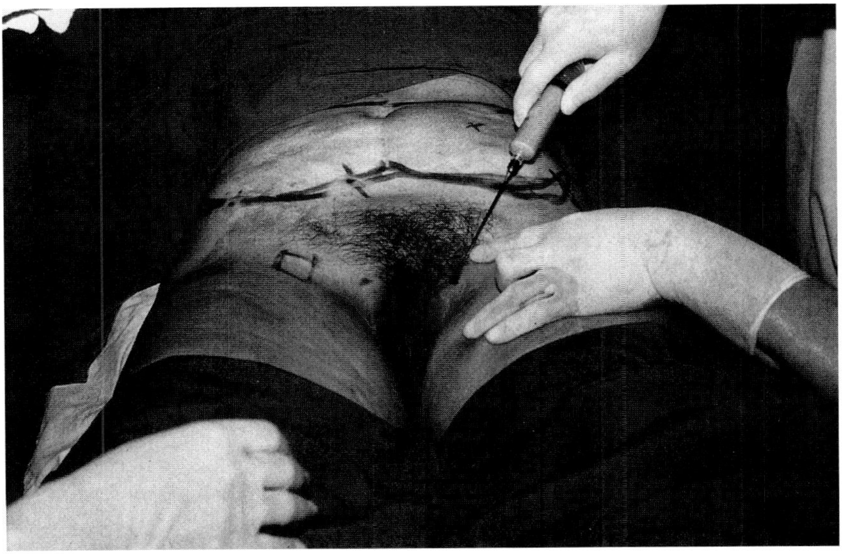

Figure 7 Fat being injected into the labium majora.

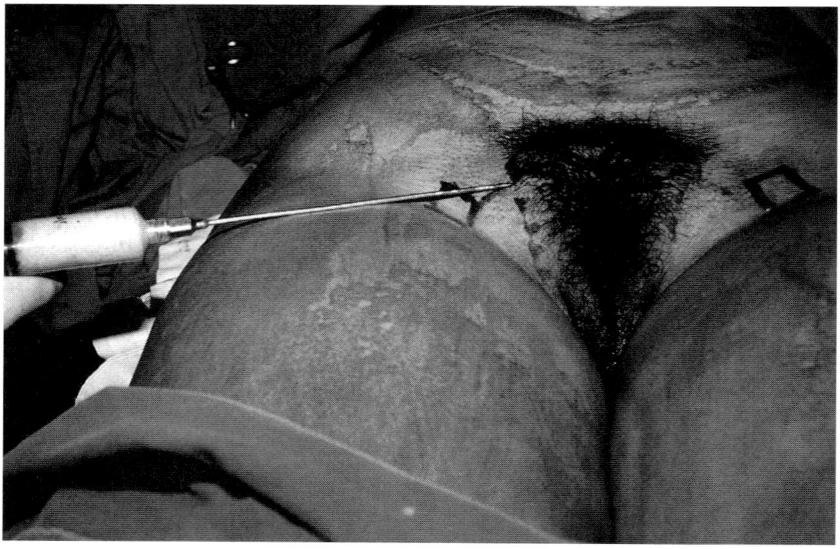

Figure 8 Injection in the mons pubis.

absence of subcutaneous fat and the presence of the tunica albuginea) accounts for this particular behavior of the injected fat. In women, a decrease of 20–30% of the transplanted fat has been observed.

VI. POSTOPERATIVE CARE

The patient is ambulatory with occasional minimal pain that can be relieved by using acetaminophen. To decrease the inflammation, tenoxican (a nonsteroidal anti-inflammatory drug, NSAID) 20 mg is given intramuscularly (IM) the day after surgery and at the same time prednisone is started using 20 mg orally a day for 5 days. Cold compresses are used four or more times a day in the first 48 h. After that, warm compresses are used two to three times daily as necessary (for swelling and discomfort). Patients are able to resume their normal activities (job, studies) the day after surgery.

Sexual activities are permitted after 4 days postoperatively, since there is minimal inflammation and lack of pain.

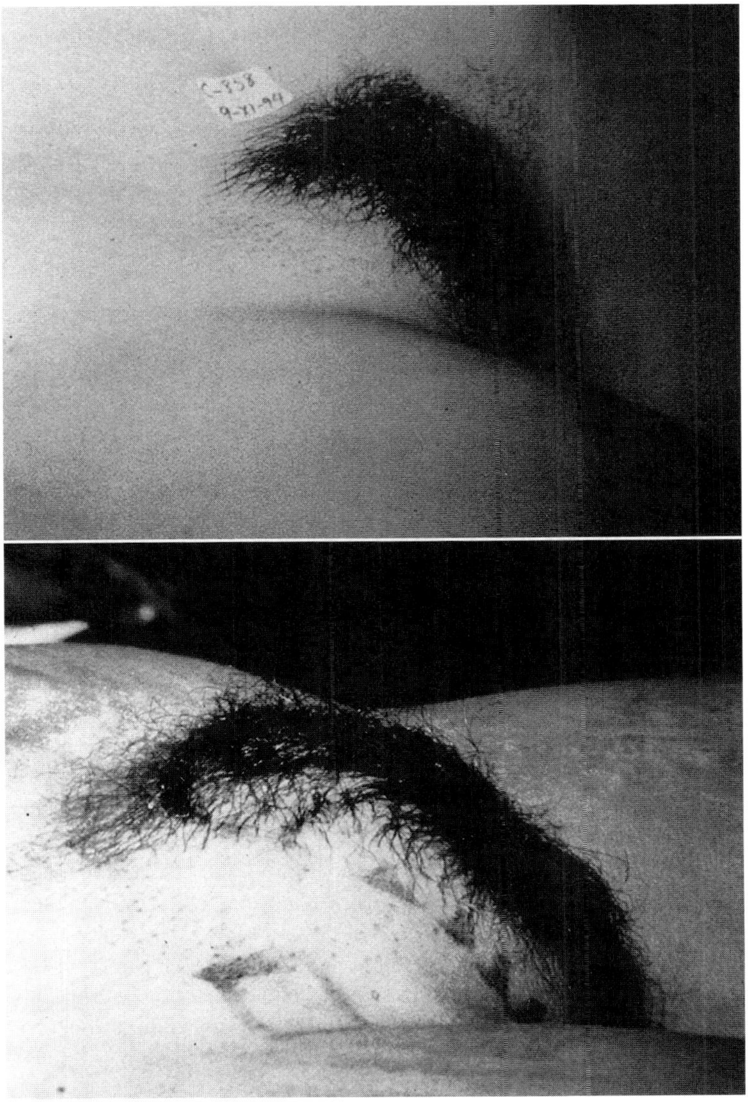

Figure 9 Preoperative and postoperative fat transplant in female genitals.

VII. COMPLICATIONS

Absorption, infection, and embolism are the most common complications mentioned in general in the cases of fat injection. Owing to the manipulation of the fat, some adipocytes do not survive the trauma of the transfer. Because of this and the lack of adequate blood support, some degree of fat absorption takes place. The percentage of absorption can be decreased by manipulating fat very gently and avoiding unnecessary manipulation and desiccation. Fat must always be injected in small segments in different layers.[15,16]

When using tumescent anesthesia in the donor area, bleeding is not usually seen. Blood losses are less than 1% of the total aspirated.[10]

Cyst formation and calcification can occur, but they are generally small and self-limited.[8] If a cyst does not disappear quickly, it may be treated with intralesional triamcinolone injections. To decrease the possibility of cyst formation, never inject fat in the penis in large amounts. Transitory deformities (penile distortion) and cysts[16] are better prevented by not exceeding 60 mL of fat per session in male genitals.[6] In any case, it should be possible and preferable to perform another injection some months later.

To avoid embolism, never inject fat in a centripetal direction but inject it when the cannula is being withdrawn.

Infection, although not seen by the authors, can be a potential major problem. Strict sterile technique is essential.

VIII. DISCUSSION

Fat grafting in male and female genitals with the goal of enhancing these organs is a simple procedure. It must be done very carefully to avoid complications and to have satisfied patients. In men, the injection should not exceed 60 mL of fat per session. In women, the recommended amount is 100–120 mL.[6] Injection of larger amounts of fat results in more frequent penile distortion, cysts, or even necrosis.[16]

For getting better proportion, in some cases it may be necessary to combine fat aspiration from the pubis as well as fat injection into the penis.

Some colleagues have advocated the use of papaverine injection to provoke erection of the penis and facilitate its manipulation (1st International Symposium on Facial Rejuvenation, Feb. 1993, Lima, Peru). That procedure is not only unnecessary but risky. Cases of priapism and necrosis when using such a drug have been described.

Combining fat transfer with section of the suspensory ligament of the penis could increase the length of the organ.[17] However, techniques which are more complex and invasive may cause more significant complications. Those modifications have proved to be related to a larger incidence of bleeding, hematoma, or neural disturbances. In general, when a procedure is simple, safe, and provides good results, it is more readily acceptable to patients.

By using injectable bovine collagen with glutaraldehyde, it is possible to avoid IV sedation or local anesthesia.[16] However, the increase in diameter will occur only in the foreskin proximal to the glans, and the results will be temporary. Furthermore, there is always the risk of immunological reactions, slow resolution of granulomas, sterile abscesses, necrosis, and other allergic reactions. When this occurs in the facial skin, it constitutes a difficult problem, but if occurring in the penis, the situation could be even more disastrous.[18]

The following are principles of genital fat transplant enhancement:

1. Fat is not reabsorbed when injected into the penis. Persistence of fat in the penis possibly is related to its peculiar anatomical conformation (rich vascularization, almost total absence of subcutaneous fat, and the presence of the tunica albuginea). In women, the percentage of absorption is about 20–30%.
2. Postoperative pain is so slight that many patients have been able to have sexual intercourse 3 or 4 days after the procedure.
3. The penis injected with fat increases not only in diameter but, according to the patient, also in length. This latter increase may be due to the weight of the fat and the subsequent stretching of the tissue similar to what occurs with tissue expanders.
4. When administered properly, autologous fat transplantation in the genitals is an acceptable procedure resulting in highly satisfied patients. This is mostly due to the fact that they are using their own tissue instead of a prosthesis. Rejection or allergy is not possible if the material used was harvested from the same person.[8]
5. Finally, patients are told that this type of surgery is not only an injection of fat, it represents in reality an injection of self-esteem. It is obvious that sexual satisfaction for the patients and their mates does not depend exclusively on the size of the genital organ. To paraphrase a distinguished Latin American writer: "Love is the music, and sex is the instrument."[19]

REFERENCES

1. Hernandez-Perez E. Implantes de Gasa Autologa. Lipo Implante. In: F Camacho, F Dulanto, eds. Cirugia Dermatologica, Madrid: Biblioteca, F. Aula Medica, 1995, pp. 563–569

2. Bircoll M. A nine-year experience with autologous fat transplantation. Am J Cosm Surg 1992;9:55–59
3. Newman J, Ftaiha Z. The biografical history of fat transplant surgery. Am J Cosm Surg 1987;4:85–87
4. Hernandez-Perez E, Marroquin-Burgos R. Fat injection of the penis: a personal technique. Am J Cosm Surg 1993;10:135–138
5. Fournier PF. Fat transfer. In LC Parish, GP Lask, eds. Aesthetic Dermatology. New York: McGraw-Hill, 1991, pp 267–275
6. Hernandez-Perez E, Machado A. Fat transplants in male and female genitals. Am J Cosm Surg 1996;13:109–111
7. Klein JA, Kassardjian N. Lidocaine toxicity with tumescent liposuction. Dermatol Surg 1997;23:1169–1174
8. Hernandez-Perez E. Practice perspectives: fat injection in different parts of the body. Dermatol Nurs 1998;10:135–138
9. Klein JA. The tumescent technique for liposuction surgery. Am J Cosm Surg 1987;4:263–267
10. Hernandez-Perez E, Henriquez A, Gutierrez J. Clarifying concepts in modern liposuction. Int J Aesth Restor Surg 1994;2:65–67
11. Jones J, Lyles M. The viability of human adipocytes after closed-syringe liposuction harvest. Am J Cosm Surg 1997;14:275–280
12. Johnson G. Autologous fat graft by injection: Ten years experience. Am J Cosm Surg 1992;9:61–65
13. Fulton J, Suarez M, Kimball S, Barnes T. Small volume fat transfer. Dermatol Surg 1998;24:857–865
14. Jackson R. Frozen fat—does it work? Am J Cosm Surg 1997;14:339–343
15. Hernandez-Perez E. Bi-level lipoinjection for facial wrinkles. Am J Cosm Surg 1992;4:73–75
16. Brandow K, Newman J. Facial multilayered micro lipoaugmentation. Int J Aesth Restor Surg 1996;4:95–110
17. Reed HM. Augmentation phaloplasty with girth enhancement employing autologous fat transplantation: A preliminary report. Am J Cosm Surg 1994:11;85–90
18. Sito G, Sorrentino L. The mushroom technique for penile enhancement. Am J Cosm Surg 1998;15:165–166
19. Allende I. El plan inifinito. Diana Literaria, Mexico, 1992

20

Autologous Fat for Liposuction Defects During and After Surgery

Pierre Fournier
Private Practice, Aesthetic Plastic Surgery, Paris, France

Liposuction is a blind and tactile operation and even in the best hands defects may occur. Excellent training and experience are needed to avoid such defects. The liposuction surgeon not only has to think as a surgeon but also as an architect and as an artist.

A very rigorous technique may decrease or avoid such defects. The use of tumescent anesthesia, several approaches (to allow criss-crossing tunnels), superficial work, using a syringe to measure the exact amount of fat removed through the different approaches, and good preestimation of the amount of fat to be removed are required. All this is acquired with time, experience, and observing the work of many colleagues.

I. DEFECTS HAPPENING DURING THE LIPOSUCTION PROCEDURE

The pinch test that should be routinely performed during the operation is the best sign warning the liposuction surgeon that removal of fat has been uneven. Liposculpturing with small cannulae (2 or 3 mm in diameter) in a thick area has to be performed to obtain the same thickness as in a thinner neighboring area using several approaches. The operation proceeds with the pinch test being used frequently to avoid a new uneven area.

Should the liposuction surgeon be unhappy with the correction, he or she should reinject a variable amount of the patient's own fat, 60–100 mL or more, which will give a more homogenous result. The reinjected fat should be harvested at the beginning of the operation using the syringe technique, and this precaution has been recommended by many authors. Even if no defect has to be feared, reinjection of a certain amount of fat after a liposuction procedure has always been recommended by Giorgio Fischer. This is a routine procedure for many.

II. LIPOSUCTION DEFECTS AFTER SURGERY

Postliposuction defects may be seen shortly after the operation if they are significant or 4 or 6 months after if they are not. The areas involved most of the time are the outer and inner thighs or the abdomen, but defects can occur in any treated area on the body (Figs. 1 and 2). They are visible and palpable, and they are seen more often when only one approach has been used. The skin may be very adherent to the aponeurosis in some places or a thin layer of fat may be present. Skin depressions and skin elevations are alternating, sometimes irregularly displayed, or linear. The only possibility in such cases is to remove fat tissue in the elevations where too much has been left if there is an excess which does not correspond to the normal layer and to reinject the harvested fat in the depressed places.

The surgeon must follow a very careful and thorough technique (Figs. 3–6):

1. Marking the elevations and depressions very accurately.
2. Use either local anesthesia with or without associated general anesthesia.
3. A mesh undermining of all the defects should be done with small cannulae using several approaches before harvesting the excess fat in the elevated places.
4. Liposculpting in the elevated places should be performed very carefully using two or three approaches using the syringe technique, and the extracted fat should be reinjected in the depressed places, which were carefully mesh undermined previously, using two or three approaches.

In certain cases, the whole uneven areas can be treated together. In others cases, each elevation and each depression is treated separately. When there is not enough fat to remove for grafting in the uneven area, the fat is taken from another place on the patient's body very carefully.

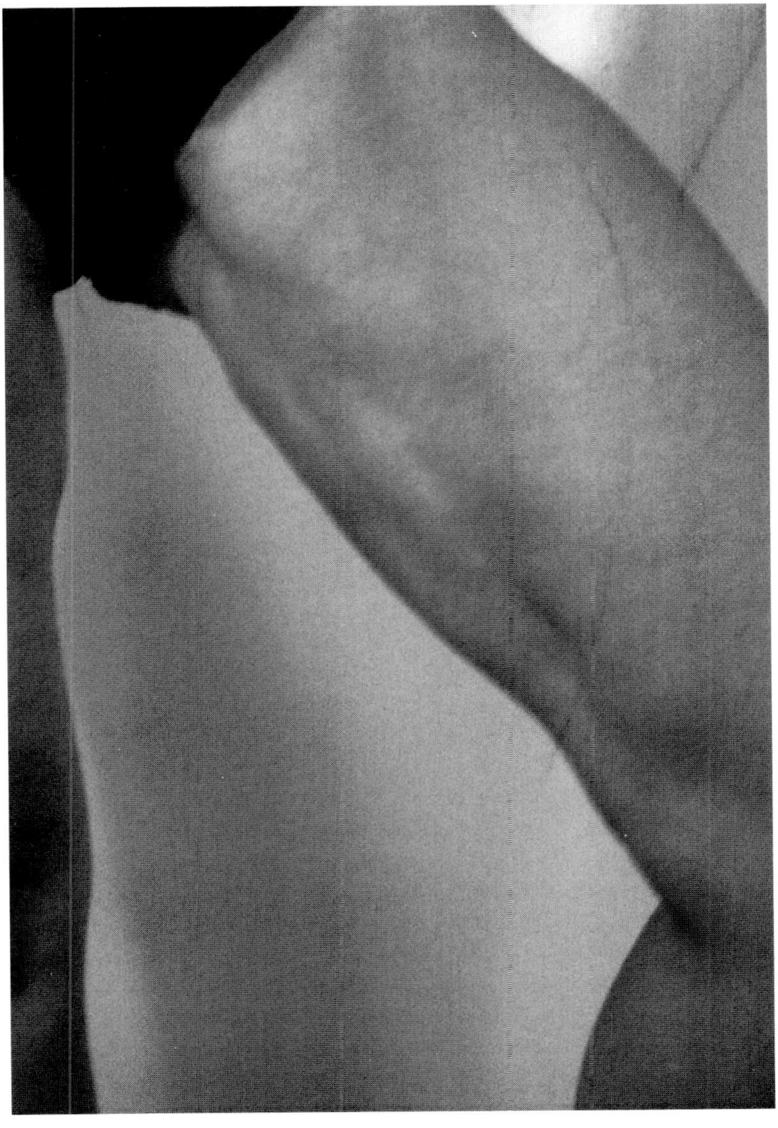

Figure 1 Defect of the inner thigh that may need correction.

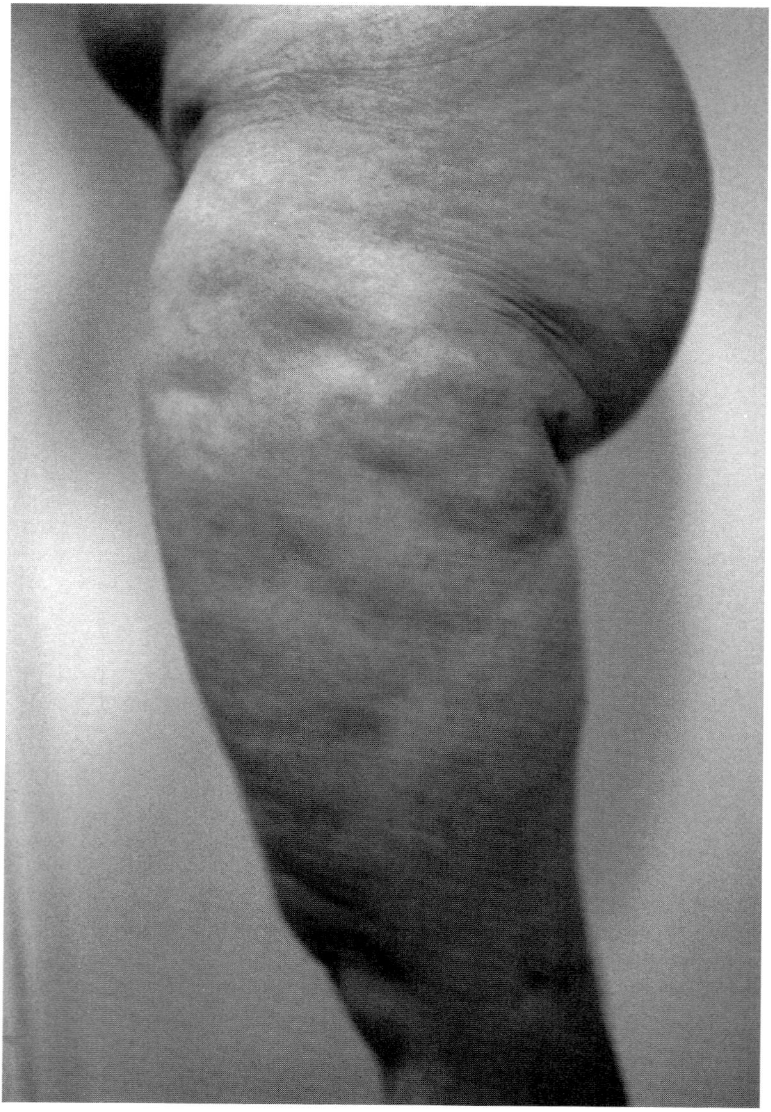

Figure 2 Defect of the outer thigh that may need correction.

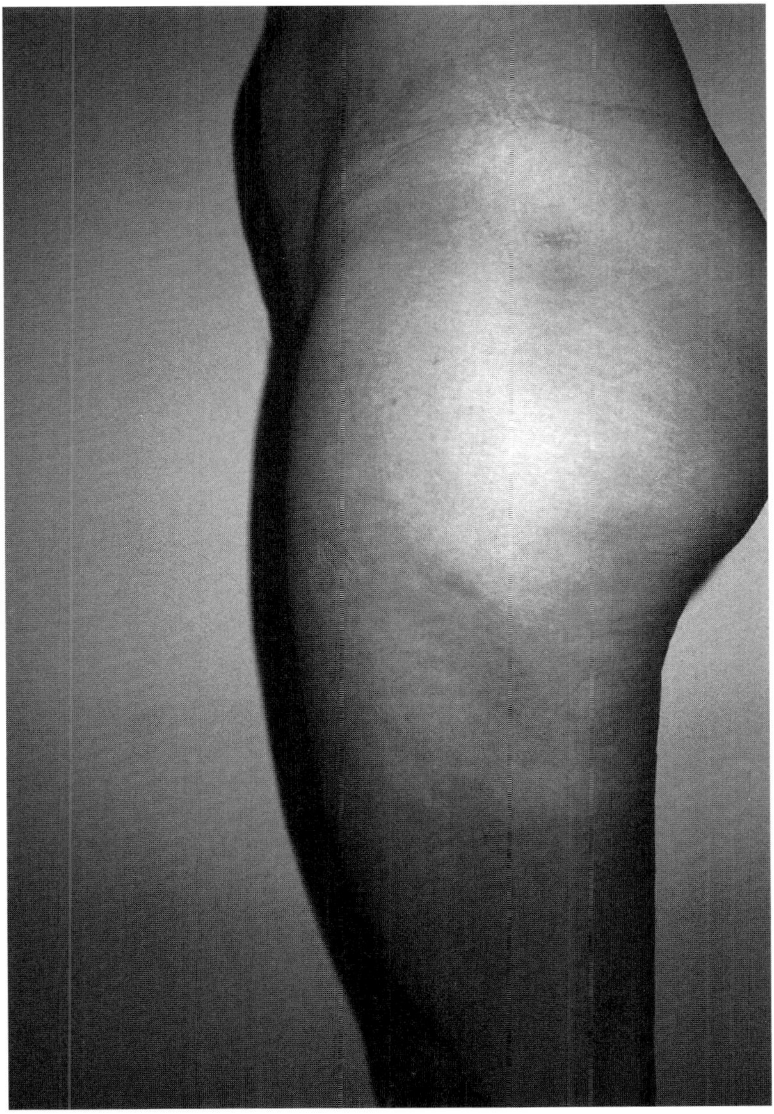

Figure 3 Depression of outer thigh with neighboring elevation.

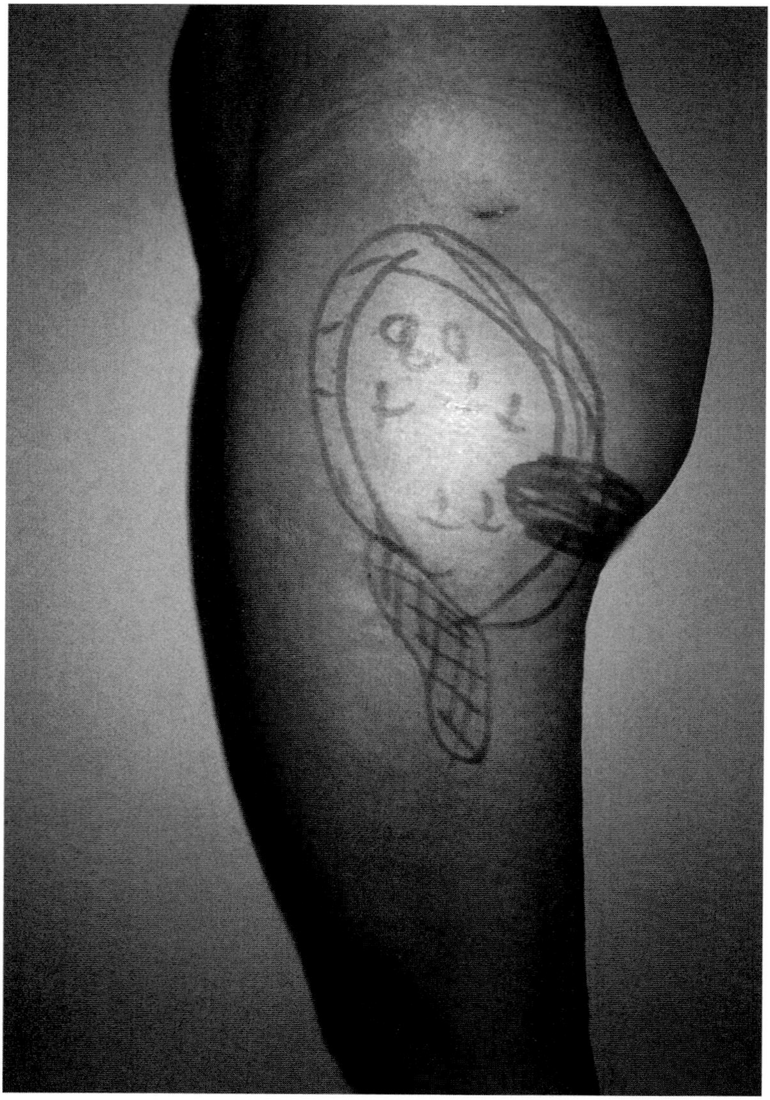

Figure 4 Marking the patient.

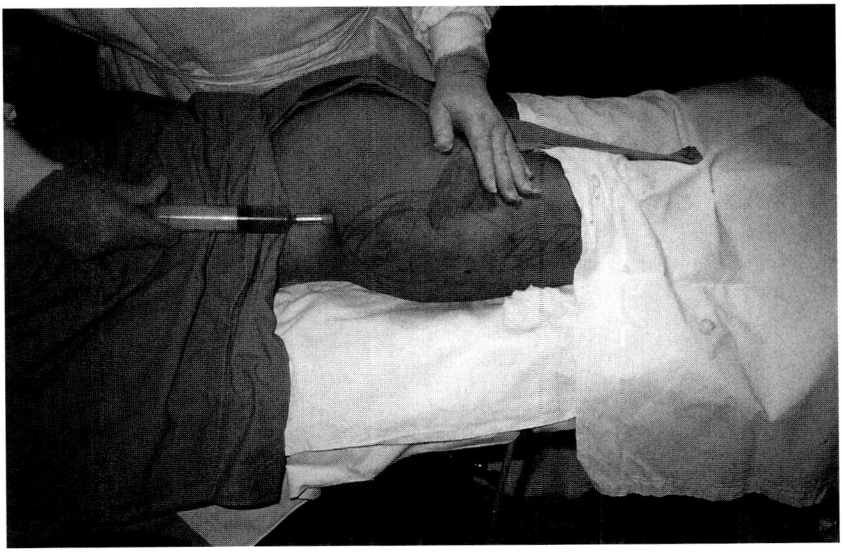

Figure 5 Reduction liposculpture in the elevated area.

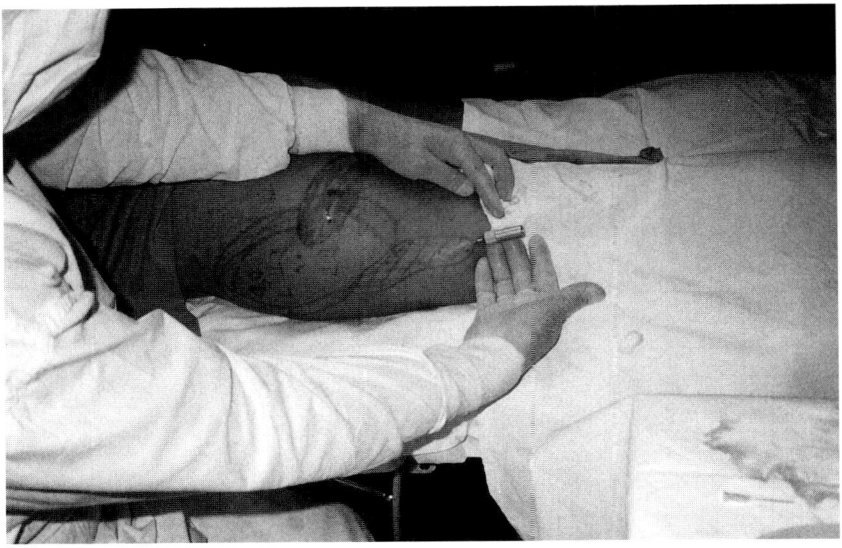

Figure 6 Syringe fat grafting after mesh undermining.

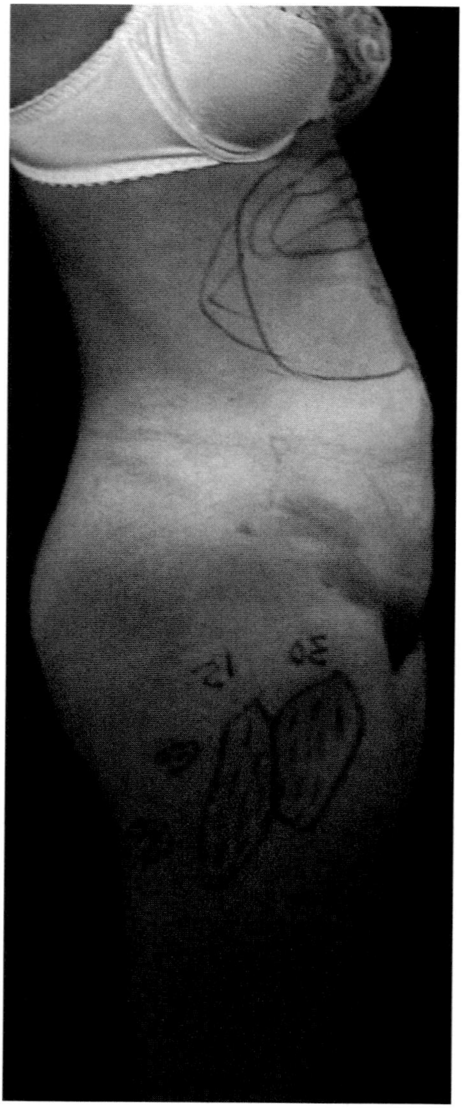

Figure 7a Preoperative depression of outer thigh.

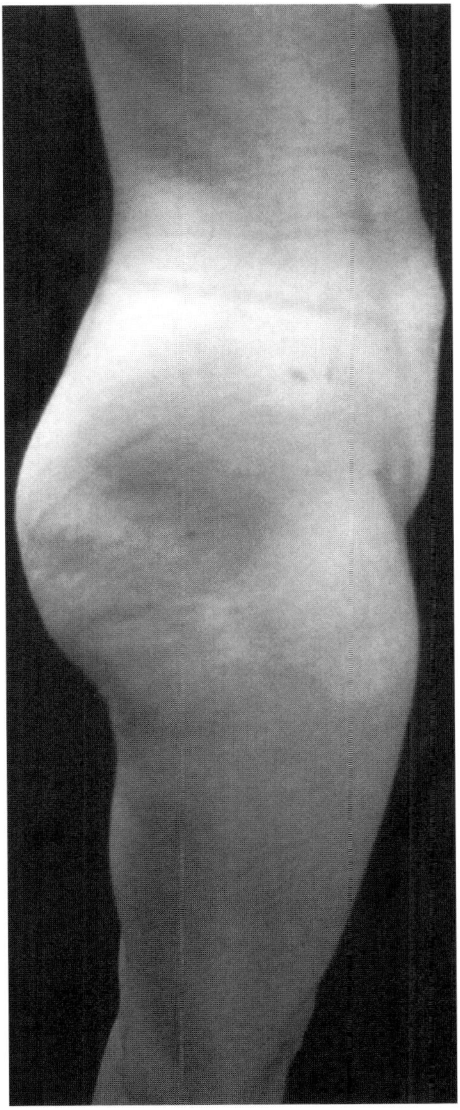

Figure 7b Result 1 year after two syringe fat grafting sessions 6 months apart.

Uneven results after a liposuction are rather frequent and may need several operations 3 or 6 months apart to correct each of them (Fig. 7). Before liposuction, the patient should be warned that there is always a risk of an uneven result according to the amount of fat tissue that has to be removed and the quality of skin. The culprit is not always the liposuction surgeon. When the surface defect is minimal, superficial liposculpturing can be performed and may be efficient. Several procedures may also be necessary in such cases.

III. CONCLUSIONS

Liposuction defects during and after surgery are rather frequent and very often a touch-up is necessary. This touch-up should be performed after 6 months when the final results of the liposculpture are really obtained. Sometimes if the defect is very visible, restoring a normal or subnormal surface can be performed after 3 months. Syringe fat grafting is the best technique to correct such defects. Repeated procedures are often necessary.

21
Lipofilling Acne Scars

David S. Alkek
University of Texas, Southwestern Medical School at Dallas, Dallas, Texas

Autologous fat has been used for over 100 years as soft tissue grafts for plastic reconstructive and cosmetic procedures. In 1893, Neuber[1] transplanted fat from the arm to reconstruct a defect in the face caused by tuberculosis. Over the next 50 odd years, surgical specialists investigated the use of free fat grafts in thoracic surgery, neurosurgery, orthopedics, and finally in cosmetic and reconstructive breast surgery. Peer[2,3] caused a rebirth of interest in 1950 when he showed that fat grafts retained about 50% of their volume.

In the 1960s, dermal fat grafts were used by Sawhney[4] and Boering[5] in the face. The development of liposuction surgery in the 1980s renewed interest in fat grafting. Illouz[6] first began reinjecting fat obtained during liposuction. At the 1986 meeting of the American Society for Dermatologic Surgery, Fournier presented a modification of the Illouz technique. In his "micro-lipoextraction and injection," he showed that small amounts of fat can be harvested through a sharp 13-gauge needle with a syringe and reinjected. This relatively easy technique revolutionized soft tissue augmentation. It initiated the use of an autologous material to correct small skin defects without the problems associated with artificial materials.

Since then several refinements have been made. The best donor areas were determined to be the buttock or outer thighs. The suctioned fat was washed in saline or Ringer's lactate. Coleman suggested that the fat be centrifuged for concentration.[7] Numerous devices were developed for harvesting and injecting the fat. Klein[8] greatly reduced the morbidity and facilited office-based liposuction by developing the tumescent anesthesia technique. Each cosmetic surgeon refined his or her own technique based on the general principles set down by Fournier. Jackson[9] and others discovered that harvested, unused fat may be frozen, thawed at a later time, and reinjected, thus saving reharvesting for a touch-up procedure. He even felt that frozen fat was better than fresh fat.

I. TYPES OF ACNE SCARS AND BEST TREATMENT FOR EACH

Postacne scars may occasionally be hypertrophied, especially at the angles of the jaw. Atrophic acne scars may be of the icepick type, saucer-shaped, linear, or irregular in shape. Icepick scars are small, sharp holes that look as if they were made with a pin or icepick in clay. They have hard, squared-off edges and cannot be flattened out by pinching or stretching the skin around them. Saucer-shaped or thumbprint scars are valleys or depressions with gradual, soft edges. IF these are pinched or stretched, they will flatten out. Acne scars may be linear, stellar, or irregular in shape. There may be valleys with soft shoulders or canyons with hard edges. Some may flatten with distention, whereas others may not.

It is important to do a simple test on scars to determine if they are amenable to correction by filling substances. The scar should be pinched or stretched with the fingers. Scars that have hard, squared-off shoulders such as icepick scars will not flatten out. Some saucer-shaped scars will completely flatten and disappear with the maneuver. This demonstration shows that an injectable substance will fill in the defect. If a hard shouldered scar is injected with a soft tissue filler, it will merely raise the defect to the top of the mound of filler, which will look like a volcano. Scars with hard edges must be resurfaced. Laser resurfacing will improve shallow scars; however, dermabrasion is the best treatment for deeper acne scars of all types.

There are many soft tissue fillers that can be injected into the dermis. Fat must be injected into the fatty subcutaneous layer of the skin, not the dermis. Fat goes where fat is. The deeper atrophic scars that are amenable to fat injection are on the middle and lower cheeks, sides of the chin, and occasionally the temples. These areas have deeps sebaceous glands that can form large cystic acne nodules that destroy subcutaneous tissue. Sometimes the lower cheek or sides of the chin appear sunken from multiple healed acne lesions. Especially in men, there may be fibrous bands underlining the scar. These may be subcised as described by Orentrich[10] to release the skin from its underlying tissue prior to injecting with fat.

The majority of patients with acne scars usually have a combination of icepick, saucer-shaped, and/or irregular scars. Therefore, a combination of treatment modalities may be the best. Each individual is different and therefore one treatment is not suitable for everyone. Deep acne or irregular scars are excised either with a punch or scalpel, or punch grafts are done. Dermabrasion is done 6–8 weeks later before the collagen is matured. Remaining valleys may be filled with soft-tissue fillers 3–6 months after the dermabrasion. Fulton[11] described a combination procedure in which he excises scars, does subcision and fat injection, dermabrades some areas, and uses a CO_2 laser on other areas of the face in the same settings.

II. PATIENT CONSULTATION AND SELECTION

After careful assessment of the types of scars present, the best procedure or combination should be described to the patient. Alternative procedures and the benefits and risks of each should be fully discussed with the patient. This is particularly important with fat injection, since two or three sessions may be required to achieve the desired result. It is important to stress this fact so that the patient does not expect perfect and lasting correction after the first treatment. The patient must be told that there will be an overcorrection up to 50% after the procedure, and that the overcorrection and some of the desired correction may dissipate over 6 months. There may be bruising, redness, or infection. If all of these are discussed before the procedure, the patient will consider them to be acceptable problems. If they are not discussed until after the fact, the patient may consider them to be complications, and this can undermine the surgeon's rapport with the patient.

The patient must have realistic expectations, since perfection is unlikely to be attained. The patient must also have an adequate donor site. Patients who are very thin will have difficulty with harvesting and retention. The patient should be warned that the transferred fat is made up of living fat cells that will also expand if the patient gains more weight. Thus, the patient may get more than 100% correction.

III. DONOR SITES

Fat may be used from the abdomen, buttocks, medial and lateral thighs, and any other site that has adequate fat. Hudson and others[12,13] prefer the outer thigh. This fat is resistant to loss with weight reduction, especially in females. This area may easily be reached through the buttock crease, which leaves a minimal scar. The patient is told to shower and particularly wash the donor area with Hibiclens the night before the procedure.

IV. EQUIPMENT

The equipment for microlipoextraction and filling is uncomplicated (Table 1; Fig. 1). No Klein pump for anesthesia or large cannulae and suction apparatus are necessary. For facial lipoinjection, especially for acne scars, small amounts of fat are required and may easily be obtained with a 10 mL syringe. Various microcannulae have been designed for microlipoextraction, some of which are sharp, whereas others are blunt with openings on the tip or on the sides (Fig. 2). The microcannula or 14-gauge needle is attached to a Luer-Lock syringe.

V. ANESTHESIA

Lidocaine 1% with epinephrine should be used to anesthetize the entry point for the cannula as well as the areas to be injected in the face. The scars are areas to be lipoinjected should be carefully outlined with gentian violet before the patient is anesthetized. The lidocaine anesthesia may distort the tissue and obliterate the defects to be corrected. The tumescent lidocaine mixture (100–300 mL) described by Klein[8] is injected into the donor area. After waiting 15 min for vasoconstriction, the donor fat is harvested. No oral or parenteral sedation is necessary.

VI. PROCEDURE

The patient is placed lying on the side with the back to the surgeon (Fig. 3). The donor site is cleaned, marked, and draped. After local anesthesia to the entry site, a small stab incision is made with a #11 blade in the lateral buttock crease.

Table 1 Equipment for Microlipoinjection

Tumescent anesthetic solution (Klein's) for donor area lidocaine with epinephrine for local anesthesia
Spinal needle
20-mL syringe for tumescent anesthesia
3-mL syringe for local anesthesia
Three to five 3-mL syringes for lipoinjection
Closed Luer-Lock for transferring fat to 3-mL syringes
Microliposuction cannula
14- or 16-gauge needle for lipoinjection
Saline or Ringer's solution
Rack to hold syringes upright

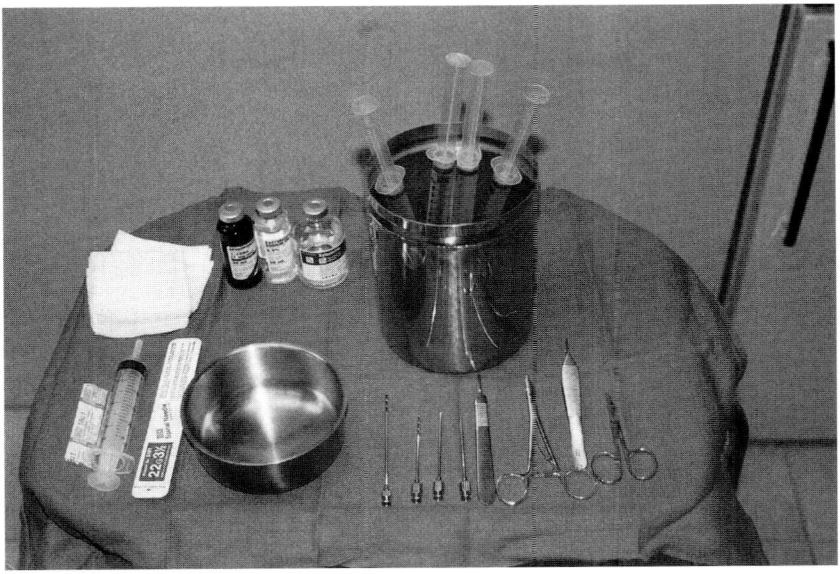

Figure 1 Equipment used for lipotransfer.

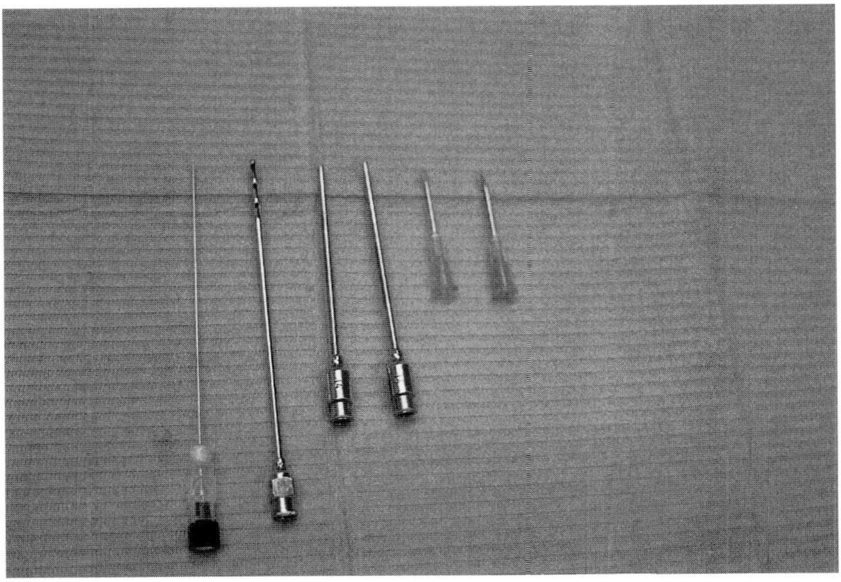

Figure 2 Microcannulae for lipoextraction. Some are sharp and some blunt. Openings can be on the tip or on the sides. On the right are 14- and 16-gauge needles.

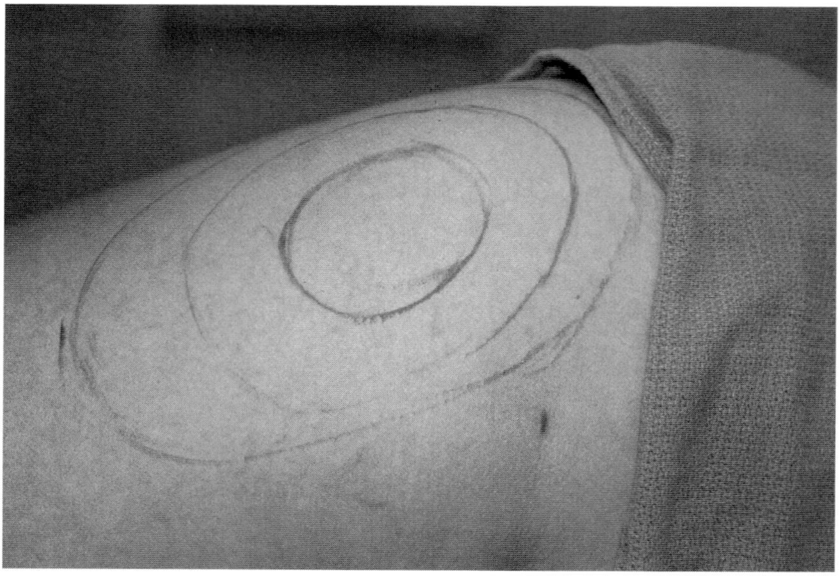

Figure 3 Donor site on lateral thigh with patient on the side and back to the surgeon.

A long 18-gauge needle is used to tumesce the donor area rapidly. After waiting the appropriate time, the cannula attached to the syringe is inserted all the way to the donor site before negative pressure is started. The thumb and index finger of the dominant hand is sufficient to maintain suction (Fig. 4). The other hand is placed over the donor area to feel and guide the cannula, much as in a large liposuction.

The cannula and syringe are moved in and out like a piston, maintaining suction on the syringe. Care should be made not to withdraw too far and suck air into the syringe. Stroking rapidly will give several milliliters of fat and tumescent mixture in a short time. Obtain four to five syringes of fat, but do not attempt to fill each syringe completely. The first ones should have only yellow fat and the last ones may be blood tinged. The syringes are removed from the cannula as they are filled, capped, and stood upright with the tips down (Fig. 5). The fat will rise, whereas the mixture of cloudy tumescent fluid and serum will settle the bottom (Fig. 6). This infranate is carefully ejected, and the remaining fat cells are washed two to four times with normal saline or lactated Ringer's solution until the washings are clear. Occasionally there is yellow clear liquid

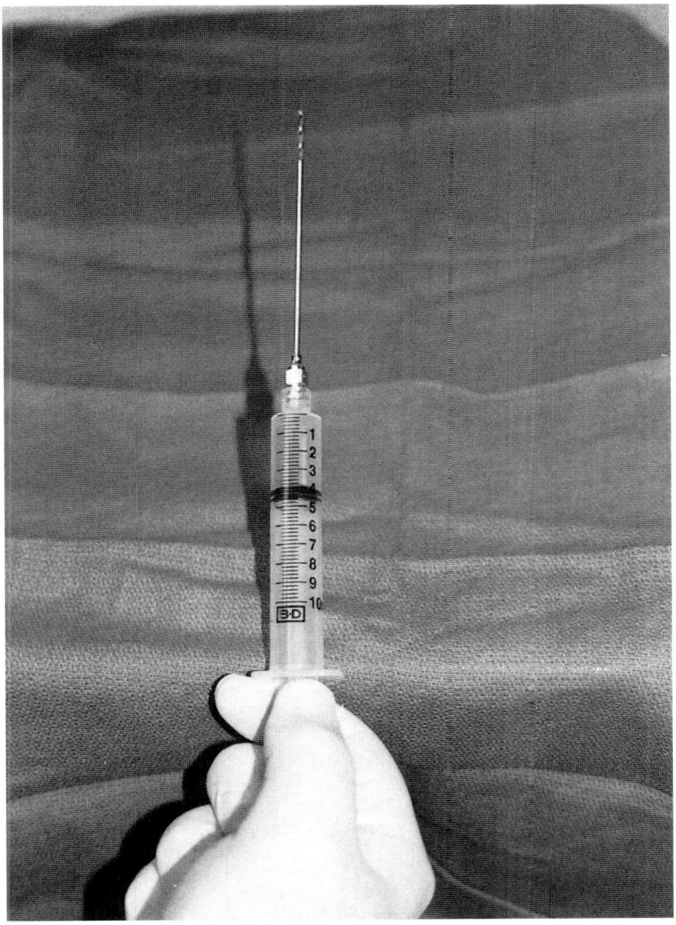

Figure 4 Suction is maintained with the thumb and index finger of dominant hand.

at the top which consists of free lipid supernatant and should not be injected into the patient. The syringe may be inverted and this part ejected. The cleaned fat is transferred to 3 mL syringes using a closed Luer-Lock method so that they may be easily centrifuged for 3 mins. The 3-mL syringe may be used more accurately and with less pressure in placing the lipoinjection and facilitates freezing and rethawing for subsequent touch-ups (Fig. 7).

Figure 5 The filled syringe is removed from the cannula, capped, and stood upright with the tip down in the rack.

While the fat is being washed and prepared for injection, the donor incision is cleaned and a small pressure bandage applied. No suture is necessary, since the incision should not be more than 2 mm long. The patient is tuned on the back and the face is cleaned. Lidocaine anesthesia, local or by nerve block, should have been done before the liposuction to allow the swelling to subside and for maximal vasoconstriction. The 16-gauge needle should be inserted in natural facial lines where possible, although in treating defects in the central

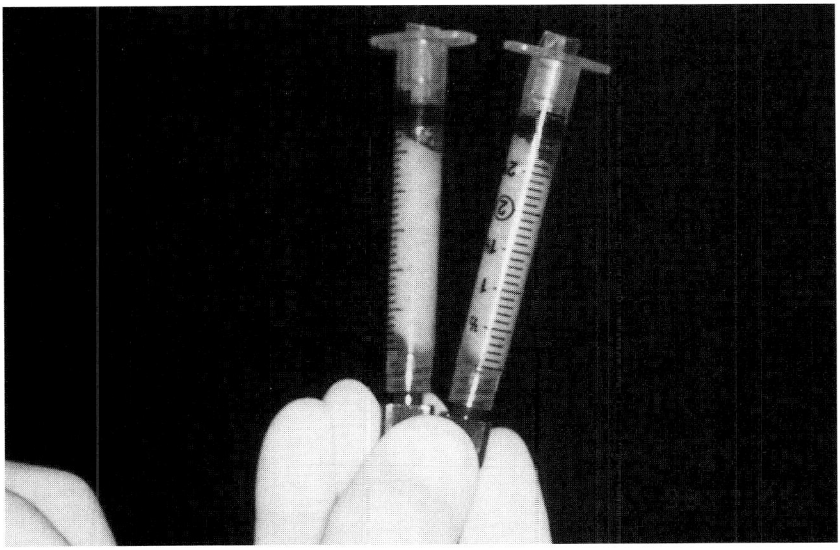

Figure 6 The fat rises and the cloudy tumescent fluid and serum settle to the bottom.

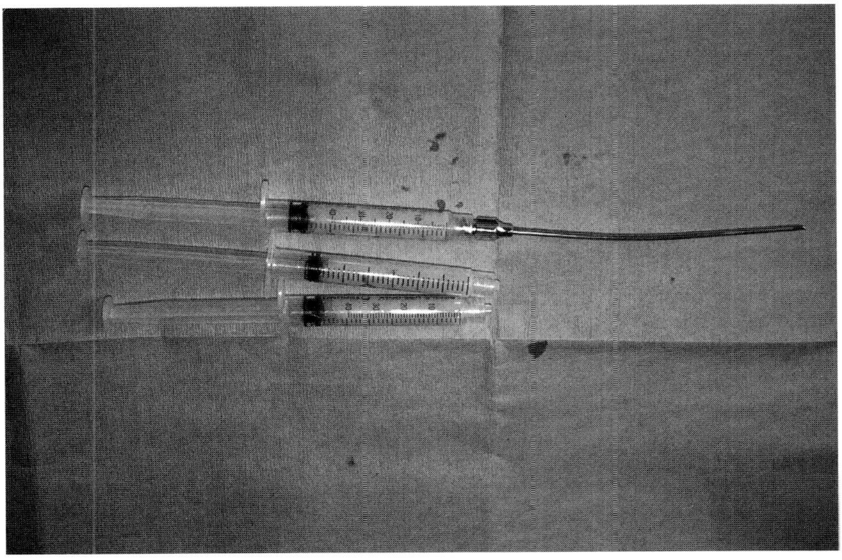

Figure 7 The fat is transferred to 3-mL syringes for lipoinjection or for freezing for subsequent touch-ups.

cheeks or temples, this may not be possible. Many beginning and some experienced physicians prefer a pistol-like device that holds the syringe and delivers a measured amount of fat governed by a ratchet. With experience it is possible to give accurate correction with a hand-held syringe. The operator should inject fat as the needle is withdrawn, attempting to deposit small aliquots of fat along several lines and planes. Remember to remain below the dermis. An overcorrection of 50–100% should be made to account for any nonfat solution and nonviable cells. The deposited fat is malleable after placement and can be manipulated into the desired areas and configuration. If the insertion is made at the oral commissures, it is closed with 5-0 fast-absorbing gut, whereas other sites are left unsutured.

VII. POSTOPERATIVE CARE

Once milliliter of betamethasone (Celestone) is given immediately postoperatively or 20 mg of oral prednisone daily for 5 days. Cephalexin 500 mg or minocycline 50 mg is given twice daily for 7 days. The patient is sent home with an ice pack to the face and told to keep it on until bedtime. No dressings are necessary for the face other than bacitracin to the injection sites. A pressure dressing to the donor area should remain for 24 h. It is changed by the patient and another gauze with bacitracin applied. Showers may be taken, but the donor incision should not be soaked for 4 days. The patient should be warned that the incision may drain for 3–4 days and that they should return for follow-up in 1 week and approximately monthly. The patient should be warned to return immediately if there is increased swelling, redness, pain, or bleeding. The essential points of lipotransfer are summarized in Table 2.

VIII. FAT SURVIVAL AND REINJECTION

Fulton[14] used volumetric and caliper measurements to show that there was an initial increase in augmentation for 15–30 days, presumably due to inflammatory response. Then there is gradual loss of volume until 60 days. After that, the graft volume stayed the same. The best areas for fat retention are the chin and cheeks, whereas it is retained the shortest time in the lips and nasolabial folds. This is also true with other injectable fillers and is probably because these areas are more animated and "wear out" the implants. The patient should

Table 2 Summary for Lipotransfer

1. Tell the patient that several sessions may be required and not to take aspirin or nonsteroidal anti-inflammatory drugs (NSAIDs) before the procedure.
2. Tumesce the donor area, such as the outer thigh or buttock.
3. Collect several syringes of fat and stand them point down in a rack.
4. Wash the fat gently and transfer to 3-mL syringes or injection gun. Let the fat concentrate by gravity or centrifuge.
5. Inject the fat evenly in small amounts as the needle is withdrawn.
6. Carefully mold the fat into the desired configuration.
7. Place pressure dressing over the donor site.
8. Use steroids, antibiotics, and ice pack postoperatively.
9. Wait at least 3 months between injection touch-ups.

be reevaluated at 2 months and reinjected if necessary. It may require three to four reinjections to achieve optimal results.

IX. COMPLICATIONS

Compared to allografts, complications with microlipotransfer are very few. The biggest problem, although not truly a complication, is lack of graft survival. If the patient is told that only 50% is expected to survive and two or more sessions are often required for desired results, the initial disappointment can be accepted. Bruising or slight erythema may persist for a week or so. Hematoma is also a possibility. Infection is rare and usually prevented with prophylactic antibiotics. Divots may also be a problem. These complications are more likely in the donor area where there is more surgical "trauma." Overfilling may occur if not enough fat is resorbed. It may also be a problem much later if the individual gains weight and the injected fat bulges. Either of these problems may be corrected with microextraction or triamcinolone injection.

X. ALTERNATIVE METHODS FOR FILLING DEPRESSED ACNE SCARS

Besides fat, there are numerous other soft tissue fillers (Table 3). Th ideal filling substance has yet to be developed (Table 4). Among the autologous materials, fat has an advantage as a filler in that it is easily obtainable and long lasting if it

Table 3 Types of Soft Tissue Fillers

Autologous	Semisynthetic	Synthetic fillers
Fat	Zyderm/Zyplast	Goretex
Dermis	Dermalogen	SoftForm
Collagen	Alloderm	Artecoll
Isolagen	Hylaform Gel	Liquid silicone
Autolagen	Restylane	
Dermalogen		

survives. Its main disadvantage is that it must be injected into the subcutaneous layer and cannot treat very small or superficial defects. Deepithelialized dermis has been described as a filler. Punch grafts of skin posterior to the ear are taken, the epidermis removed, and pieces of this dermis are placed into pockets subcised under depressed facial defects. Fournier[15] separated connective tissue from extracted fat and injected this intradermally and subcutaneously.

Isolagen[16–18] is a recently approved autologous process in which a punch biopsy of skin is sent to the Isolagen laboratories where it is cultured and fibroblasts and type I collagen are harvested. This material is returned to the physician for injection. Its advantage is that it is the patient's own living fibroblasts, which continue to produce collagen. The disadvantages are its expense and the logistics involved with scheduling the injection of living cells within 48 h of shipping.

Autologen,[19] like Isolagen, is processed from the patient's own collagen except that a large piece of excised skin, such as from a facelift, is sent for processing. The dermis is pulverized to produce a dispersion of intact collagen fibers for injection. The advantage is that the intact collagen with its natural cross links should promote collagenesis. The disadvantage is that it requires 3 cm of skin to produce 1 mL of Autologen.

Dermalogen,[20] introduced in 1998, is a suspension of intact (not fragmented) human collagen fibrils. It is processed from human skin procured from accredited tissue banks and is processed to ensure safety and freedom for human immunodeficiency virus (HIV), hepatitis, and other infectious agents. It does not

Table 4 The Ideal Filling Substance

Easily obtainable	Safe, nontoxic
Easy to use	Nonallergenic
Easy to set up	Long lasting
Predictable results	Inexpensive
Versatile	

require skin testing and can be injected like Zyplast (Collagen Corp.) into the mid dermis. Typically, three injections are necessary at 2-week intervals. This produces a progressive improvement that lasts longer than Zyplast. Its advantage is that it may be used "off the shelf" in a patient on the first visit. The disadvantage is the need to reassure the patient of the safety of human tissue from a tissue bank.

Semisynthetic fillers are derived from living tissue but not from the same patients. Zyderm (Collagen Corp.) and Zyplast are produced from bovine skin. They are readily available and easy to use. The disadvantages are possible allergenicity (3% of the general population) and lack of longevity. Bovine collagen is slowly degraded by human collagenase over several months. Its correction is of shorter duration in very active areas such as the lips and expression lines. It may last up to 18 months in depressed scars. The author, in an unreported observation, has noted that after repeated injections of bovine collagen, less and less is needed, until the correction is permanent. This is probably because repeated injection with a foreign body stimulates the patient's own collagen formation. This of course varies from patient to patient.

Alloderm is decellurized human skin. Skin from the tissue bank is reduced to its noncellular and noncollagenous content and dehydrated. This is sent as a solid wafer to the physician who rehydrates it, trims it, and surgically inserts it into larger skin defects and the lips. It serves as a matrix for fibroblast regrowth. The disadvantage is that it cannot be used for small defects such as acne scars.

Hylaform Gel is hyaluronic acid, a collagen precursor, derived from cock's comb. It is nonallergenic; however, it is not permanent. Eighty percent of patients report moderate or greater improvement at 12 weeks.

Restylane is a higher concentration of hyaluronic acid derived from bacteria. It can be used in small defects and 60% of patients are satisfied at 1 year. However, 50% have immediate swelling after injection, and a smaller number have intermittent swelling over several weeks.

Synthetic fillers are manufactured substances designed for soft tissue implantation. They are made from inert nonallergic materials. Goretex (Collagen Corp.) is expanded polytetrafluoroethylene (ePTFE) and is produced as threads, sheets, or hollow tubes. Its manufacturer introduced it in hollow tubes as SoftForm. These solid materials are threaded with a needle or, in the case of SoftForm, with a trochar that is included in a kit. This material is not suitable for small defects such as acne scars.

Artecoll is a suspension of microscopic methacrylate beads in a collagen matrix. After injection, the collagen matrix is degraded leaving the Plexiglas beads which are encapsulated as foreign bodies by tissue fibroblasts. This can be injected and is permanent, but it is unforgiving. Placement must be precise because it is very difficult to impossible to remove without scarring.

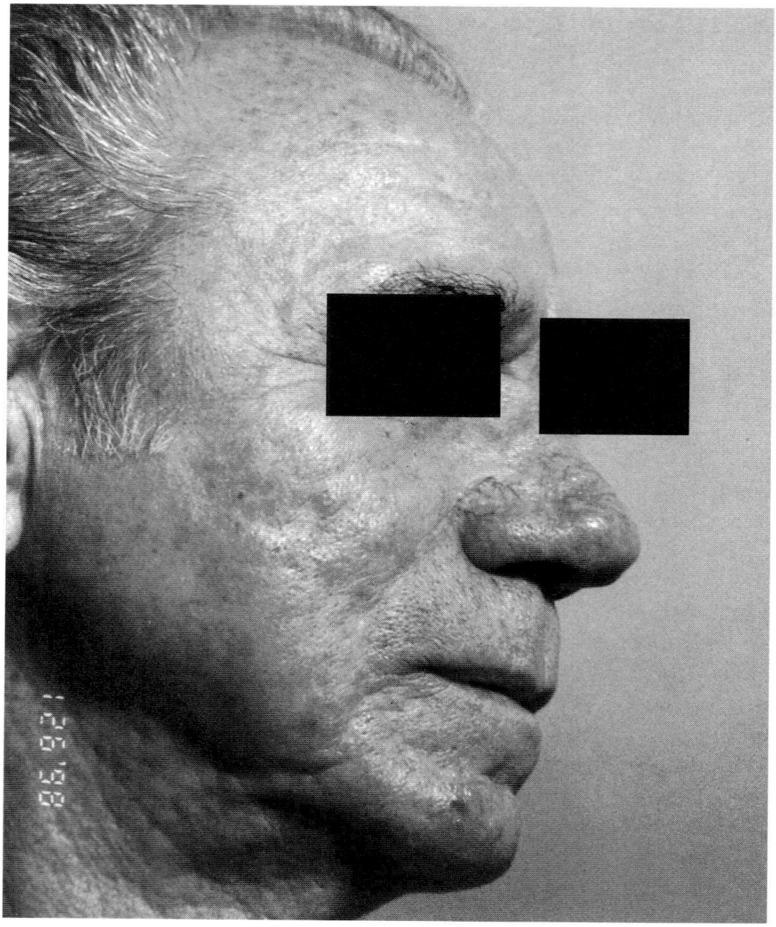

Figure 8a Preoperative patient with acne.

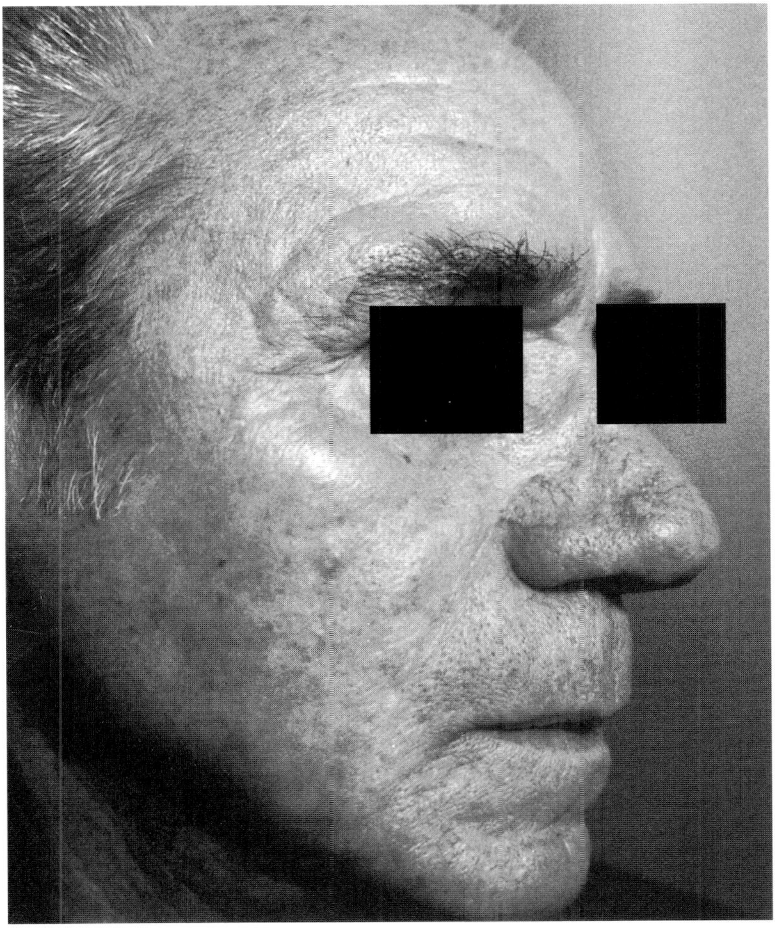

Figure 8b Six months postoperative lipoinjection.

Liquid silicone ushered in soft tissue augmentation of small defects such as scars and wrinkle lines in the 1950s. Several dermatologists and plastic surgeons skillfully corrected saucer-shaped acne scars; depressed areas of the face due to disease, trauma or surgery; and facial wrinkles and furrows. Orentreich[21] was a pioneer in this field and developed the microinjections technique. Alkek[22] described the use of medical-grade silicone in lip augmentation. The original clinical study of injectable silicone was never submitted to the US Food and Drug Administration (FDA), and because of problems with silicone-filled breast implants, the FDA forbade the use of silicone breast implants and injectable silicone. This was deplorable, since pure medical-grade injectable silicone met more of the criteria of the ideal filling substance than any other material up to this time.

XI. CONCLUSIONS

Autologous fat transplantation has a definite place in correcting facial defects such as deep furrows, sunken areas, and atrophic scars with soft, distensible edges, such as saucer-shaped acne scars (Fig. 8). The extraction technique is rather simple, but care must be used in injecting the fat deep into the subcutaneous layer. There are many newer autologous, semisynthetic, and synthetic materials that can be used for dermal filling of scars and wrinkles. The physician must carefully assess the scars, because not all scars may be improved with filling substances.

REFERENCES

1. Neuber F. Fettransplantation. Chir Kongr Verhandl Deusch Gesellsch Chir 1893; 2:66
2. Peer LA. Los of weight and volume in human fat grafts with postulation of “cell survival theory.” Plast Reconstr Surg 1950;5:217
3. Peer LA. The neglected “free fat graft,” its behavior and clinical use. Am J Surg 1956;92:40
4. Sawhney CP, Banerjee TN, Chakavarti RN. Behavior of dermal fat transplants. Br J Plast Surg 1969;22:169
5. Boering G, Huffstadt AJ. The use of derma-fat grafts in the face. Br J Plast Surg 1968;20(2):172
6. Illouz YG. The fat cell “graft”: A new technique to fill depressions. Plast Reconstr Surg 1986;78:122

7. Coleman SR. Long term survival of fat transplants: controlled demonstration. Adv Plast Surg 1995;19:421–425
8. Klein JA. The tumescent technique for liposuction surgery. Am J Cosm Surg 1987;263–267
9. Jackson RF. Frozen fat—Does it work? Am J Cosm Surg 1997;14(3):339–343
10. Orentreich DS, Orentreich N. Subcutaneous incisionless (subcision) surgery for the correction of depressed scars and wrinkles. Dermatol Surg 1995;21:543–549
11. Fulton J, Silverton K. Rejuvenation of the acne-scarred face. Am J Cosm Surg 1998;15:281–289
12. Hudson DA, Lambert EV, Block C. Site selection for auto transplantation. Aesthet Plast Surg 1990;14:194–199
13. Pinski JB, Roenick HH Jr. Autologous fat transplantation: long term follow up. J Dermatol Surg Oncol 1992;18:179–184
14. Fulton J, Suarez M, Silverton K, Barnes T. Small volume fat transfer. Dermatol Surg 1998;24:857–865
15. Fournier P. Facial recontouring with fat grafting. Dermatol Clin 1989;8:523
16. West T, Aster T. Autologous human collagen and dermal fibroblasts for soft tissue augmentation. Dermatol Surg 1998;24:510–512
17. Alkek DS. Isolagen, a new autologous collagen. Cosm Dermatol 1998;11:30–32
18. Alkek DS. A new autologous collagen technique. Am Cosm Surg 1998;15:325–328
19. Alster T, West T. New options for Soft tissue augmentation. Skin and Aging 1998;32–36
20. Elson M. Soft tissue augmentation with allogenic human tissue matrix. Cosm Dermatol 1998;24–28
21. Orentreich D, Orentreich N. Injectable fluid silicone. In: HH Roenigk, HH Roenigk, Jr, eds. Dermalologic Surgery. New York: Marcel Dekker, 1989
22. Alkek DS. Lip augmentation with silicone. Dermatol Surg Oncol 1991;17:747–748

22
Autologous Frozen Fat for Transplant

Ziya Saylan
Saylan Cosmetic Surgery Center,
Düsseldorf, Germany

Long-term satisfactory results for tissue augmentation can be obtained with even large amounts of autologous fat. However, swelling in the recipient site persists at times for 3–4 weeks until the water is absorbed

Using frozen fat over the past 4 years, the author has noted less swelling and bruising in the injected site. Swelling persists for only 2–3 days. Less swelling occurs because of water crystallization from freezing and reduction of water during the thawing period.

I. HARVESTING

The harvesting is done with low-pressure suctioning using large-diameter cannulae and without centrifuging. The parts of the body chosen must be a low-tension area such as the gluteal region and the abdomen which enable the fat cells to survive. The fat cells obtained from parts of the body where a fluctuation of volume occurs reduces the amount of cells as the patient fasts or diets. After waiting about 30 min, using a metal strainer (open technique) to allow fluid to drain, 5-mL syringes are filled with fat and marked with the name and the birthday of the patient. Afterwards it is placed in a box made especially for this purpose and catalogued.

II. FREEZING

The freezing of fat is done in a normal commercial refrigerator deep freeze. Shock freezing (immediate, quick freezing) is not desirable and is believed to destroy the fat cells. A constant temperature is maintained, and there are safeguards against power outages with a backup power supply. Any drastic increase in temperature will result in totally damaging the stored fat tissues and rendering them useless. Tissue frozen at lower temperatures have 6–8 h to refrigerate in case of loss of power (P. Oellig, Pathology Institute of Mulheim, Germany, personal communication). The fat is frozen at –30°C, and a built-in warning system raises and alarm with any temperature changes. Labeling and cataloging of the frozen fat samples is critical.

III. WARMING THE FAT

The syringe filled with frozen fat is taken out of the deep freeze as the patient arrives at the office, and by means of holding the frozen syringe for 20 mins in the hand, the patient warms up her or his own fat.

The thawing process releases water, making the fat denser, which makes it more vital. During the warming-up procedure, water collects at the bottom of the syringe when it is held in a vertical position, and this is drained.

IV. LIPOINJECTION

The injection of the frozen fat does not differ from conventional lipoinjection. The amount injected in one part of the face must be less than 4 mL. Low-pressure filling is necessary, which requires a thick needle and a small syringe (smaller than 5 mL). The larger syringes produce too much pressure at the time of injection and kill the fat cells.

Small pearls of fat are injected in different layers. An injection close to periosteum or to the vessels facilitates a better blood supply, which will make it possible for the fat to survive longer. Subcision may be necessary to free up skin attachments to deeper tissue. The fat is distributed in different layers homogeneously and evenly to the facial contours where the blood supply is optimal. Molding the fat is done immediately.

It is best to administer the desired amount in two or three separate sessions. By injecting the total fat desired at one time, the body is unable to nourish the newly introduced fat and a large percentage of fat cells diminishes owing to insufficient blood supply. This results in uneven distribution. The first amount of fat is injected 8 days after harvesting, the second amount after 3–6 weeks, and the third and final amount after 6–12 months.

V. HISTOLOGY

Histological study of freshly harvested fat cells has shown that the cells are swollen because of the injected tumescent solution (Fig. 1). The proportion of cells to collagen septa has been changed showing more fat (swelling) than the collagen. The normal histological appearance is restored after freezing and then thawing (Fig. 2).

The histological study of injected frozen adipose tissue has shown totally vital adipose tissues with a normal amount of collagen septa and some blood vessels, which is evidence of revascularization (Fig. 3).

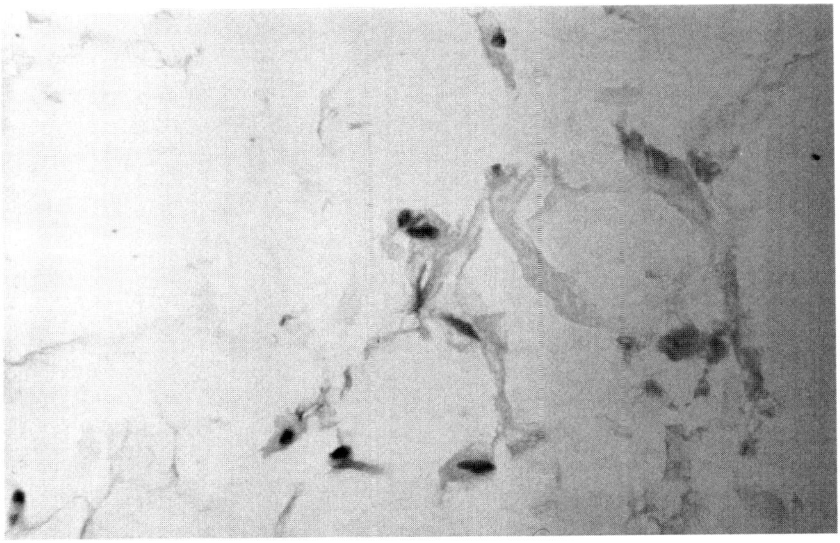

Figure 1 Swelling of fat cells freshly harvested.

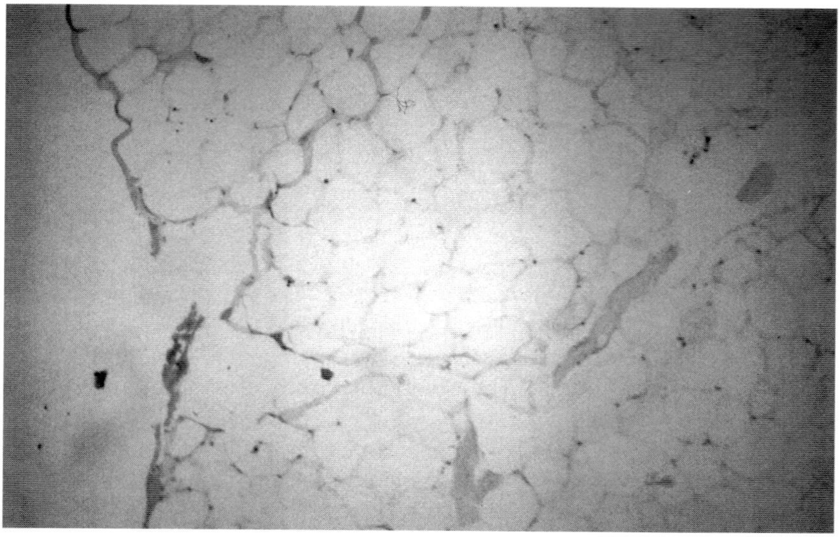

Figure 2 Normal histologic appearance of fat follows thawing of frozen fat.

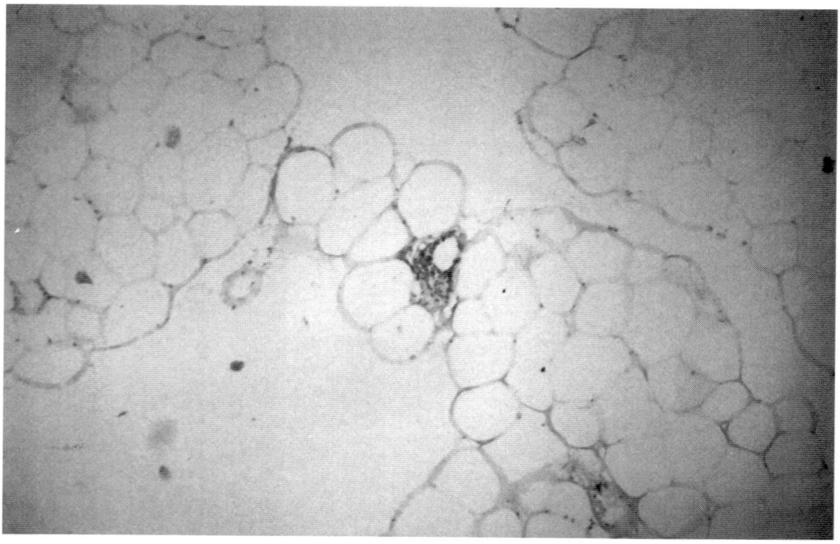

Figure 3 Histologic appearance of implanted frozen fat shows live adipose tissue with collagen septa and small blood vessels.

A. Advantages of Interval Frozen Fat Injections

1. Controlled results
2. Better possibility of contouring and replacement
3. Less swelling and no bruising
4. Long-lasting results
5. Easy to obtain and perform
6. Less pain for the patient

VI. DISCUSSION

The technique of using frozen fat will assist the surgeon in sculpturing and contouring the face and body. The desired results can be obtained without major swelling and bruising. The frozen fat survives if it is obtained and stored properly.

23
Long-Term Frozen Fat Transplant

Katsuya Takasu and Shizu Takasu, M.D.
Takasu Clinic, Nagoya, Japan

Fournier[1] has shown the possibility of adipose tissue conservation by refrigeration. The purpose of preserving fat tissue is to save the excess material extracted, thereby avoiding a new harvest when repeated injections are anticipated. Autocollagen can be separated from adipose tissue with an ultrasonic generator to break up fat cells and then centrifuging the tissue.[2] This can then be used as a dermal filler. Fat can be frozen up to 7 years without losing its potential to be a permanent soft tissue filler.

I. TECHNIQUE

Fat is harvested from the convenient sites, such as the abdomen, buttocks, and thighs, using a syringe or liposuction machine with a 2- to 3-mm cannula. Tumescent technique is used to fill out the donor areas.[3] The aspirate is rinsed repeatedly with saline solution to remove the blood component. A fine metal sieve is used to drain out the excess fluid. For immediate transplantation, the collected fat is packed into multiple 1-mL syringes with an 18-gauge needle. The rest of the adipose tissue is then placed in a 15-mL sterile test tube (Fig. 1a) and put into a freezer in which the adipose tissue can be maintained at a temperature of –70°C (Fig. 1b). When reutilizing the frozen fat, it is sufficient to leave the test tube containing the fat at room temperature for a few hours. The implantation of the thawed materials is performed in the same manner as for fresh adipose tissue. The fat is injected in a retrograde fashion while withdrawing the needle. It is important to transplant only a small quantity (less than 0.5 mL) of fat in the same place, so that each is in continuity with a vascular supply.[1,4] For

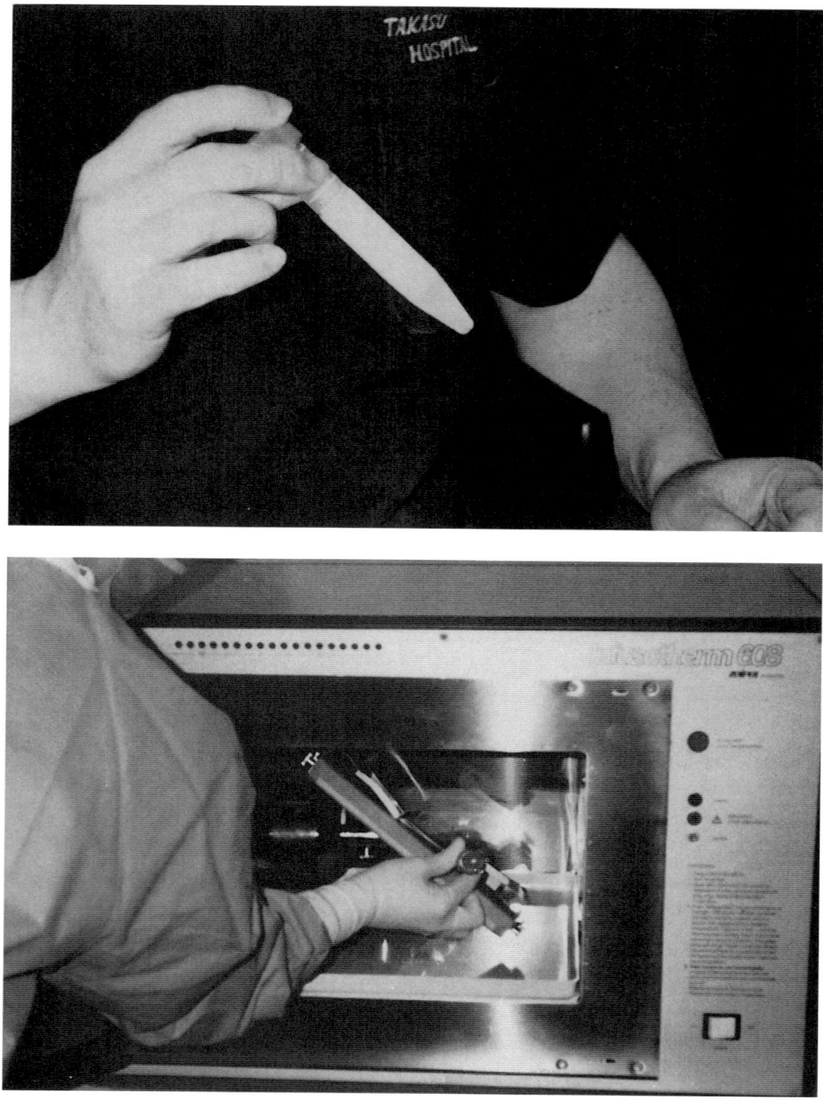

a

b

Figure 1 (**a**) Fat is placed in a test tube for freezing. (**b**) Freezer used for fat preservation.

cheek augmentation, multiple puncture sites are used and adipose tissue is injected in a crisscross manner and in several layers. The frozen fat is also used to correct deep facial wrinkles, such as vertical glabellar creases, transverse forehead lines, and nasolabial folds. Wrinkles are undermined subcutaneously prior to injection.[5] No attempt is made to overcorrect facial depressions. This way the patient can return to a normal appearance sooner. Augmentation volume required can be achieved easily with repeated injection of preserved fat (Figs. 2–4).

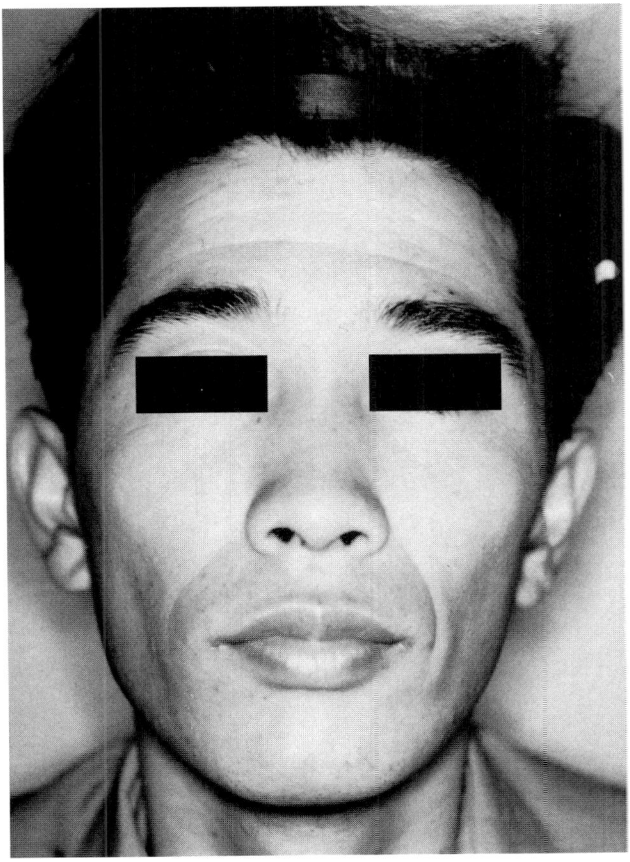

Figure 2a Preoperative 36-year-old man with volume defect in the cheek and upper eyelids.

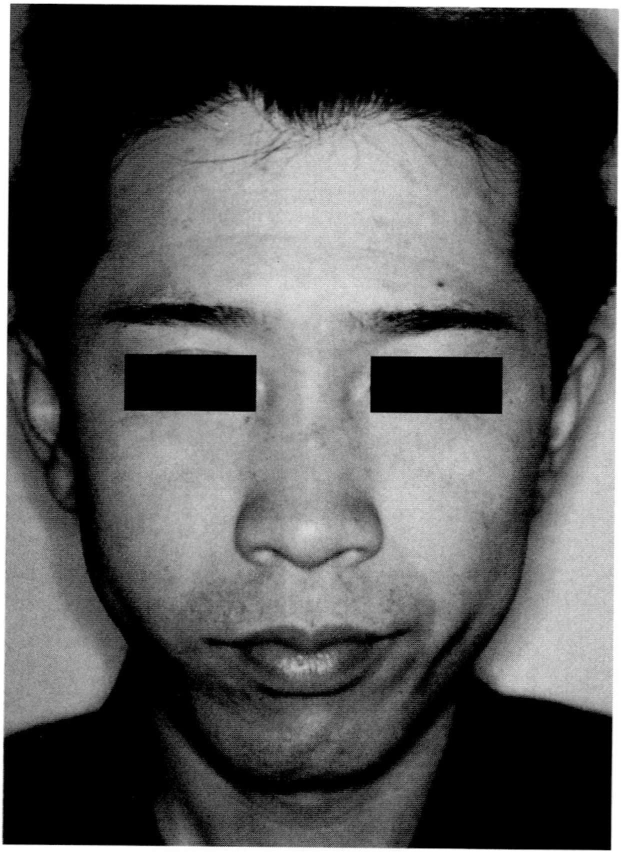

Figure 2b Patient 9 months after fresh fat injection in the cheek (16 ml in each side) and upper eyelids (1.4 ml in each side). Note undercorrection in the lower part of the cheek.

II. DISCUSSION

The purpose of preserving fat is to conserve the excess material extracted and to avoid an additional harvest procedure when repeated injections are anticipated. Before determining the efficacy of this technique, a few questions need to be answered. Can adipocytes maintain their viability after freezing and thawing? How long can they be stored frozen without losing viability?

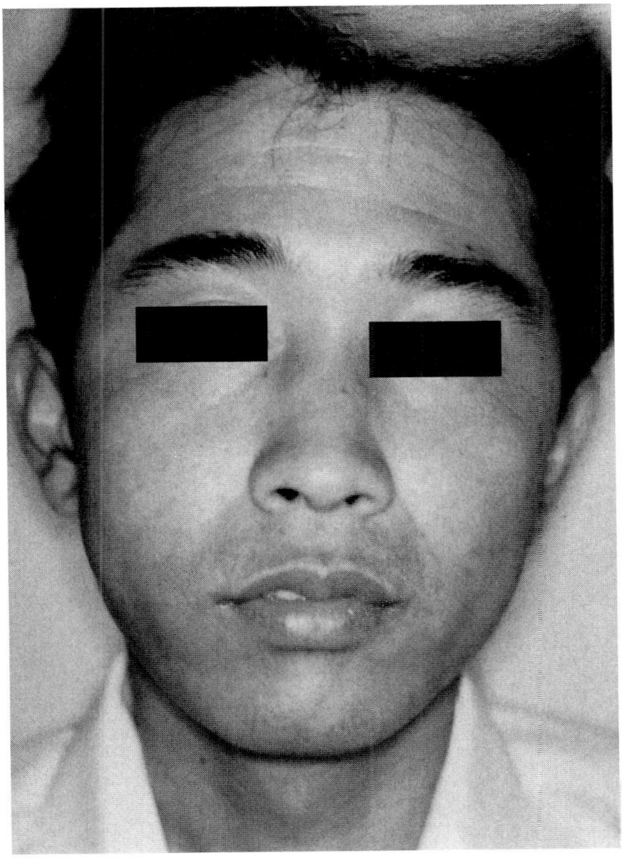

Figure 2c Satisfactory fullness in the cheek one year after frozen fat injection (10 ml in each side).

Fournier[1] described the possibility of conservation of fat in a freezer. Although he mentioned that the results obtained with frozen fat were no different than those obtained with the fresh implants normally used, he did not clearly show the viability of frozen fat in any experimental studies or clinical cases. He did not state the possible storage period. Sattler and Sommer, in 1997[6], suggested from their clinical experiences that it was safe to store the fat in a freezer for reinjection for up to 2 years. However, they did not discuss any clinical cases or describe their results with injection of frozen fat. It was not clear how they drew the conclusion that the safe preservation period was at least 2 years.

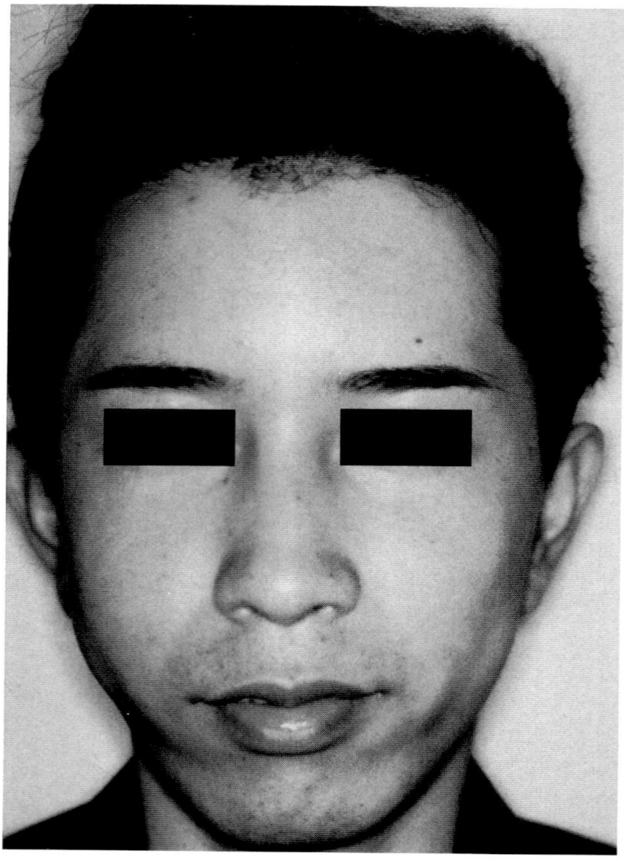

Figure 2d Six months after injection of 2.5 ml frozen fat in the horizontal forehead lines.

Jackson, in 1997[7], presented his 4-year experience with the use of frozen fat as an adjunct to fat transplantation after liposuction. He described that fat cells did remain viable in a frozen state for 6–9 months and possibly longer.

Although the viability of store frozen adipocytes is unknown, clinical results undoubtedly show that the frozen fat does maintain some of the volume transplanted in the facial regions. It is also suggested that fat can be stored in a freezer for up to 7 years without losing its potential to be a permanent soft tissue filler. Further experimental and clinical studies need to be done regarding the viability of frozen fat.

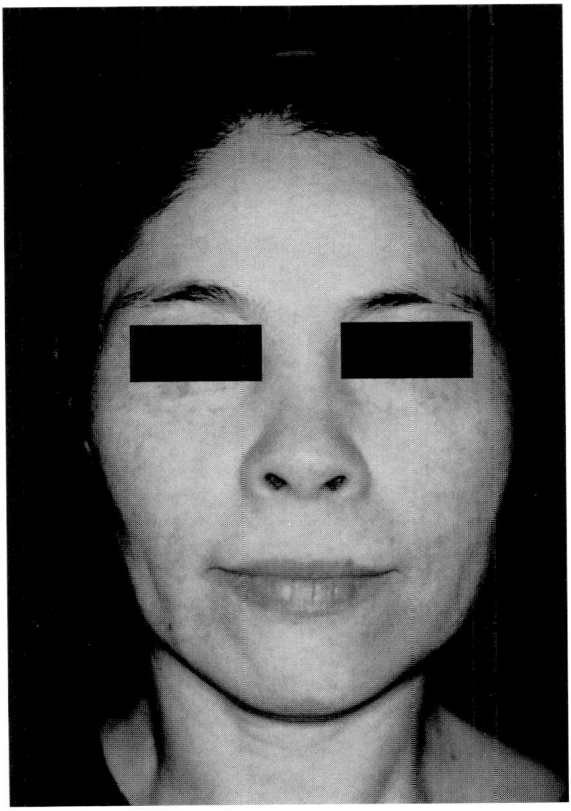

Figure 3a A 46-year-old woman with moderate atrophy of the subcutaneous fat in the cheek regions.

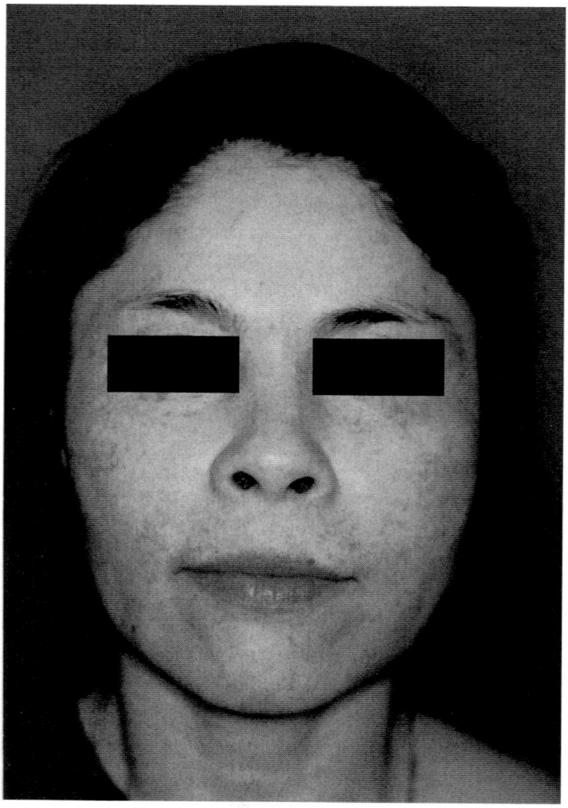

Figure 3b Three months after transplantation of 16 ml of fresh fat to each cheek. The patient has had some improvement, but more filling was needed.

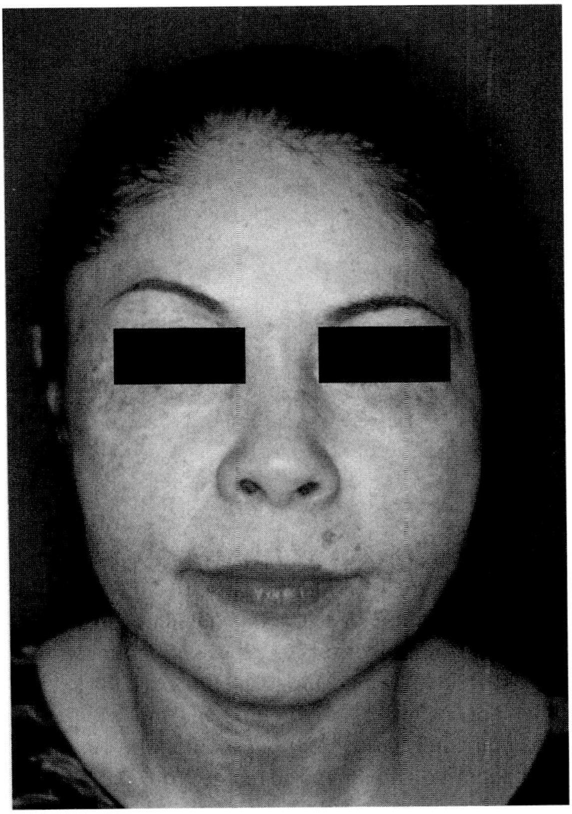

Figure 3c Satisfactory results 2 years after injection of 15 ml of frozen fat in each cheek.

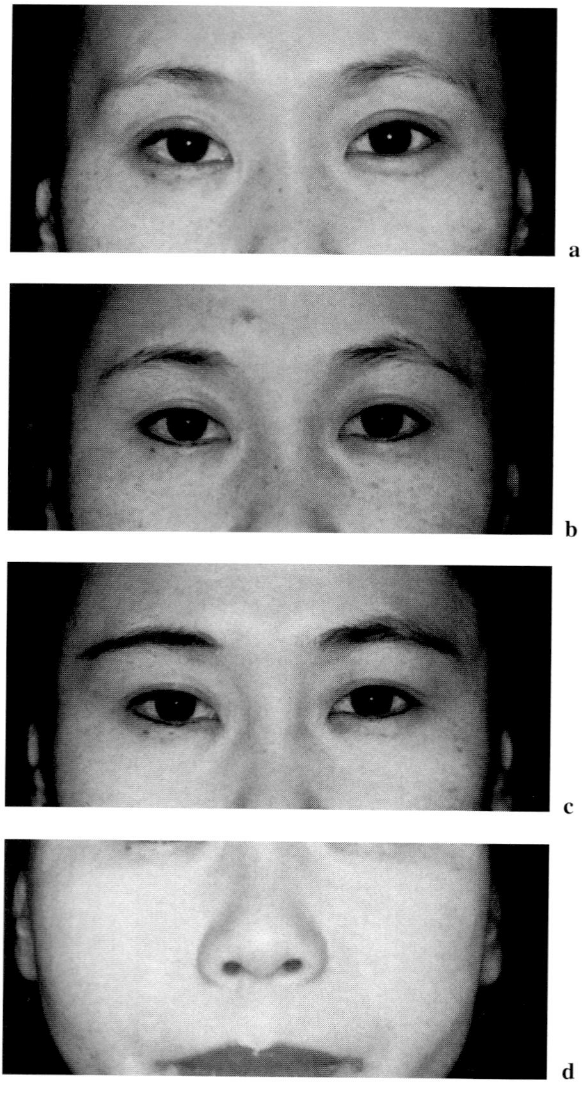

Figure 4 (**a**) A 28-year-old woman with depressions in the upper eyelids. (**b**) Five months after injection of 1.4 ml fresh fat in each upper eyelid. The fullness of the upper eyelids was not sufficient and required transplantation of 1.0 ml of frozen fat. (**c**) Fourteen months after the injection of frozen fat to upper eyelids (1.0 ml for each) and nasolabial folds (1.5 ml for each). The improvement of the nasolabial fold was not sufficient and required additional filling with frozen fat (3.0 ml for each). (**d**) Nine months after injection of 3.0 ml frozen fat in the nasolabial fold, each side.

REFERENCES

1. Fournier PF. Facial recontouring with fat grafting. Dermatol Clin 1990;8:523–537
2. Takasu K. A method for separating collagen. In: PF Fournier, ed. Liposculpture: The Syringe Technique. Paris: Arnette; 1991, pp 409–412
3. Klein J. Tumescent technique for regional anesthesia permits lidocaine doses of 35 mg/kg for liposuction. J Dermatol Surg Oncol 1990;16:248–262
4. Niechajev I, Sevchuk O. Long-term results of fat transplantation: clinical and histologic studies. Plast Reconstr Surg 1994;94:496–505
5. Takasu K, Takasu S. Endoscope assisted tunneling for fat injection to correct facial wrinkles. Int J Aesth Restor Surg 1997;5:87–90
6. Sattler G, Sommer B. Liporecycling: Immediate and delayed. Am J Cosm Surg 1997;14:311–16
7. Jackson RF. Frozen fat: does it work? Am J Cosm Surg 1997 14:339–343

24

Autologous Fat Mixed with Polymethylmethacrylate

Marcelo Benedetti and Adriana Ponti
Clinica Estetica Charcas, Buenos Aires, Argentina

Injectable fillers for the correction of contour defects through soft tissue augmentation have not been completely successful. Autologous fat alone tends to reabsorb, as does collagen and fibril. Silicone, which is an excellent filler when using midcrodrops, has been taken off the market by the US Food Drug Administration (FDA).

The ideal implant should be[1,2]:

1. Inexpensive
2. Easy to use
3. Inert
4. Nontoxic
5. Nonallergenic
6. Nonmigrating
7. Not absorbed by surrounding tissue (i.e., permanent results)
8. Not carcinogenic
9. Noninflammatroy

The combination of autologous fat and microparticulate alloplastic combination appears to be superior to other methods.[4] The mixture of autologous fat and polymethylmethacrylate (PMMA) microspheres has been studied by the authors over the past few years and has been found to be an excellent filler.

I. INDICATIONS FOR MIXED LIPOFILLING

The mixture of PMMA and autologous fat has been implanted in multiple areas, including

1. Nasolabial folds
2. Cheek bones
3. Lips
4. Mandibular area
5. Chin

II. TECHNIQUE

Autologous fat is harvested under tumescent anesthesia[3] with the syringe technique,[4] using a cannula with a blunt or sharp-edged suctioned hole. The internal diameter of the cannula should be 3 mm. The amount of fat extracted is 10 mL at a time, which is then washed with normal saline. It is essential that the harvesting site be as far away as possible from the recipient site.

Polymethylmethacrylate microspheres are biocompatible,[5–7] spherical, transparent, smooth surfaced, and free from impurities as well as having characteristics that allow selection by size (between 40 and 60 μg). The microspheres can be sterilized with ethylene oxide followed by prolonged period of degassing (2 weeks).[6] The mixture of the substances is obtained by dispersing 3 mL of microparticles with 7 mL of fat and transferring the mixture from one syringe to another until a complete and homogeneous suspension is achieved (Fig. 1). The resultant mixture is injected subdermally at the junction between the dermis and subcutaneous fat using 1-mm diameter cannula with a single side opening. The entrance site must be placed slightly away from the area to be corrected (Fig. 2). The object is to spread the injected material homogeneously through the tunnels made by the cannula, thus avoiding its accumulation in a localized area and enhancing the possibility of fat cell survival.[8] The quantity of material to be injected must be predetermined so as to avoid overcorrection (Fig. 3).[6,9]

III. COMPLICATIONS

All biological procedures have associated complications. In using PMMA and autologous fat there are the typical complications associated with liposuction and reinjection procedures.

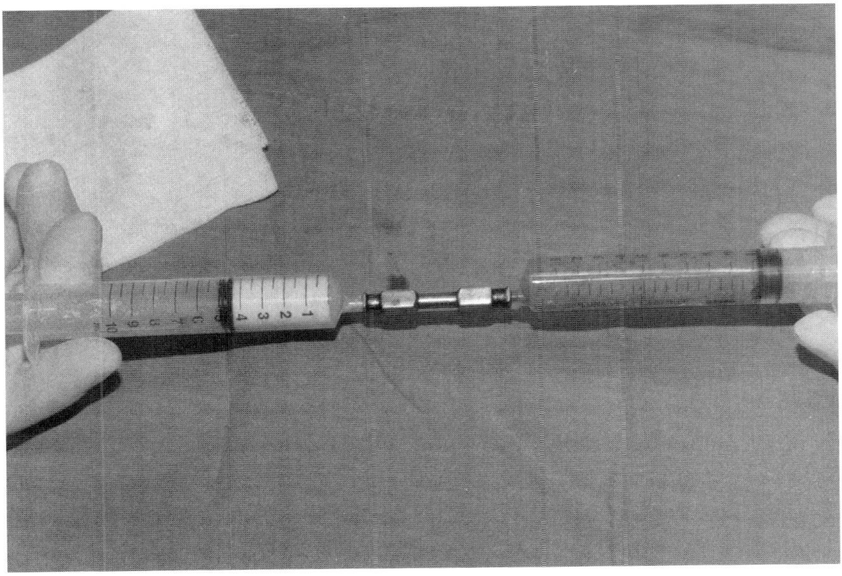

Figure 1 Mixing PMMA microparticles with autologous fat by transferring from one syringe to another.

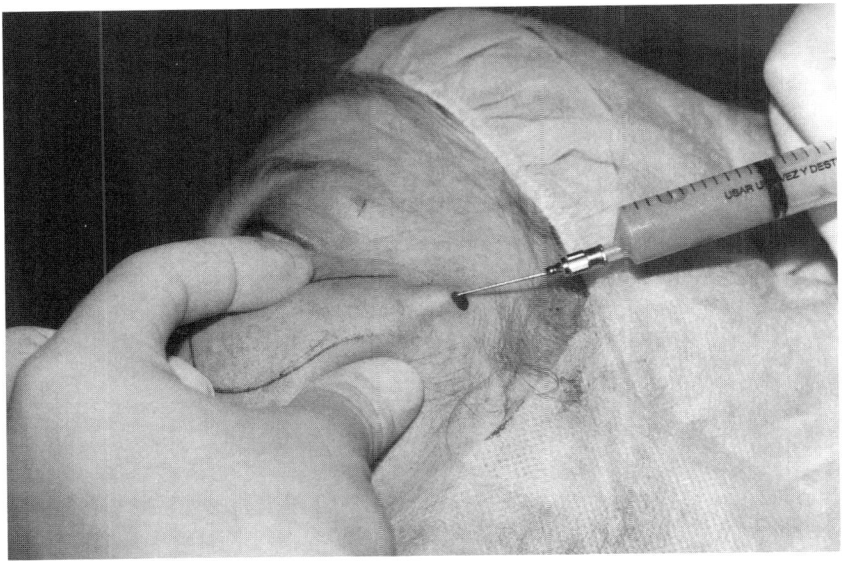

Figure 2 Needle entrance site should be slightly away from the area to be corrected.

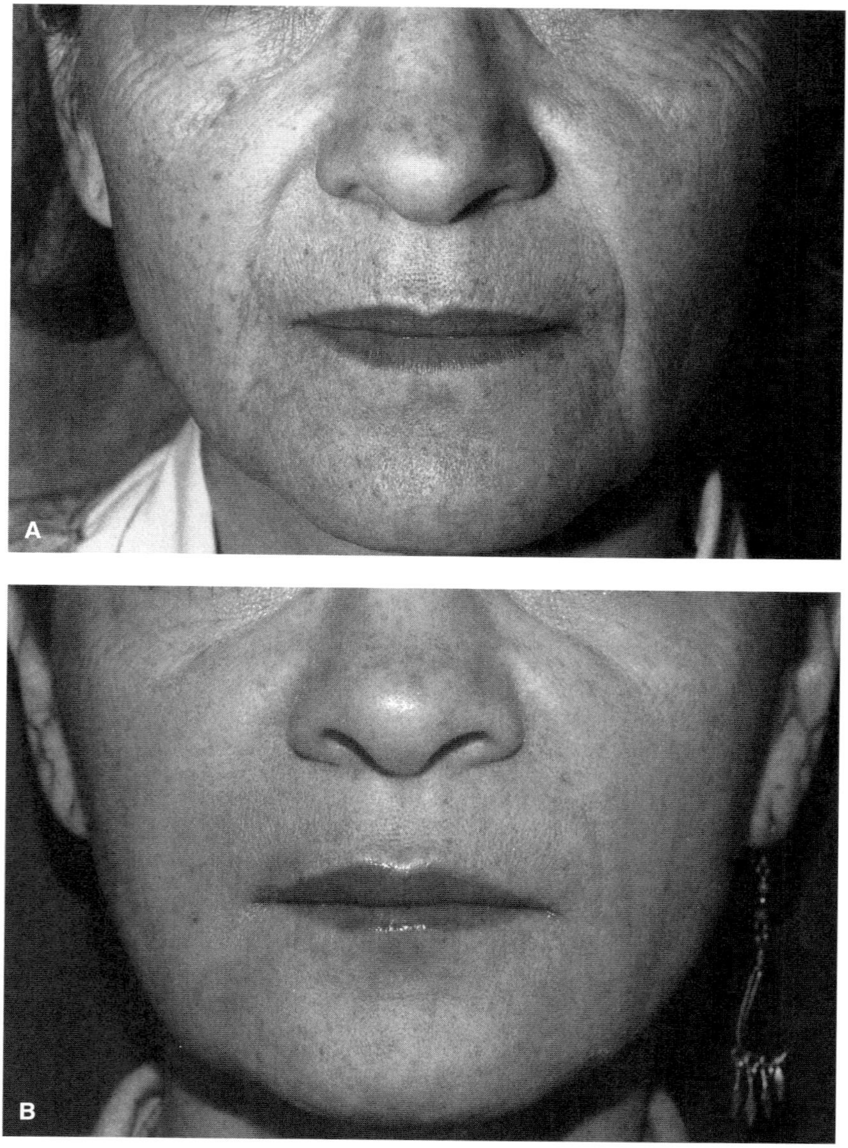

Figure 3 (**A**) Preoperative photograph. (**B**) Postoperative injection into lips. There is very little reabsorption of the implant.

A. Liposuction—Harvesting Site

The most frequent complications are related to the harvesting area. After the extraction of fat tissue, the patient may undergo mild bruising, slight edema, and hypesthesia.[10,11]

There are two different possibilities which may affect the complications: (1) The implant may be indicated together with a liposculpture procedure, in which case the harvested area is part of liposculpture procedures and will undergo the typical complications of such operations. (2) If the extraction of fat tissue is performed only to be used in the implant and the anatomical area involved is very small, thus reducing the complications.

B. Implantation—Recipient Site

The complications observed are slight swelling, minor hematoma, redness, and discomfort with pressure, all of which disappear within 48–72 h. This is similar to complications observed by other authors using the autologous and heterologous implants.[6,12,13]

Of all the areas treated, the lips have been associated with the most discomfort. The application of ice on the recipient site postoperatively is recommended.

IV. DISCUSSION

The major disadvantage of autologous fat transplantation is the loss of volume, which is quite variable.[8,12,14,15]

Large collections of fat are subject to liquefaction, central necrosis, and cyst formation. This, coupled with technical variations, are the reasons for the loss of volume and the unpredictable results.[8] To counteract these complications, autologous fat tissue is mixed with PMMA microparticles. PMMA is biocompatible, inert, and already in use in medicine both in opthalmology[16,17] and orthopedics.[18–20] Recently, PMMA has been used in commercial preparations mixed with collagen or with other synthetic polymers.[5–7,21] The loss of volume due to fat resorption can be offset by these solid microspheres.

Histopathological studies were done on patients implanted with autologous fat and PMMA microspheres in the abdominal wall subcutaneous tissues. Ten weeks after implantation, inflammatory cells were present and fibrous connective tissue surrounded the microspheres and the fat cells. The fat cells tend to be replaced by fibrous connective tissue, which also surrounded the microparticles.

There are advantages of this mixed implant over bovine collagen and PMMA. One of the benefits resides in the autologous character of the carrier (fat), which avoids hypersensitivity and allergic reactions. In volume retention, the permanency of some of the fat cells constitutes a positive element in comparison with the total resorption of the collagen.[22,23]

Another implant which gives good results in Bioplastique (methyl-methylpolysiloxane microparticles, Uroplasty, Inc., Minneapolis, MN),[2,24,25] but indications for the use of this substance are limited to the nose, cheek bones, and chin. This product is not easily implanted, since it should not be injected into superficial tissue owing to the large particle size, which might cause its extrusion or a hard mass can be palpated if too much material is injected.[2,24,26]

Goretex (extended polytetrofluroethylane) (Collagen Corp., Palo Alto, CA) has been in use for more than 20 years for vascular grafts, abdominal wall repair, and facial implants.[27,28] No allergic reactions to the product have been reported so far. In the case of facial implants, Goretex is used in the form of sheets or threads. For the implantation of sheets, a surgical procedure is required, which demands more time and may result in more complications. The threads, which are easy to implant, are indicated for lips and nasolabial folds, but the main disadvantage of this type of implant is that it is palpable. The edges of the sheet are not sharp, but there may be discomfort and palpability. Contraction may occur, and if the implantation is superficial, there is the risk of infection and later extrusion.[27,28] Another of its drawbacks is the difficulty in correcting the implant position if there should be patient dissatisfaction.

V. CONCLUSIONS

Although there is a wide range of injectable fillers available for the surgeon, in our experience, mixed lipofilling using PMMA with autologous fat is an excellent option which meets requirements for the ideal injectable filler. This is especially true if the correction is associated with a liposculpture procedure where autologous fat is readily available. Mixed lipofilling requires less necessity for reinjection than lipofilling alone.

REFERENCES

1. Grabb WC, Smith JW. Plastic Surgery. 3rd ed. New York: Little Brown, 1984, pp 93–107
2. Mak K, Toriumi DM. Injectable filler materials for soft-tissue augmentation. Otolaryngol Clin N Am 1994;27(1):211–222

3. Klein JA. The tumescent technique for liposuction surgery. Am J Cosm Surg 1987; 4(4):263–267
4. Fournier PF. Microlipoextraction et microlipoinjection. Rev Chir Esthet Langue Francaise 1985;41:36–40
5. Lemperle G, Ott P, Charrier J, Hecker J, Lemperle M. PMMA microspheres for intradermal implantation: Part I, Animal research. Ann Plast Surg 1991;26:57–63
6. Lemperle G, Hazan-Gautier N, Lemperle M. PMMA microspheres (Artecoll) For skin and soft tissue augmentation: Part II: Clinical investigations. Plast Reconstr Surg 1995;96(3):627–634
7. McClelland M, Egbert B, Hanko V, Berg RA, DeLustro F. Evaluation of Artecoll polymethylmethacrilate implant for soft tissue augmentation: biocompatibility and chemical characterization. Plast Reconstr Surg 1997;100(6):1466–1474
8. Neichajev I, Sevchuk O. Long term results of fat transplantation: clinical histological studies. Plast Reconstr Surg 1994;94(3):496–506
9. Ersek RA, Beisang AA. Bioplastique: a new biphasic polymer for minimally invasive injection implantation. Aesth Plast Surg 1992:16(1):59–65
10. Ferreira J. Complications in liposuction: Am J Cosm Surg 1993:10(2):139–142
11. Vila Rovira L, Serra Renom JM. Liposuccion en Cirugia Plastica y Estetica. Barcelona: Salvat, 1987
12. Chajchir A, Benzaquen I. Fat grafting injection for soft tissue augmentation. Plast Reconstr 1989;84:921–934
13. Fournier PF. Autocollagen. Conference in the Simposio International de Lipodistrofias, Vth Congreso Internacional de Flebologia y Linfolgia, Buenos Aires, Argentina, Oct 1996
14. Guerrerosantos J. Autologous fat grafting for body contouring. Clin Plast Surg 1996;23(4):619–631
15. Ferreira J. Lipofilling. Meeting de Cirugia Cosmetica y Medicina Estetica Masculinas, Buenos Aires, Argentina, April 1998
16. Ravalico G, Baccara F, Lovisato A, Tognetto D. Postoperative cellular reaction on various intraocular lens materials. Opthalmology 1997;104(7):1084–1091
17. Strenn K, Menapace R, Vans C. Capsular bag shrinkage after implantation of an open-loop silicone lens and a PMMA capsule tension ring. J Catar Refract Surg 1997;23(10):1543–1547
18. Jensen ME, Evans AJ, Mathis JM, Kallmes DF, Cloft HJ, Dion JE. Percutaneous PMMA vertebroplasty in the treatment of osteoporotic vertebral body compression fractures: Technical aspects. Am J Neuroradiol 1997;18(10):1897–1904
19. Deb S, Vazquez B, Bonfield W. Effect of crosslinking agents on acrylic bone cements based on PMMA. J Biomed Mater Res 1997;37(4):465–473
20. Cho SH, Song HR, Koo KH, Park YJ. Antibiotic impregnated cement beads in the treatment of chronic osteomyelitis. Bull Hosp Joint Dis 1997;56(3):140–144
21. Szuchmacher A. Metacrilato—Curso Internacional Multidisciplinar, Face No Seculo XXI, XXXIV Congresso Brasileiro de Ciugia Plastica, Sao Paulo, Brazil, Dec 1997

22. Kononas TC, Bucky LP, Hurley C, May JW Jr. The fate of suctioned and surgically removed fat after reimplantation for soft-tissue augmentation: a volumetric and histologic study in the rabbit. Plast Reconstr Surg 1993;91(5):763–768
23. Marques A, Brenda E, Saldiva PH, Amarante M, Ferreira M. Soc Bras Cir Plast, Sao Paulo 1998;13:37–44
24. Ersek RA. Bioplastique: Specific technical advice on its use and possible complications. Aesth Plast Surg 1992;16:67–68
25. Ersek RA, Beisang AA. Bioplastique: a new textured copolymer microparticle promises permanence in soft-tissue augmentation. Plast Reconstr Surg 1991;87: 693–702
26. Nacul AM, Nacul MP. Uso de Bioplastique na correcao de contorno dos tecidos moles. Arquivos Catarinenses de Medicina 1994;23:114–115
27. Vinal M. El uso de PTFE—Experiencia personal. I1 Simposio Internacional Hacia el Siglo XXI de la Sociedad de Cirugia Plastica, Buenos Aires, Argentina, May 1997
28. Morretti E. Goretex: Curso Internacional Multidisciplinario: “Face no Seculo XXI” XXXIV Congresso Brasileiro de Cirugia Plastica Sao Paulo, Brazil, Dec 1997

25
Autologous Collagen and Fat Transfer

Julio A. Ferreira
South American Academy of Cosmetic Surgery and Gendarmeria Hospital, Buenos Aires, Argentina

The techniques of autologous collagen tissue filler and autologous fat transfer may be used alone or in combination. When used together, the transplantation is performed in two layers.

1. Autologous collagen is injected into the dermis.
2. Autologous fat is injected into the subcutaneous tissue.

Indications for autologous collagen and lipotransfer include:

1. Filling of defects
2. Rhytids
3. Rejuvenation of aging (i.e., hands, face)

These techniques may also be combined with dermabrasion or skin peel.[1]

I. FAT TRANSFER

A. Technique

Traditional liposuction procedures are used for fat extraction with the exception that minimum local anesthetics drugs are utilized to avoid adipocyte lysis (Fig. 1). The fat is retrieved with blunt-tipped, single-hole microcannulae or a 14-gauge needle. Prior to extraction the 10-, 20-, or 50-mL syringe is filled with 2 or 5 mL of saline. The fat is removed from the deep subcutaneous layer in an area close to the proposed injection site to prevent differences in the adipocytes.

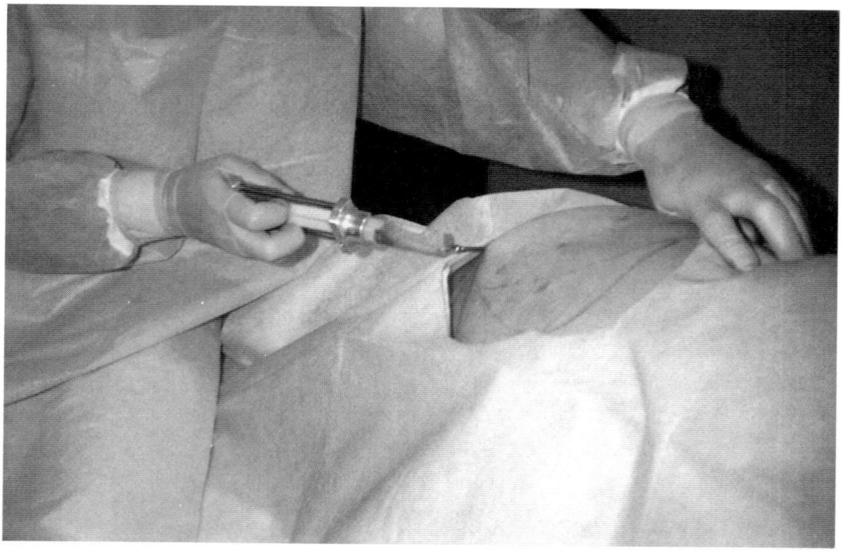

Figure 1 Traditional liposculpture techniques are used for lipoextraction except for the use of minimal local anesthetic.

The filled syringe is placed in the grid facing downward so that it can be decanted (Fig. 2). Several minutes later, the remains of blood products and saline are in the bottom and the fat is on the surface (Fig. 3). The liquid content is emptied and the syringe is filled again with saline solution, placed tip downward, and then decanted. This is repeated 8 to 10 times until the fat is free of blood products and the fat is pure yellow.

The fat is transferred to a 5-mL syringe using a traversing tip (Fig. 4). With a smaller syringe, the pressure produced on the plunger during injection is lower (Poiseuille's law). The needle for injection should be the same size as the one used for extraction. The bevel of the tip is face downward during injection (Fig. 5). After injection, the fat is minimally manipulated to leave the area smooth and even. The quantity of fat injected is the surgeon's decision, but 50% overcorrection can easily be handled. The author prefers to avoid overcorrection.

B. Fat Survival

Fat survival depends on vascularization of the area, the patient's characteristics, and preoperative and postoperative care. Most authors find 40–60% of the transplanted fat survives a year after transplant.

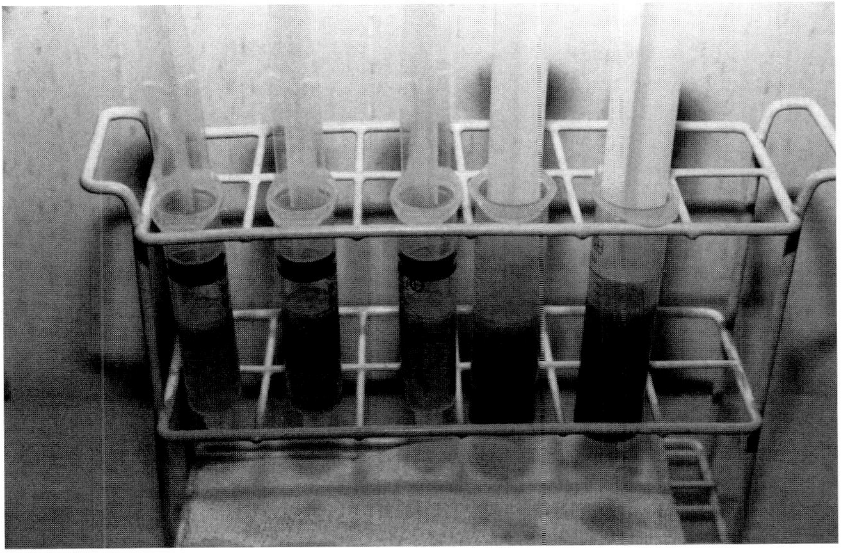

Figure 2 The filled syringe is placed in the grid with the tip facing downward.

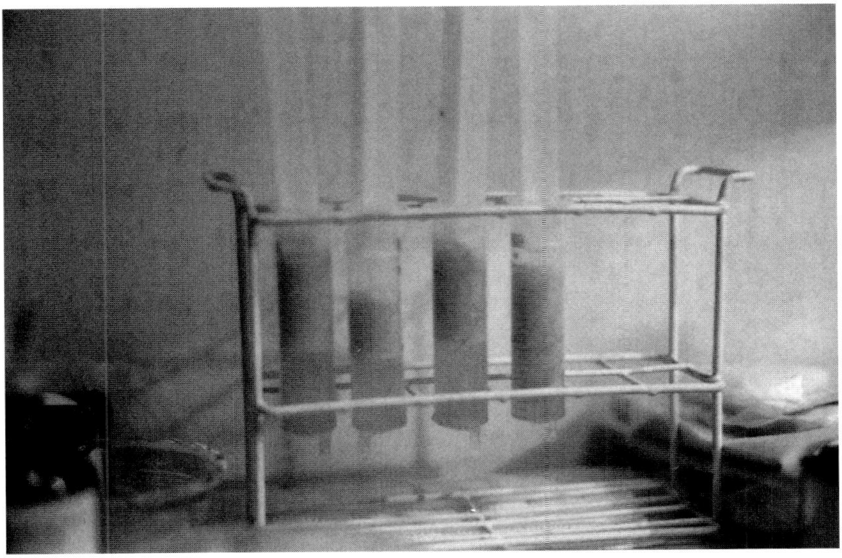

Figure 3 After several minutes, the blood products and saline are in the bottom and the fat is on the surface.

Figure 4 Fat is transferred to a 5-mL syringe so that the plunger pressure is lower during injection.

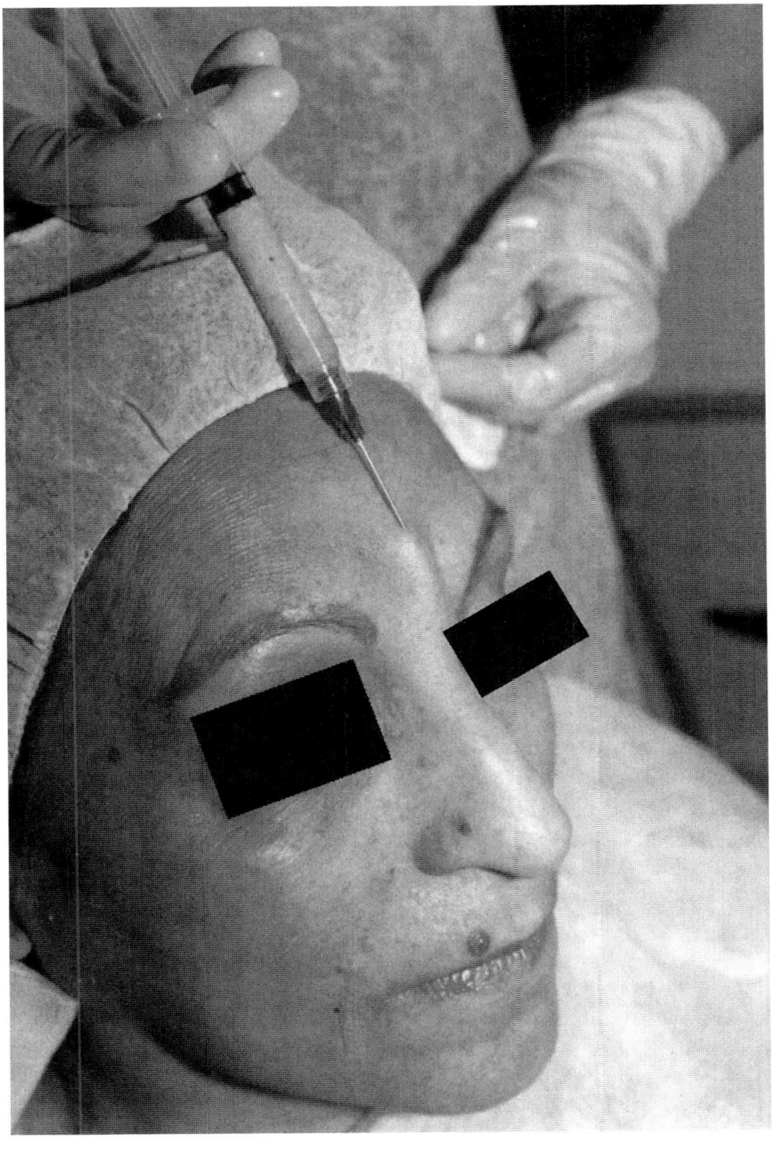

Figure 5 Injection of fat is performed with the bevel of the tip facing downward.

II. AUTOLOGOUS COLLAGEN FILLING

When extracted adipose tissue is subjected to true lipolysis, there is cell wall degradation, empty adipocytes, lack of vessels, and residual connective tissue. This inert tissue (collagen) can be obtained by ultrasound-assisted liposuction and/or ultracentrifugation.[2]

A. Technique

The author extracts autologous fat with a syringe using conventional liposculpture techniques. The tissue is washed with distilled water, which breaks the cell wall owing to different concentrations of sodium and potassium inside and outside the cell. Lavage is repeated 8 to 10 times until the tissue is free of blood products.

The tissue is treated with ultrasound (Fig. 6) to lyse the cell walls through the cavitation effect. Depending upon the power of the ultrasound instrument, the treatment lasts 3–45 min until there is a dispersed pasty whitish material[2] (Fig. 7).

The syringe is placed in a 5000-rpm centrifuge for 5 min. Layers separate out with centrifugation.

1. Upper layer is greenish yellow in color and contains fatty acids in suspension.
2. Middle layer is whitish in color and containing the autologous collagen.
3. Lower layer is reddish in color and contains blood products, distilled water, and residual anesthetic.

The upper and lower layers are discarded and the middle layer is transferred to a 2-mL syringe.

The material is injected using a small needle in the dermis. If the correct plane is used, the skin turns white and has an orange peel appearance. Residual tissue in the syringe may be frozen for use at a later date. Frozen collagen that darkens in color should not be used, since oxidation has occurred and the tissue is rancid.

III. ADVANTAGES

The use of autologous fat and collagen cannot result in immune reactions, has no known harmful effects, and has a relatively endless supply without cost. The injections can be repeated as many times as needed, and the tissue can be preserved in the freezer.

IV. COMPLICATIONS

In case of hematoma and bruising (Fig. 8), the author uses external ultrasound to the area and 0.5% vitamin K cream with 0.3 g polysulfate of mucopolysaccharide gel are applied.

Figure 6 Ultrasound treatment of extracted autologous fat.

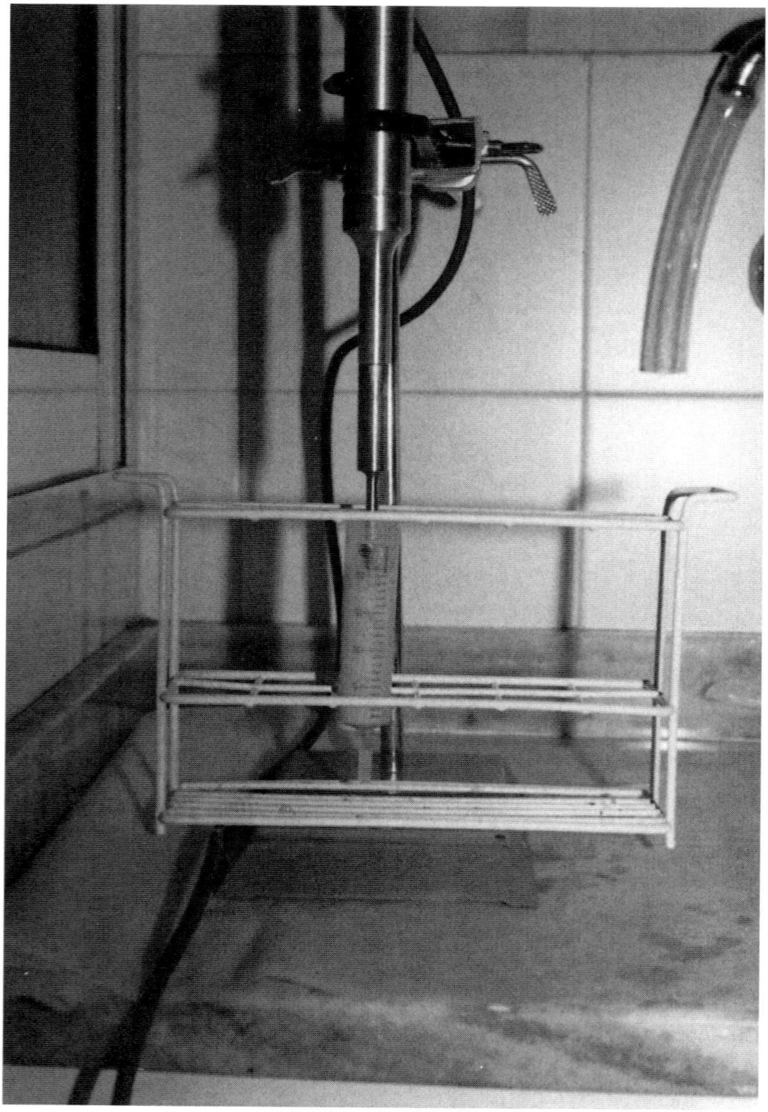

Figure 7 Pasty whitish material at completion of ultrasound treatment.

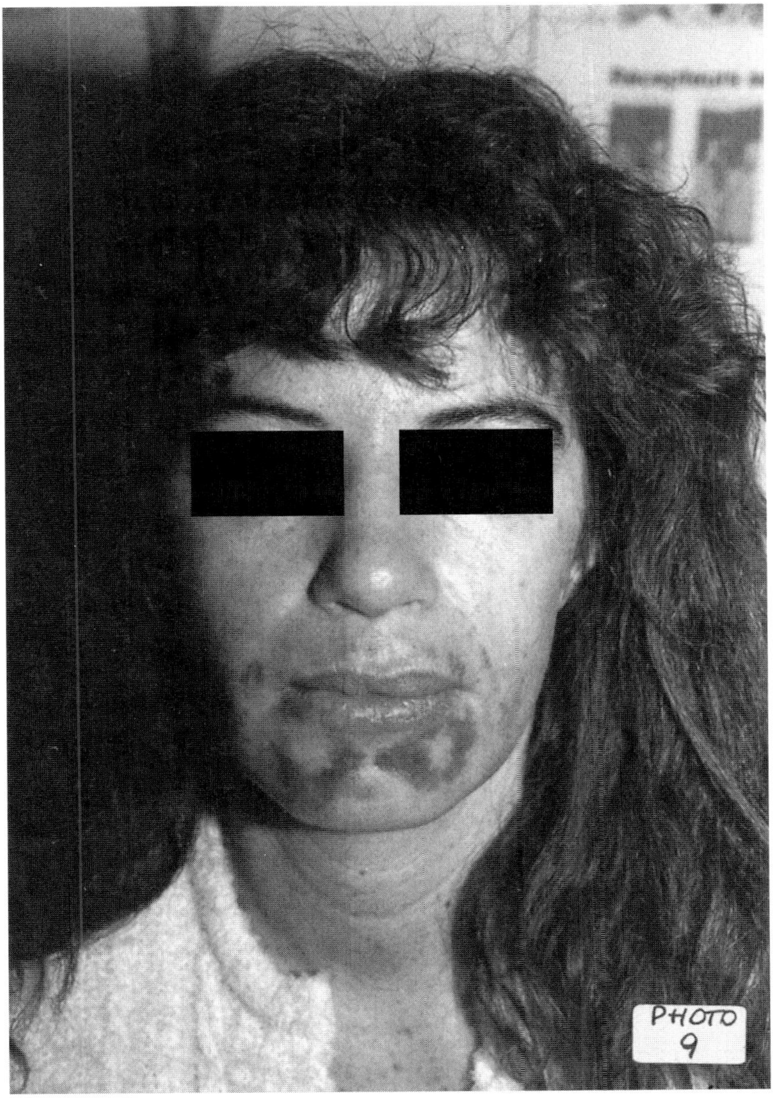

Figure 8 Postoperative bruising.

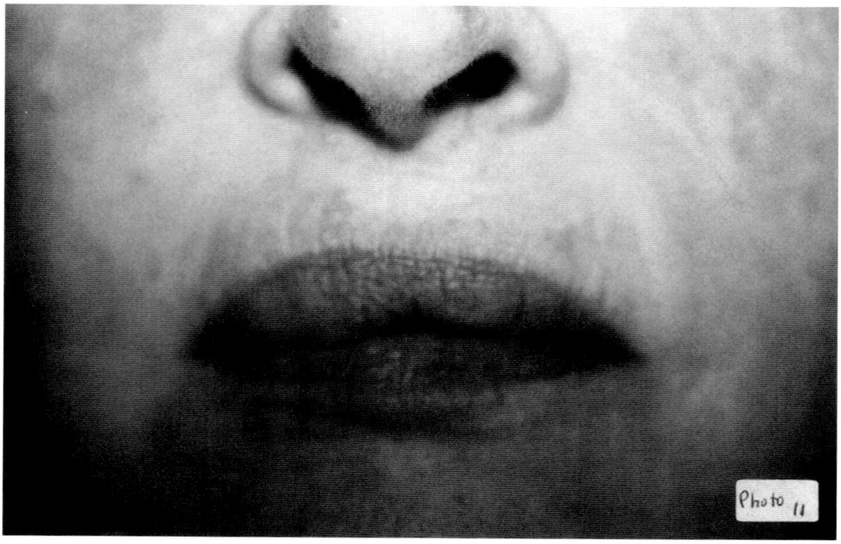

Figure 9 Persistent nodules in lips causing distortion.

Persistent nodules (Fig. 9) are treated with 3 Mhz ultrasound for 5 min daily for 1 week.

The author has not encountered any infections, abscess, persistent edema, cyst, or large hematomas. To prevent loss of vision from lipoinjection on the glabellar area, a blunt-tip cannula should be used to avoid intravascular injection.

V. DISCUSSION

The combined use of autologous fat and autologous collagen allows better refining of esthetic corrections than either method alone (Fig. 10). Each type of tissue has the problem of resorption, but reinjection allows long-term satisfactory results.

REFERENCES

1. Fournier PF. Auto-lipo-implants de collagene. Rev Chir Esth Langue Francaise 1990;15(59):39–43
2. Zucchi M. Ultrasonic liposculpturing. Aesth Plast Surg 1992;16(4):287–298

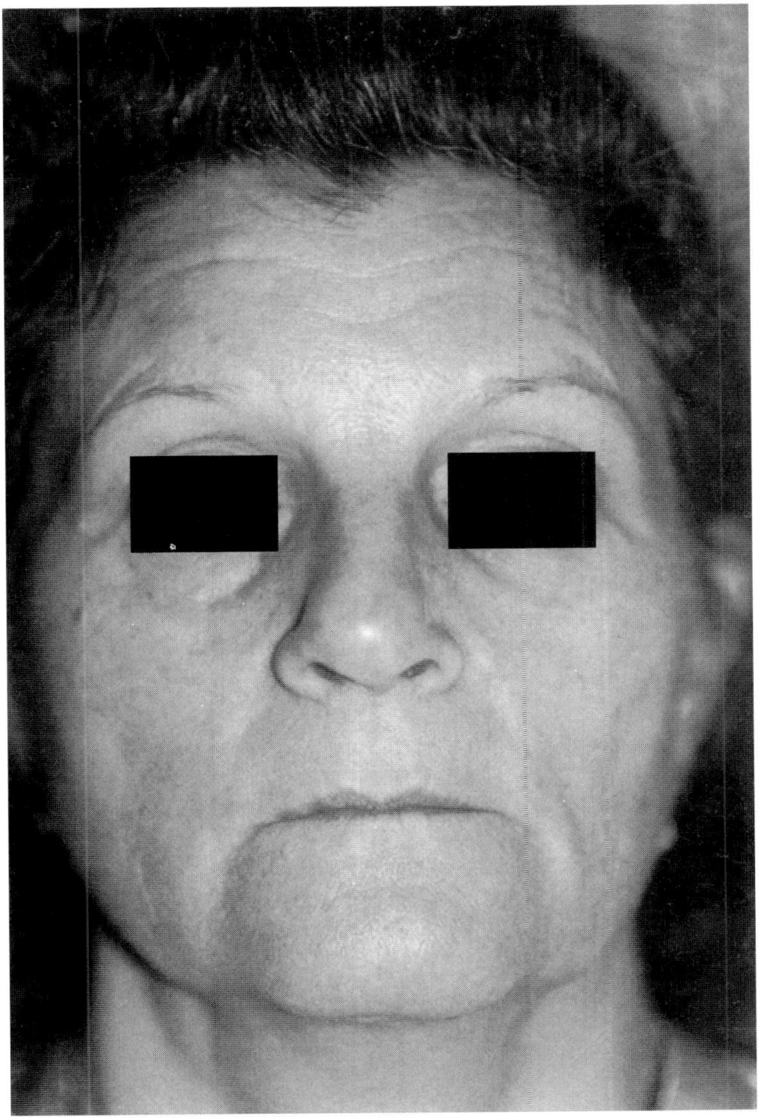

Figure 10a 55-year-old female prior to treatment.

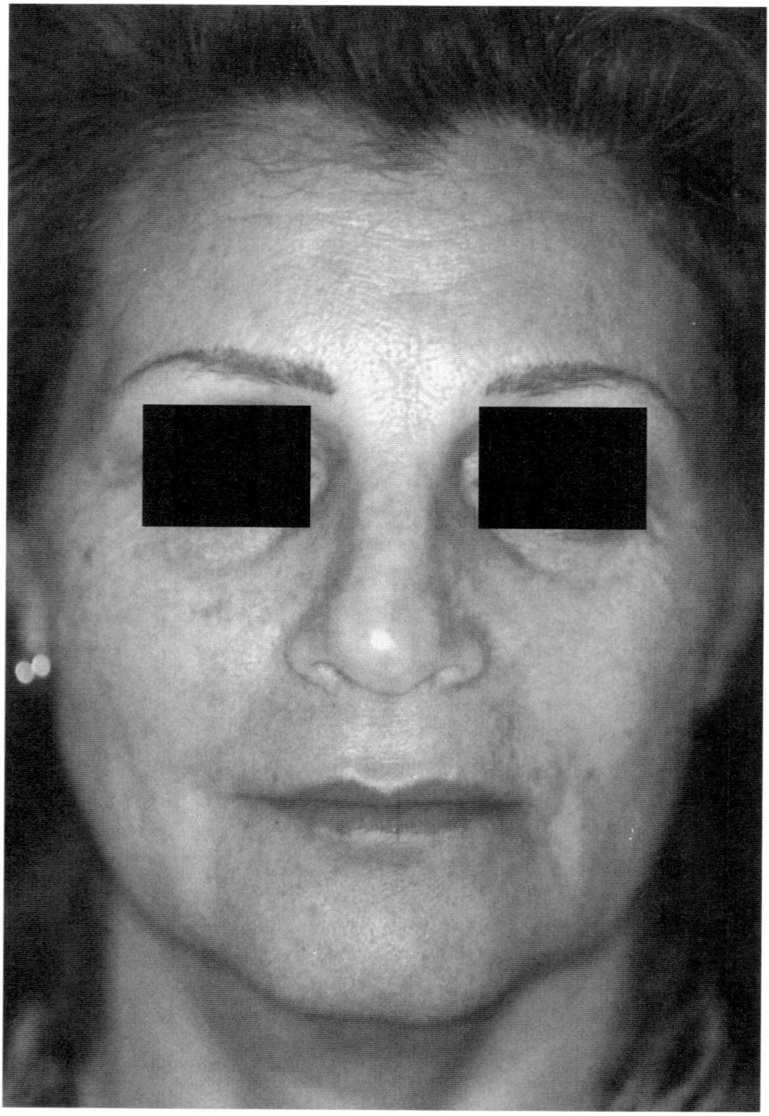

Figure 10b Six months after fat transfer, autologous collagen filling, and phenol peel.

26

New Consideration in Fat Transfer: A Possible Role for Maintaining Interstitial Protein to Reduce Shrinkage of Transferred Volume

Mitchell V. Kaminski, Jr.
North Shore Cosmetic Surgery and Wellness Center, Northfield, Illinois

James E. Fulton
Newport Point Medical Center, Newport Beach, California

John J. Wolosewick
University of Illinois, Chicago, Illinois

Cosmetic surgeons continue to search for a soft tissue filler that is safe, economical, and reliable to enhance body features esthetically. This substance must be nonallergenic. Autologous fat would be ideal, but it has not proven itself to be reliable enough because of significant problems with reabsorption. Other problems associated with fat transfer include oil cysts and abscesses secondary to fat cell necrosis. Various fat transfer techniques have been evolving from the later nineteenth century to the present day.[1–5] Research into methodologies to prevent absorption of transplanted volume slowed in the 1960s as prosthetic implants became widely used, but research increased again in the 1990s in the wake of the silicone controversy.[6] In the 1980s, liposuction was popularized. In the 1990s, liposuction was made safe following publication by Klein of the tumescent technique for regional anesthesia.[7,8] These methodologies make autologous fat readily available. With the work of Coleman and more recently Fulton's work, interest in fat grafting has been rekindled.[9]

There are several considerations regarding graft survival that are established. For example, it is accepted that harvesting must be aseptic and low pressure and only small pearls of transplanted fat can be successfully nourished by recipient tissues. It is also helpful to withdraw the injection cannula during extrusion to maximize the fat's exposure to the host bed. Layering each homograft between healthy tissue is necessary. The diameter of the cannula used to harvest and implant the fat must, therefore, be 4 mm or less and hemostasis meticulous.[10]

When used for large-volume transfers, such as breast augmentation, there is the added late problem of "eggshell" calcifications seen on subsequent mammograms.[11] Although distinct from the fleck and chip-like microcalcifications associated with tumors suspicious for cancer, they nevertheless are a clinical annoyance and another reason why cosmetic surgeons have not enthusiastically promoted autologous fat transfer for breast augmentation. Trauma to cells during harvesting and processing should be minimized. Potentially traumatic events include washing and filling techniques. Despite the annoying shrinkage problem, interest has remained high in autologous fat transfer, because occasionally the transplanted tissues take beautifully. Further, improvement in cell survival has been observed when fat has been frozen/thawed and used at a later date or centrifuged as described by Coleman.[9] Finally, when plasma and platelets are added back to fat prior to large-volume transfer, survival improved from an expected 50 to 60% of enhancement volume to 70–80%. Although not obvious, these two observations may be consistent and could point to a heretofore unappreciated factor in fat transfer survival; namely, the presence of harmful accumulations of intercellular/intracellular water due to wash down of intercellular proteins by current processing protocols.

I. WATER BETWEEN AND WITHIN FAT CELLS

Increasing the distance between cells caused by processing tissues with protein-free solutions impairs oxygenation and nutrition at a critical time. Freezing and thawing will remove water from and between the cells. Restoring protein to the fat transfer solutions used to harvest cells will keep fluid from accumulating in the interstitial space and from entering the cell. Could it be that the harvest and wash phases currently used for fat transfer are associated with overhydration of the tissues?

With regard to freezing, each time a sample is frozen and thawed, more subnatant fluid develops. This fluid can only be from lipocytes or their interstitial spaces. This dehydration by freezing phenomenon can be observed when one compares a package of lunch meat that has never been frozen to one that has. The meat from the unfrozen package is plump and there are no water droplets in the sealed serving bag. However, frozen/thawed lunch meat is rubbery and the package contains droplets of water. It has been observed by one of the authors that when autologous soluble protein obtained by plasmapheresis and platelets are added to the fat transfer sample, lipocytes quickly gather toward the supranatant and the size of the infranatant increases (see Chap. 18). The original hypothesis was that there would be an improvement in graft survival if autologous plasma containing growth factors and platelets are added back to the fat prior to injection. Early experience demonstrated that instead of the expected 50–60% loss in volume at 90 days, there is instead a 70–80% survival.

Rather than being a series of metabolic events related to growth factors, it appears that freezing, adding protein, or centrifugation might accomplish similar things. Namely, harvested fat tissues exposed to an extremely low-colloid osmotic pressure (COP) due to the harvesting and washing techniques, imbibe and accumulate interstitial and intracellular water. A fair amount is known about the physiological principles that govern the movement of fluid from one space to another within the body.[12] Over 100 years ago, a formula known as Starling's equation was devised.[13] Kaminski[14,15] has used Starling's equation to understand the effect and repair of edema secondary to profound hypoproteinemia. After reviewing Fulton's[15a] work, it became apparent that maintaining colloid pressure might be added to the list of steps taken to promote graft survival in fat transfer. If successful, the need for plasmapheresis to obtain soluble protein might be eliminated. Also, instead of resuscitating the sample by restoring COP with plasmapheresis protein at the last step prior to transplantation, COP can be maintained by simply adding soluble protein in the form of human serum albumin (HSA) to each step from infiltration with Klein's solution through the wash steps.

II. SOLUBLE PROTEIN IS ABUNDANT IN THE INTERSTITIAL SPACE

What is potentially wrong with tumescing and then washing fat with isotonic solutions? Normally, the lipocyte is bathed in interstitial fluid which is rich in soluble protein. It is the gel of the gel-sol matrix. Normal circulating total protein is between 6.2 and 9.0 g%. In a healthy individual, 60% of total body albumin is in the interstitial space. The total protein of interstitial fluid is 4 g% and is dynamic. Every 24–48 h, each molecule of albumin makes its way from the heart, out of circulation through a pore in the capillary membrane, across the interstitial space, and then back to circulation through lymphatic channels. The presence of soluble protein in circulation and in interstitial fluid is profoundly important.[16,17]

First, large molecules on one side of a semipermeable membrane that cannot easily cross will create an osmotic gradient. Thus, a –7 mm Hg pressure exists for lipocytes and all other cells of the body packing them tightly together. Second, soluble protein in the interstitial space causes series resistance to interstitial fluid flow. Further, if this protein is washed out, there is no negative pressure surrounding cells. Water will disperse the cells even if it contains sodium and chloride ions at isotonic levels. Intracellular protein constituents are a dominant osmotic force. Nevertheless, some fluid could be drawn across the cell wall and the cell increases in size. These cells would be more fragile. Following fat transfer, absorption of interstitial fluid will occur and a portion of the cells subjected to this trauma will rupture as evidenced by shrinkage and calcification within the breast tissue of patients that have had breast augmentation using autologous fat as it is presently harvested and prepared.

III. DIFFERENCE BETWEEN OSMOTIC PRESSURE AND ONCOTIC PRESSURE

The reason 5% dextrose or salt is added to an intravenous infusion solution is not to provide sodium and chloride or dextrose. Rather it is to increase osmolarity toward isotonicity to prevent red cells from taking up water and exploding. Osmolarity/isotonicity is different than oncotic pressure created in solutions containing colloid. When a molecule is greater than 10,000 D, it is called a colloid. It is capable of generating an oncotic pressure if it is restricted to one side of a semipermeable membrane because of its size. Very small molecules and ions such as sodium, potassium, glucose, and urea easily cross the capillary membrane. Therefore, they cannot create an oncotic gradient. Osmolarity is measured by freeze point depression. The greater the number of particles in solution (not their size), the colder the solution must be before it will freeze. On the other hand, colloid restricted to one side of a semipermeable membrane creates gradients that are measured in millimeters of mercury. However, COP stands for colloid osmotic pressure, not colloid oncotic pressure. This can be a source of confusion and is worth a brief explanation.

A. Why Colloid Oncotic Pressure Becomes Colloid Osmotic Pressure

The Landis and Pappenheimer equation for determining colloid osmotic pressure (COP) takes into account the fact that all soluble proteins, whether albumin, globulin, or fibrinogen, are highly negatively charged.[12]

$$COP = 2.1(TP) + (0.16\ TP^2) + 0.009\ TP_3$$

In so being, positively charged sodium ions surround the core protein. Sodium then attracts and holds water. The effect of accumulating more fluid on one side of a semipermeable membrane has two parts. First, the oncotic force generated by the large protein or colloid molecule. Second, these colloid molecules also hold sodium against its tendency to distribute on each side of the membrane equally. Therefore, there is a boost in the pressure gradient. This is why the name is changed from colloid oncotic pressure to colloid osmotic pressure.

Total protein (TP) which is the sum of albumin, globulin, and fibrinogen generates COP. Each colloid molecule, however, has slightly different properties. This is described by the Gibbs-Donnan equation.[12]

$$[Na^+]_A \times [Cl^-]_A = [Na^+]_B \times [Cl^-]_B$$

Globulin and fibrinogen molecules are very large: 150,000 and 400,000 daltons, respectively. Albumin is only 69,000 D. Here too, it is the number of molecules in solution, but not their size, that are held on one side of the semipermeable membrane that generate COP. Albumin is the most common colloid molecule and in addition, gram for gram, it creates the most pressure. That is, 1 g of albumin will have twice as many molecules as 1 g of globulin and five times as many molecules as 1 g of fibrinogen. This is a small but worthwhile point to make when selecting a soluble protein solution to add to the steps in harvesting and washing cells for fat transfer. An albumin solution will give the greatest oncotic pull when compared to a more expensive globulin solution. Starch molecules, as found in hetastarch or dextran, should never be used for this purpose. They are on average very large and cannot be evacuated through the lymphatics. They are effective when added to the circulatory volume but would cause sustained localized edema if placed in the interstitial space.

B. Starling's Equation

At first glance, Starling's equation can appear daunting if not discouraging.[14,18] It nevertheless is straightforward and is a formula that represents little more than hydrostatic pressure pushing fluid through the capillary pore (P_c through δ)

versus COP forces (π_c) holding fluid in circulation, and the rate of fluid flow across the gel-sol matrix (Kf is inversely proportional to π_{if}) and back into circulation via lymphatic channels (Q_{lymph}) (Fig. 1). When all of these factors are combined, the entire equation is written as:

$$J_v = Kf\,[(P_c - P_{if}) - \delta\,(\pi_c - \pi_{if})] - Q_{lymph}$$

Where J_v is interstitial fluid flow.

Histological features of adipose cells and similar to the schematic drawing in Figure 1 as connective tissue in a thick-spread preparation of pig subcutaneous tissue are seen in Figure 2.

1. Components of Starling's Equation

a. $P_c - P_{if}$*: Hydrostatic Pressure*

The pressure in the capillary minus the opposing pressure in the interstitial space is known as the hydrostatic pressure. Central venous pressure (P_c) is the pressure pushing fluid across the endothelial membrane throughout the body. In the lung, the pulmonary capillary wedge pressure describes the hydrostatic

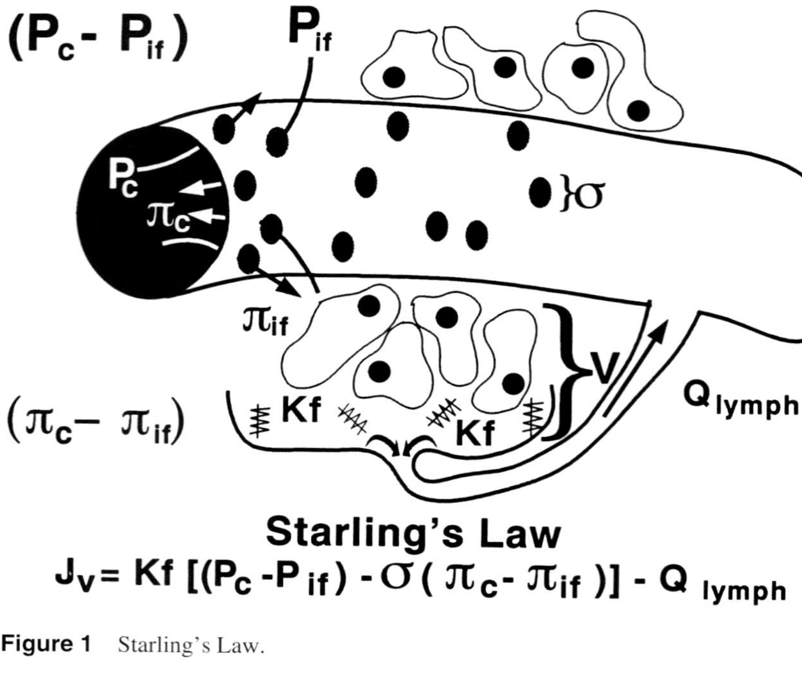

Figure 1 Starling's Law.

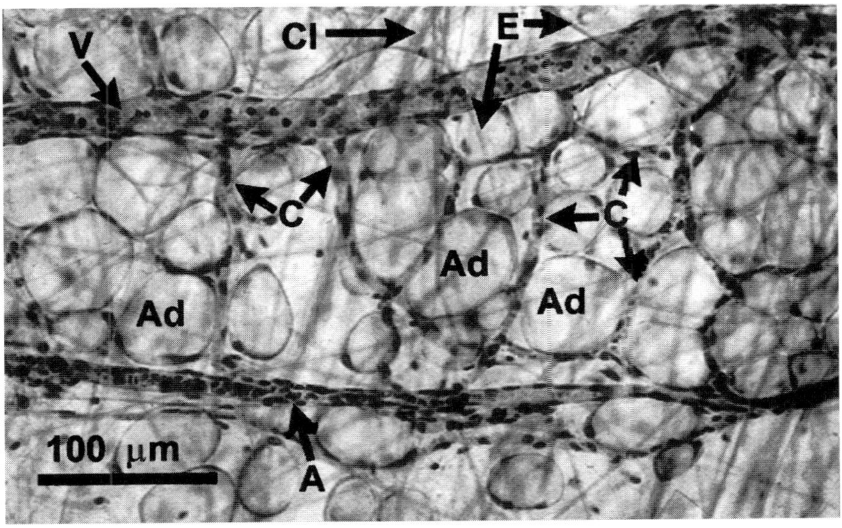

Figure 2 Pig subcutaneous tissue stained with Frankel's orcein method for connective tissue elements. Even though a spread preparation disrupts the in vivo architecture, numerous features are noteworthy. The microvasculature associated with adipose cells is illustrated with an arteriole and venule seen coursing through the extracellular matrix composed largely of collagen and elastic fibers. Of particular interest are the capillary networks (outlined by the arrowheads) that encompass or border much of each adipocyte. V, venule; C, capillary; Ad, adipocyte; A, arteriole; E, elastin. (The images were photographed from slides in the Tissue Biology slide collection of the Department of Anatomy and Cell Biology, College of Medicine, at the University of Illinois at Chicago.)

pressure forcing fluid across the capillary membrane into pulmonary interstitial spaces. These hydrostatic forces are opposed by a rather low but present pressure in the interstitial space (P_{if}). The hydrostatic pressure then is $P_c - P_{if}$.

b. $\pi_c - \pi_{if}$: *Oncotic Pressure and COP*

The total protein in circulation creates colloid osmotic pressure (π_c). At any given time at any given pore in the vascular endothelium, there is more protein concentrated in circulation than there is at that site in the interstitial space. Thus π_c creates a constant negative force within the body, holding fluid in circulation and keeping interstitial fluid flow to a minimum, packing cells together, and preventing edema. π_c Must always be higher than P_c. This is called the COP/wedge gradient in the lung. If P_c increases, as it does in congestive heart failure or fluid overload, π_c must increase or there will be a massive flow of fluid from circulation into the interstitial space. This increased interstitial fluid flow will wash down the protein in the interstitial space, causing it to translocate back in circulation via lymphatics and raise π_c in an attempt to maintain a favorable COP/wedge gradient and avoid pulmonary edema.[19] The protein in the interstitial space (π_{if}) therefore, has two purposes. The first purpose is to act as a reserve. In a healthy individual, 60% of all the albumin in the body is is in the interstitial space. If that individual has a catastrophic ongoing blood loss and is resuscitated with only packed red blood cells and saltwater without colloids, we have the equivalent of an ongoing plasmapheresis. This can be tolerated for some time, because as stated above, π_{if} moves through the lymphatic system back into circulation in an attempt to restore circulating colloid (π_c). If the bleeding and colloid-free resuscitation continues, π_{if} reserves are depleted, decreasing the gel of the gel-sol matrix. As fluid now freely leaves circulation and rapidly traverse the interstitial space with less gel, lymphatic clearance cannot keep up and edema develops. Edema is mobilized from muscle by contraction. Edema that develops in dependent fatty tissues is mobilized by change of position. It has been observed that neither hydrostatic pressure nor low colloid osmotic pressure will produce edema if interstitial protein is normal or near normal. That is, greater than 1.8 g% with normal π_{if} being between 2 and 4 g%. Therefore, it appears that the target total protein in harvested and washed specimens of fat for transfer should be equal to or greater than 1.8 g%.

c. Q_{lymph}: Lymph Flow

Lymphatic channels are the back door for fluid from across the capillary pore through the interstitial space and then back into circulation. Obstruction to lymphatic clearance creates lymphedema.

d. Kf: Filtration Coefficient

Kf is the filtration coefficient and is related to π_{if}. That is, how much gel there is in the gel-sol matrix. As π_{if} decreases, Kf will increase. The net effect for either profound hypoproteinemia or an elevated central venous pressure (CVP) is less gel in the interstitial fluid. This results in edema and separation of cells with loss of structural integrity of that organ. Accumulation of edema in the interstitial space secondary to profound hypoproteinemia in vivo is called pseudoleaky membrane.[14] This is because there is no derangement in the pore membrane size. Rather edema accumulates because of the decrease in series resistance to transinterstitial fluid flow due to the decrease in π_{if}.

e. Sigma (σ): Pore Membrane Size

The size of the aperture in the semipermeable membrane is termed σ. It is different for every tissue in the body. For example, the blood-brain barrier is so impervious that many antibiotics will not cross it. On the other hand, the liver's σ is very large; large enough for albumin synthesis. Albumin must be unimpeded as it enters circulation and vice versa as it returns carrying noxious molecules for hepatic detoxification.

IV. CONCLUSIONS

There is potentially a significant difference between isotonicity/osmolarity and oncotic forces. Osmolarity is determined by the number of molecules in solution and generally refers to the sum of sodium, potassium, glucose, and urea molecules. These are very small and can traverse the vascular endothelium with a speed near that of water. Oncotic pressure, on the other hand, refers to much larger molecules that comprise total protein. There are albumin plus globulin plus fibrinogen. Albumin is the most common molecule and, gram for gram, the most potent in generating oncotic force. Oncotic pressure is due to the fact that the molecule is large and cannot easily cross a semipermeable membrane, thus setting up a gradient. Because proteins are highly negatively charged and draw sodium, which holds water, the abbreviation COP therefore stands for colloid osmotic pressure, not oncotic pressure alone.

After tumescing with classic Klein's solution, the subnatant of the harvested fat contains 1.1–1.2 g% protein. This is a significant drop from the normal levels of 2% to 4 g%. When one ampule of concentrated human serum albumin (12.5 g in 50 mL) is added to 1 L of Klein's solution, the harvested fat subnatant contains 2.6 g% of protein. Alternately, between 5 and 10 mL of human serum albumin can be added to a 60-mL harvesting syringe. Three washes of a sample that is harvested and treated in the standard manner will produce a subnatant protein of 2–0 g%. Thus, the protocol to avoid hypo-oncotic trauma due to removing COP from the harvested specimen is to add 12.5 g of human serum albumin (1 amp) per liter of Klein's solution and 18.75 g/L (11/2 amps) to each liter used to wash the specimen. This is in addition to attention to other recognized factors associated with graft survival. That is, the bed into which the fat is transferred must have a good blood supply. The size of the globule transferred is a function of both the size of the cannula that harvested and injected the fat, as well as the size of the allograft dispensed by the injecting gun. Adequate time must be allowed between each wash for the cells to pack above the subnatant. This process can be accelerated by centrifugation. Finally, any oil that is observed above the fat must be wicked off.

Attention to detail in minimizing trauma to tissue harvested for fat transfer is critical. It has not previously been recognized that the lipocyte is usually surrounded by a protein-rich solution producing a colloid osmotic pressure. This pressure should be maintained with the same diligence that one maintains isotonicity. This is accomplished by adding human serum albumin to each step of the harvesting and wash phases of the graft material.

REFERENCES

1. Shiffman MA. Principles of liposuction. Am J Cosm Surg 1997;14:331–334
2. Peer LA. Loss of weight and volume in human fat graft. With postulation of a "cell survival theory." Plast Reconstr Surg 1950;5:217–230
3. Agris J. Autologous fat transplantation: a three year study. Am J Cosm Surg 1987;4:95
4. Asken S. Autologous fat transplantation: micro and macro techniques. Am J Cosm Surg 1987;4:111
5. Alexander RW. Liposculpture in the superficial plane: closed syringe system for improvements in fat removal and free fat transfer. Am J Cosm Surg 1994;11:127
6. Connell EB. The exploitation of autoimmune disease—breast implant litigation and its dire implications for women's health (Editorial). J Women's Health 1998;7(3):329–338
7. Klein JA. The tumescent technique for liposuction surgery. Am J Cosm Surg 1987;4:263–267
8. Klein JA. Tumescent technique for regional anesthesia permits lidocaine doses of 35 mg/kg for liposuction: peak plasma concentrations are diminished and delayed 12 hours. J Dermatol Surg Oncol 1990;16:248–263
9. Coleman WP III. The history of liposuction surgery. Dermatol Clin North Am 1990;8:381–383
10. Jones JK, Lyles ME. the viability of human adipocytes after closed-syringe liposuction harvest. Am J Cosm Surg 1997;14:275–280
11. Bircoll M. Autologous fat transplantation: an evaluation of microcalcification and fat cell survivability following (AFT) cosmetic breast augmentation. Am J Cosm Surg 1988;5:283
11a. Coleman S. Long-term survival of fat transplants: Controlled demonstrations. Acstnet Plast Surg 1945;19:421–425
12. Guyton AC. Capillary dynamics and exchange of fluid between the blood and the interstitial fluid, In: A Guyton, ed. Text Book of Medical Physiology. 7th ed. Philadelphia: Saunders, 1986, p 348
13. Kaminski M, Haase T. Albumin and colloid pressure: Implications for fluid resuscitation, In: B Kaufman, ed. Critical Care Clinics: Fluid Resuscitation of the Critically Ill. Philadelphia, Saunders 1992:311–322

14. Kaminski M, Haase T. Use of albumin in total parenteral nutrition solutions: Understanding Starling's law and the resolution of hypo-oncotic edema. In: C Van Way, ed. Handbook of Surgical Nutrition. Philadelphia: Lippincott, 1992:272–282
15. Rosas M, Kaminski M, Ninos N, et al. Intestinal edema and decreased tensile strength during hypoproteinemia (abst). JPEN 1991;15(Suppl):28
15a. Fulton E Jr. Breast contouring by autologous fat transfer. Am J Cosm Surg 1992;9:273–279
16. Kramer G, Harms B, Gunther R. The effects of hypoproteinemia on blood-to-lymph fluid transport in sheep lung. Circ Res 1981;49:1173–1180
17. Kaminski M, Williams S. Review of the rapid normalization of serum albumin with modified total parenteral nutrition solution. Crit Care Med 1990;18:327
18. Civetta J. A new look at the Starling equation. Crit Care Med 1979;7:84–91
19. Rackow E, Fein A, Siegel J. The relationship of colloid osmotic pressure gradient to pulmonary edema and mortality in critically ill patients. Chest 1982;82:4

27

Medical Legal Aspects of Autologous Fat Transplantation

Melvin A. Shiffman
Private Practice, Cosmetic and Reconstructive Surgery, Tustin, California

The few complications encountered in autologous fat transplantation make this procedure low risk for litigation. Autologous tissue has less potential problems than most of the presently available injectable materials.

There is need for a complete explanation to the patient of the procedure and its material risks and complications as well as any viable alternatives and their risks and complications.

When injecting the glabellar area, a blunt-tipped needle or cannula of sufficient size should be utilized.[1] The patient should be informed of the four reported cases of unilateral blindness and one case of several central nervous system damage following injection of fat into the glabellar area.[2–4]

Fat transplantation to the breast should be undertaken with extreme caution. Complications such as oil cyst(s) and microcalcifications may occur and require a thorough explanation to the patient preoperatively. There are surgeons who will testify to the "fact" that it is an unwarranted and dangerous procedure because of the possibility of calcifications in the breast which cannot be distinguished from possible malignancy. They will also testify that the results of augmentation are minimal. There have been at least four claims filed against physicians performing fat augmentation of the breast.[5] One was lost at trial, two others settled out of court, and one tried and won. The Doctor's Company will not cover this procedure under its medical malpractice insurance.

There is a need for accurate and honest research in evaluating the complications of augmentation of the breast with autologous fat. Specifically, the study should include the actual risk of microcalcifications, the procedures to distinguish this from possible breast malignancy, and the long-term effect of these calcifications (both progression of calcifications and psychological effect to the patient).

REFERENCES

1. Skouge JW. Autologous fat transplantation in facial surgery. In: WP Coleman III, CW Hanke, TH Alt, S Asken, eds. Cosmetic Surgery of the Skin: Principles and Techniques. Philadelphia: Decker, 1991, pp 239–249
2. Coiffman F. Lipoinjection complications. In: V Henderer, ed. Plastic Surgery 1992. Vol 2, Amsterdam: Excerpta Medica, 1992, pp 759–760
3. Niechajev F, Sevcuk O. Long term results of fat transplantation: clinical and histologic studies. Plast Reconstr Surg 1994;94:496–506
4. Teimourian B. Blindness following fat injections. Plast Reconstr Surg 1988;82:361
5. Gorney M. Fat injection to breast compromises safety. Commenting on calcifications. Cosm Surg Times 1998;2(6):3

28

Editor's Analysis of Fat Transfer Techniques

Melvin A. Shiffman
Private Practice, Cosmetic and Reconstructive Surgery, Tustin, California

This book shows the varieties of techniques used by various cosmetic surgeons in performing autologous fat transfer. In order to dissipate some of the confusion encountered by surgeons wishing to start performing fat transfer or by surgeons who have had unsuccessful attempts at fat transfer, the editor has analyzed these techniques to see which are most useful in obtaining a high percentage of fat retention.

1. *Donor Site*: There is a consensus in the literature that fat cells with alpha receptors (least likely to respond to weight change) are in fat deposits which are preferred as donor sites. Those areas that are easily accessible, tend to have excess fat, and are not fibrous include the abdomen, outer thighs, and hips.

2. *Tumescent Solution*: The donor site is best infiltrated with tumescent solution containing epinephrine in a physiological fluid, saline, or Ringer's lactate. Lidocaine in the solution will depend upon the type of anesthesia being used. In patients having liposuction of the body under local tumescent anesthesia, 500 mg lidocaine, 12.5 mg sodium bicarbonate, and 1.0 mg epinephrine in each 1000 mL is necessary. With general anesthesia, no sodium bicarbonate and minimal to no lidocaine need be used.

An interesting new concept is the use of 12.5 g albumin in each 1000 mL of physiological solution in order to maintain a normal oncotic pressure external to the fat cells when injecting larger amounts of fluid into the tissues. Thus, oncotic pressure is maintained when the fat is withdrawn from the body by suctioning or aspiration. The use of a similar amount of albumin in the fat to be injected would dilute the transplant and would require overfilling the recipient area.

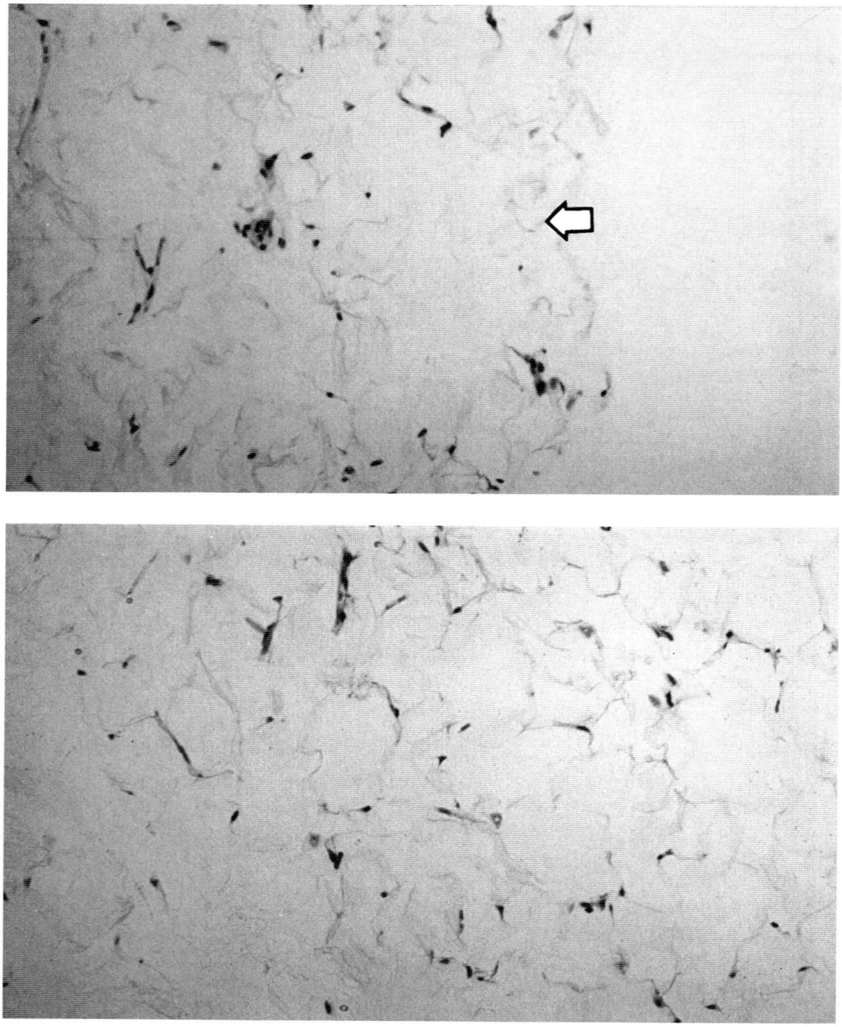

A

B

Figure 1 (**A**) Destruction of fat cells in the periphery of the fat pearl (arrow) with liposuction at –700 mm Hg using a 3-mm cannula. Hematoxylin-eosin, 200×. (**B**) Center of pearl removed at –700 mm Hg using a 3-mm cannula showing intact fat cells. Hematoxylin-eosin, 200×.

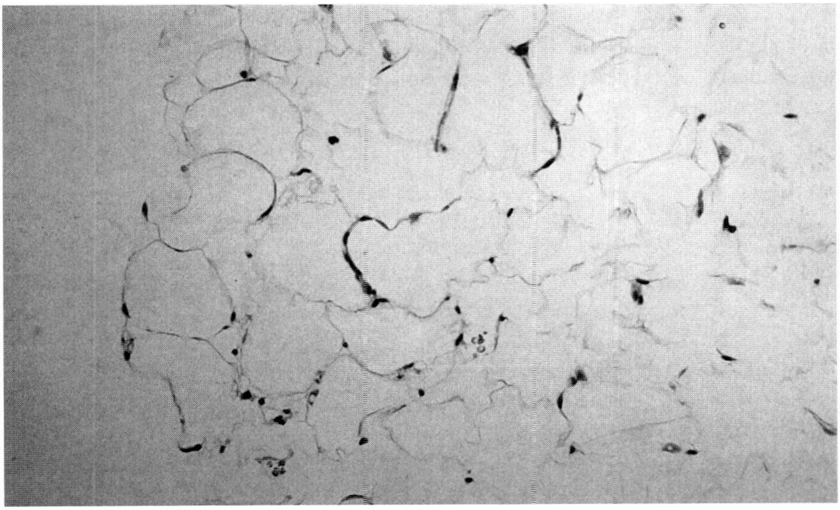

C

Figure 1 (**C**) Periphery of fat pearl removed at –500 mm Hg using a 3-mm cannula showing intact fat cells. Hematoxylin-eosin, 200×.

3. *Cannula with Pump Versus Needle with Syringe*: Fat aspirated by cannula with a suction machine or a needle with a syringe is viable as long as the pressure on the vacuum pump is less than –700 mm Hg, at the most –500 mm Hg, or with the larger syringe (20–60 mL) mounting the plunger halfway or two thirds of the way up the syringe. The negative pressure in the syringe rapidly reduces as the fluid and fat is pulled into the syringe. Vacuum in the pump at –700 mm Hg results in some damage to the fat cells which are in the periphery of the fat "pearl," which is not seen at –500 mm Hg suctioning (Fig. 1).

4. *Fat Trap*: A fat trap should be used when collecting fat by means of a cannula and suction machine.

5. *Washing*: Gentle handling is essential. The fat should be washed to remove blood products and residual lidocaine, epinephrine, and sodium bicarbonate. The wash solution should be physiological, normal saline, or Ringer's lactate. Gentle tilting to mix the solution and fat followed by decanting is important. Avoid *vigorous* mixing, shaking, or straining, and do not let the fat be exposed to dehydration. Use only fat which is yellow and blood free.

A promising beneficial addition to the fat following washing is 12.5 g albumin to each 1000 mL. This will maintain a normal oncotic pressure outside the fat cells so that cells will not expand from external fluids being drawn into the cell and disrupting it.

6. *Injection into Recipient Area*

a. *Subcision*: Areas with skin attachment to underlying tissues (scar, muscle, fascia, or bone) must be subcised before fat is injected.
b. *Anesthesia*: The injection of local anesthesia into the recipient site tissues (other than the skin) is not physiological and may result in damage to the fat cells being transferred (oncotic pressure is decreased allowing fluid to enter and expand the fat cells).
c. *Location of Injection Site*: The needle insertion should be far enough from the area being filled to prevent extrusion of fat through the entry site.
d. *Injection Needle*: The smallest needle used for injecting fat should be 18 gauge when fat retrieval has been through a needle or cannula that is at least 18 gauge in size. A sharp needle does not usually require a skin incision but may cause undesirable bleeding. A small blunt cannula requires an incision, which should be placed in an inconspicuous place.
e. *Multilayering*: Fat injected should be placed in small amounts in any single spot (under 0.33 mL). Therefore, the fat has to be inserted in various layers and from multiple injection sites (when possible).
f. *Ratchet Gun Versus Syringe*: When injecting fat with a syringe using finger or hand pressure, there is usually some point where the fat gets stuck in the needle and increasing pressure is applied. This will result in too much fat being placed in a single spot. The chances of fat survival, therefore, are decreased. A ratchet gun is a simple solution to the problem.

7. *Dressings:* Avoid compression dressings to prevent shifting of the fat transplant and loss of fat cells from compression of the blood supply needed for neovascularization.

8. *Frozen Fat:* There is some evidence that frozen fat survives. It is, therefore, reasonable to freeze fat (slowly) to limit the need for fat retrieval a second or more times. The freezing must maintain quality control with careful distinct labeling of the fat container, strict sterile precautions, functioning temperature gauge with an alarm system if the temperature elevates, and a means of rapid action by the health care personnel to respond to the alarm and reestablish the proper temperature within 8 hours. Fat not adequately protected by these means should not be used.

Index

Absorption (*see* Transplantation, complications: resorption)
Acyclovir 141, 146
Ad-Hoc Committee on New Procedures 2
Adipocytes (*see* Fat)
Agris 46
Albumin 301–303, 307, 308, 313, 316
Alexander 202
Alkek 258
Allergic reaction 284
Alloderm 103, 104, 113, 255
Alpha receptors 313
American Society of Plastic and Reconstructive Surgery (ASPRS) 2, 46, 202, 203
Antiobesity medication 65, 66
Aquafor 156, 157
Artecoll 103,104, 255
Asaadi 17, 48
Asken 16, 40, 46, 200
Asymmetry, lips 114
Autocollagen 267
Autolagen 104, 254

Bames 45,199
Beeswax 69
Berdeguer 17
Betamethasone (Celestone) 102, 157, 252
Billie 10
Billings 46
Bioplastique 284
Bircoll 2, 46, 200–202
Blindness 53, 74, 311
Blood 6
Boering 243
Botox 75, 127, 128
Brandow 17
Brandy-Yoho 201
Breast:
  augmentation 203
  calcifications 203
  reduction 203
Brown 201
Brow ptosis 135
Bruning 1

Bupivicaine 147,149,151

Calcifications 203
Campbell 6, 46
Cariel 50
Carpaneda 49
Ceftriaxone 221
Celestone (*see* Betamethasone)
Cellulite 2
Cellulitis (*see* Transplantation, complication: infection)
Cephalexin 102,157,188
Chajchir 16,17
Chemical peel 131
Chlorhexidine 220
Ciprofloxacin 214
Clobetsol 163
Coleman 49, 243, 300
Collagen 69, 70, 113, 284, 287–298
Compression 191
Congestive heart failure 306
Corticosteroids 157
Cotton 1
Courtiss 48, 49
Covermark 179
Cryoanesthesia 175,180
Cupid's bow 114
Cytochrome P450 inhibitors 65, 67
Czerny 1, 199

Davis 43
Demerol (*see* Meperidine)
Dermabrasion 171, 245
Dermalogen 104, 105, 113, 254
Dextran 48
Diazepam 157
Doctor's Company 203, 311
Duoderm 163

Edema 306
Edema, pulmonary 306
Elam 18
Eppley 48
Ersek 47

Face, hemiatrophy 192
Fagrell 49
Fat:
  atrophy 135
  centrifuge, 16, 17, 27, 60, 78, 81, 82, 117, 151, 301, 307
  characteristics 5, 6
  decant 13–15, 27, 61, 117, 151, 267, 288, 315
  diet 6
  donor sites 24, 25
  drain 261
  fat trap 315
  harvesting 2, 10, 11, 25–27
  ratchet gun 17, 18, 32–34, 77, 95, 211, 217, 316
  sieve 267
  survival 6, 7, 72, 74, 75, 116, 152, 288, 301
  trauma 44, 116, 214
  vascularity 7
  washing 6, 10, 13–15, 28, 61, 177, 181, 210, 214, 217, 248, 249, 307, 315
Fentanyl 221
Fibrinogen 303
Fischer 2
Flexzan 156, 157
Food and Drug Administration (FDA) 258
Fournier 243, 244, 254, 267, 270
Fragen 10
Fulton 17, 18, 245, 252, 300, 301

Gemellus muscle, atrophy 192, 196
Globulin 303
Goretex 69, 103, 104, 113, 255, 284
Gorney 203

Gradinger 202
Graft (*see* Transplantation)
Grazer 202
Green 45
Guerney 44

Halcion (*see* Triazolam)
Hansberger 45
Haramis 17
Harvesting:
  antibiotics 25
  bruising 283, 292
  cannula 7, 60, 96, 98, 115, 151, 210, 261, 267, 280
  edema 283
  hematoma 292
  hypesthesia 283
  infranatant 2
  pressure, low 300
  pretunneling 26
  suction machine 13, 26, 78, 261, 267, 315
  supranatant 28
  syringe 13, 26, 27, 32, 60, 96, 115, 172, 210, 267, 280, 315
  tumescent technique 10, 31, 60, 71, 96, 115, 117, 127, 134,152, 153, 188, 217, 221, 222, 246, 267, 280, 301, 307, 313
  tunnels 10, 151, 152, 280
  trap 2, 10, 78
  ultrasound, external 292
  vacuum, reduced 211
Herpes simplex virus 141, 142, 146
Hibiclens 245
Hilse 45
Hin 201
Hiragun 5, 16
Hudson 47, 245
Hyalform 104, 105, 255
Hyaluronidase 39, 70, 75, 105
Hydroquinone 140
Hydrostatic pressure 304
Hydroxyzine (Vistaril) 147
Hypersensitivity 284
Hypoproteinemia 306

Illouz 2, 39, 45, 47, 243
Injection (*see* Transplantation)
Insulin 5, 10, 14, 16, 40, 77
Isolagen 105, 254
Isotonicity 302, 307, 308

Jackson 244, 271
Johnson 2
Jones 49

Kaminski 301
Kanavel 43
Ketamine 208
Klein 243, 300
Klein's solution (*see* Harvesting, tumescent technique)
Kononas 47
Krulig 2 ,200

Laser:
  acne 164
  burn 154
  cheek ptosis 155
  cheek rhytids 155
  chin 132
  collagen 133,155
  contact dermatitis 162
  contracture 133
  ectropion 133, 155, 165
  erythema 157
  hyperpigmentation 134, 142, 146, 157, 162, 163
  hypopigmentation 134, 146, 163
  infection 161, 162
  itching 163
  jowl 155
  malar 132

[Laser]
mentolabial crease 132
milia 164
nasolabial furrows 132, 156
purpura 163, 164
scarring 134, 160, 164
subcommissural furrows 132
wrinkles 164
Lanolin 69
Levin 17
Lexer 1, 43, 44
Lidocaine, toxicity 149
Lid ptosis 135
Linder 200
Lipoinjection 2, 28, 29, 32, 55
Liposuction 2, 18, 19, 25–27, 60
Loss of vision (*see* Blindness)
Lustra cream 142, 157
Lyles 49
Lynch 203

Magnetic resonance imaging (MRI) 35–39, 41
Mammograms 213
Mann 44
Marcaine (*see* Bupivicaine)
Markman 47
Mastopexy 208
May 46
Medical record 66
Meperidine (Demerol) 147
Methylmethylpolysiloxane (*see* Bioplastique)
Metoclopramide 221
Midazolan (Versed) 147, 208, 221
Miller 201
Minocycline 102
Mitnick 201
Mucopolysaccharide 292

Neichajev 17, 18, 49
Neovascularization 62
Nerve block 62
Neuber 1, 187, 243
Neuhof 44
Neuman 17
Newman 2, 17
Nguyen 18, 47
Novack 200

Oncotic pressure 302, 303, 305, 307, 316
Orentreich 244, 258
Osmolarity 302, 307
Osmotic pressure 302, 303, 308
Overcorrection (*see* Transplantation, overfill)

Papaverine 230
Paraffin 69
Peel, skin 171
Peer 1, 45, 72, 199, 243
Penn 203
Petrolatum 69
Pinch test 233
Pitman 202
Planotome 2
Platelet gel 207–218
Poiseuille's law, 288
Polio sequelae 192, 195
Polmethylmethacralate 279–286
Povidone 220
Prednisone 102
Prochlorperazine (Compazine) 147
Propofol 208
Prosthesis, buttock 191, 192
Protein 301–303, 307
Ptosis 208, 217
Pulmonary edema (*see* Edema, pulmonary)
Pulse oximeter 153

Ratchet gun (see Fat, ratchet gun)
Reabsorption (*see* Transplantation, complications: resorption)

Reparil 179
Reston 63
Restylane 104, 105, 255
Retention (*see* Transplantation, complications: resorption)
Retin A 142
Retinoids 146
Rhytidectomy 131
Ribeiro 49
Roncari 45

Samdal 48
Satler 50, 271
Saunders 45
Sawhney 243
Schaefer 202
Schorcher 45,m199
Sertraline 220
Sevcuk 17, 18, 49
Shapiro 45
Shiffman 74
Sidman 16
Silicone 69, 103, 104, 113, 164, 173, 188, 203, 214, 215, 258, 300
Skouge 5, 6, 201
Smoker 66, 220
Softform 103, 104, 113
Sommer 50, 271
Starling's equation 301, 303–307
Straatsma 1
Subcision 135, 245, 262, 269, 315
Subcutaneous musculoaponeurotic system (SMAS) 136
Survival (*see* Transplantation, survival)

Teflon 69
Telfa 156, 157, 161
Tenoxican 228
Tobin 11
Toledo 17
Toxicity, lidocaine 65
Transfer (*see* Transplantation)
Transplantation:
- acne 136, 243–259
- breast 1, 2, 24, 45, 199–206, 207–218, 300, 311
- buttocks 2
- cheeks 12, 149, 269, 280
- chin 23, 77–93, 280
- complications:
  - abscess 295
  - blindness 9, 53, 54, 296, 311
  - bruising 9, 53, 103, 121, 179, 181
  - calcifications 9, 53, 54, 75, 204, 214, 230, 300, 311, 312
  - central nervous system damage 53
  - cyst 121, 204, 230, 283, 296
  - dermatitis, contact 159
  - desiccation 9
  - discomfort 283
  - distortion 121, 181
  - edema 181, 191, 296
  - emboli 214, 230
  - erysipelas 191
  - erythema 53, 103
  - fat embolism 74
  - fibrous capsule 53
  - hematoma 53, 74, 103, 121, 253, 283, 296
  - hemorrhage 44
  - infection 2, 9, 44, 53, 75, 117, 121, 157, 191, 214, 230, 253, 296
  - injection, intravascular 159, 160
  - liquefaction 283
  - mass 53, 204
  - mucocele 121
  - necrosis 283
  - nodule 159, 296
  - oil cysts 9, 44, 53, 75, 230, 311

[Transplantation]
overinjection 160, 253
pain 53, 181, 214, 231
pseudocyst 2
redness 283
resorption 7, 9, 35, 37, 39, 41, 44, 45, 48, 49, 53, 54, 72, 93, 102, 103, 125, 128, 182, 183, 230, 231, 300
scarring 121
sliding of graft 75
swelling 9, 103, 143, 283
tenderness 9, 53
vesicles 191
volume loss 283
compression 213, 316
contouring 63
defects 233–242, 287
eyebrow 127–130
face 287
forehead 69–76, 149, 153, 269
freezing 29, 30, 50, 72, 95, 102, 119, 120, 132, 143, 180, 181, 222, 249, 251, 261–265, 267–277, 300, 301, 316
genitals, female 219–232
glabella 9, 11, 23, 69–76, 149, 153, 269, 296, 311
guidelines 23–30
hand 12, 169–185, 287
indications 7–9, 70–71
injection 2,9
lip 11, 24, 113–125, 153, 280, 282
lysis 287
malar 23, 77–93, 149
mandible 280
marionette lines 95–112, 119
massage 119, 128, 190
mental (*see* Transplantation, chin)
multilayering 316
nasolabial fold 11, 24, 49, 50, 95–112, 136, 269, 280
overfill 11, 29, 71, 74, 78, 96, 103, 143, 227, 252, 253
penile 219–232
perioral 149
periorbital 153
postliposuction 24
pressure 2
rhytids 287
scar 24
site, recipient 23, 24
submalar 23
survival 1, 43–50, 55, 119, 120, 252, 253, 288
syringe 2, 176
technique 9–19
temple 149
tethering bands 18
thighs 2
trauma 1, 9
trochanteric 2
tunnels 128, 135, 188
ultrasound 296
Triamcinolone 230, 253
Triazolam (Halcion) 147
Tuffier 1
Tumescence (*see* Harvesting, tumescent technique)

Uebel 17
Ullmann 50
U.S. Food and Drug Administration (FDA) 113, 279

Valacyclovir 141, 142, 146
Van 45
Vein stripping 171
Verderame 43
Versed (*see* Midazolam)
Vibramycin 179
Vicodin 147
Vitamin E 5,40
Vitamin K 292

Wertheimer 45
Wetmore 72
Willi 55–59
Wrinkles 269

Yoho 201

Zyderm 12, 104, 105, 255
Zyplast 104, 105, 255